Journal of Chromatography Library – Volume 5

INSTRUMENTAL LIQUID CHROMATOGRAPHY

A Practical Manual on High-Performance Liquid Chromatographic Methods

JOURNAL OF CHROMATOGRAPHY LIBRARY

Volume 1 Chromatography of Antibiotics
by G. H. Wagman and M. J. Weinstein

Volume 2 Extraction Chromatography
edited by T. Braun and G. Ghersini

Volume 3 Liquid Column Chromatography. A Survey of Modern Techniques and Applications
edited by Z. Deyl, K. Macek and J. Janák

Volume 4 Detectors in Gas Chromatography
by J. Ševčík

Volume 5 Instrumental Liquid Chromatography. A Practical Manual on High-Performance Liquid Chromatographic Methods
by N. A. Parris

Volume 6 Isotachophoresis. Theory, Instrumentation and Applications
by F. M. Everaerts, J. L. Beckers and Th. P. E. M. Verheggen

Volume 7 Chemical Derivatization in Liquid Chromatography
by J. F. Lawrence and R. W. Frei

Volume 8 Chromatography of Steroids
by E. Heftmann

Journal of Chromatography Library – Volume 5

INSTRUMENTAL LIQUID CHROMATOGRAPHY

A Practical Manual on High-Performance Liquid Chromatographic Methods

N.A. Parris
DuPont (U.K.) Ltd., Instrument Products Division, Wilbury House, Wilbury Way, Hitchin, Herts. SG4 OUR

ELSEVIER SCIENTIFIC PUBLISHING COMPANY
AMSTERDAM – OXFORD – NEW YORK 1976

ELSEVIER SCIENTIFIC PUBLISHING COMPANY
335 Jan van Galenstraat
P.O. Box 211, Amsterdam, The Netherlands

AMERICAN ELSEVIER PUBLISHING COMPANY, INC.
52 Vanderbilt Avenue
New York, New York 10017

Library of Congress Cataloging in Publication Data

Parris, N A
Instrumental liquid chromatography.

(Journal of chromatography library ; v. 5)
Includes bibliographies and index.
1. Liquid chromatography. I. Title. II. Series.
QD79.C454P37 544'.924 76-24837
ISBN 0-444-41427-4

Printed in The Netherlands

Contents

Preface . . . IX

FUNDAMENTALS AND INSTRUMENTATION

1. Introduction and historical background . . . 3
   References . . . 5

2. Basic principles and terminology . . . 7
   General resolution equation . . . 14
   Calculation of optimum column length . . . 15
   Reference . . . 18

3. Chromatographic support and column . . . 19
   Introduction . . . 19
   Sources of band broadening . . . 20
   Role of particle size in LC columns . . . 24
   Porous layer supports . . . 27
   Totally porous (microparticulate) supports . . . 28
   Dependence of column efficiency on operational conditions . . . 30
   Columns for high-pressure LC . . . 31
   Column efficiency and internal diameter . . . 32
   Methods of packing chromatographic columns . . . 34
   References . . . 40

4. Liquid chromatographic instrumentation . . . 43
   Introduction . . . 43
   Tubing and tube fittings . . . 44
   Solvent delivery systems . . . 45
   Gradient elution devices . . . 52
   Other components of the solvent delivery system . . . 56
   Sample introduction . . . 59
   Chromatographic column and couplings . . . 66
   Detectors . . . 69
   Fraction collectors . . . 71
   Measurement of mobile phase flow-rate . . . 72
   Presentation of results . . . 73
   Availability of LC equipment . . . 74
   References . . . 74

5. Liquid chromatographic detection systems . . . 75
   Introduction . . . 75
   Principal requirements of a LC detector . . . 77
   Photometric detectors . . . 77
   Fluorescence detection . . . 81
   Refractive index detectors . . . 83
   Phase transformation detectors . . . 86
   Phase transformation to flame ionisation detector . . . 87
   Other detection devices . . . 88
   Final comments on instrument design . . . 90
   References . . . 91

FACTORS INFLUENCING CHROMATOGRAPHIC SELECTIVITY

6. Nature of the mobile phase . . . 95
Introduction . . . 95
Methods of separation in the liquid phase . . . 96
Classification of mobile phases . . . 98
Development of chromatographic methods . . . 102
Elution behaviour of complex mixtures of dissimilar compounds . . . 110
References . . . 126

7. Liquid–solid (adsorption) chromatography . . . 127
Introduction . . . 127
Range of sample applicability . . . 127
Types of adsorptive packing . . . 129
Mechanism of adsorption chromatography . . . 132
Choice of separating conditions . . . 135
Practical aspects of adsorption chromatography . . . 136
References . . . 141

8. Liquid–liquid (partition) chromatography . . . 143
Introduction . . . 143
Range of sample applicability . . . 143
General considerations . . . 145
Types of liquid–liquid phase systems . . . 147
Relative merits of the various forms of partition chromatography . . . 163
References . . . 165

9. Ion-exchange chromatography . . . 167
Introduction . . . 167
Range of sample applicability . . . 168
Mechanism of ion-exchange separations . . . 174
Structure of column packings for ion-exchange chromatography . . . 176
Commercially available ion-exchange materials . . . 181
Practical aspects of ion-exchange chromatography . . . 181
Ion-pair partition chromatography . . . 187
References . . . 188

10. Steric exclusion chromatography . . . 191
Introduction . . . 191
Range of applicability of the method . . . 191
Mechanism of separation . . . 192
Column packings for steric exclusion chromatography . . . 194
Choice of mobile phases for steric exclusion chromatography . . . 202
General scope of steric exclusion chromatography . . . 204
References . . . 215

USES OF LIQUID CHROMATOGRAPHIC PROCEDURES

11. Qualitative analysis . . . 219
Introduction . . . 219
Methods of establishing or confirming the identity of an eluting peak . . . 220
Other considerations when seeking to identify an eluted component . . . 226
References . . . 227

12. Quantitative analysis 229
Introduction 229
Sources of error in chromatographic analysis 230
Manual methods of integration made after completion of the analysis 238
Integration made during the course of the analysis 240
Normalisation of the peaks 243
Normalisation of peaks with correction factors 243
Calibration by means of an external standard 244
Calibration using an internal standard 244

13. Practical aspects of trace analysis 247
Introduction 247
Sample pretreatment 247
Sample injection 250
Chromatographic considerations 252
Detection considerations 257
Quantitation of minor components 261
References 262

14. Practical aspects of preparative liquid chromatography 263
Introduction 263
Available methods for increasing the sample throughput of chromatographic columns . . 264
Effect of column geometry on chromatographic resolution 265
Considerations on the chromatographic support 267
Practical aspects of preparative liquid chromatography 268
Applications of preparative chromatography 273
Industrial-scale chromatographic separations 275
References 276

APPLICATIONS OF LIQUID CHROMATOGRAPHY

15. Published LC applications information 279
Pharmaceutical analysis 280
Biochemical analysis 285
Food analysis 288
Pesticides and related compounds 292
Oil and petroleum analysis 293
Petrochemical and related compounds 294
Inorganic and organometallic compounds 296
Polymer analysis 297

16. The latest trends and a glimpse into the future 299
References 300

Appendix 1. Derivation of the general resolution equation 301

Appendix 2. Comparison of the U.S. (A.S.T.M.) and B.S.S. sieve sizes in relation to aperture size in micrometres 303

Appendix 3. Suppliers of liquid chromatographic instrumentation and components 305

Appendix 4. Practical aspects of using simple liquid stationary phases 309
References 311

Appendix 5. Suppliers of well characterised polymer samples for molecular weight standards . 313

List of abbreviations and symbols 315

Subject index 317

Preface

There must be few, if any, involved in organic chemical analysis who have not been impressed by the impact gas chromatography (GC) has made on their approach to analytical problems. This impact was so great that by the mid-1960's it was quite apparent that there was a real need for a complementary technique for liquid phase separations, since not all compounds were amenable to GC. Although the number of samples handled by gas–liquid chromatography can be increased significantly by derivatisation of polar functional groups, it has been suggested that only some 15% of all chemicals are capable of existing in the vapour phase. Modern, high-pressure liquid chromatography (LC) has emerged as an instrumental technique offering rapid separations with simultaneous sensitive monitoring of the course of the analysis.

Much of the development of modern LC reflects the experience gained during the growth of GC. The methodology of the two techniques is superficially quite similar, a fact which is perhaps not too surprising since some of the world's most experienced gas chromatographers have pioneered the so-called renaissance of LC.

There are, however, many detailed differences between the two techniques, giving each an important area of application. It is proper that the techniques are viewed in this manner rather than as two methods competing for the same application.

Several independent schools of thought have contributed to the rapid development of modern LC. This situation has sometimes created the impression that only one of these approaches can be right for any one application. This is certainly not the situation, for often there are several ways of achieving the result; such is the versatility of LC – albeit much to the confusion of a newcomer to the technique. It is the hope of the author that this book will combine the advantageous practical aspects of these various approaches and also point out their shortcomings in such a manner that the reader is able to decide which procedure will be best for his application and, perhaps of equal importance, suit the instrumentation available to him.

The theoretical aspects of LC are dealt with only in sufficient depth that will enable the reader to grasp the basic principles of chromatography and the terminology involved. No apologies are made for this light regard for the theoretical aspects, since it is the author's experience that many who practice chromatography do so to achieve an end result, which is not to gain a thorough understanding of how and why a separation occurs but simply to obtain a separation to isolate or assay one or many components in a sample. This statement is not meant to infer a lack of scientific interest in understanding the mechanism by which separations occur, but more an appreciation that everyday pressures in most laboratories do not allow time for a thorough grasp of the theory to be obtained. In these circumstances information which is directly applicable to the problem in hand together with some indication of the most likely sources of trouble or experimental error is often of more immediate use.

It is for these would-be, or practising chromatographers that this book is primarily intended, *i.e.*, as a practical introduction to the technique of modern LC.

The author has been fortunate to have worked for a number of years in an Applications

Laboratory of the DuPont Company, who market a range of LC equipment and column packings. The experience gained in this work – which involves studying the entire spectrum of sample types, also continuously striving to solve new separation problems as well as advising instrument users in practical matters – and the frequent exchange of information in such an environment have given the author a thorough understanding of the most common difficulties encountered while practising LC on a day-to-day basis. As far as is practicable, advice on how to avoid or overcome these trivial yet frustrating pitfalls is included in the appropriate sections of the text.

In the preparation of this text the author is indebted to a number of organisations and individuals whose advice and suggestions have proved invaluable. Particular mention should be made to the DuPont Company (U.K.) Ltd., a subsidiary of E.I. DuPont de Nemours and Company, Wilmington, Dela., U.S.A., who have made the preparation of this manuscript possible by allowing the author to use data generated in their Applications Laboratories. Additionally, as my employer, the Company should also be thanked for the opportunity to contact fellow workers in this field by way of frequent attendance and participation at symposia, discussion meetings, seminars and workshop sessions.

The co-operation of companies who have allowed the reproduction of their data in this book is also gratefully acknowledged, as are the time and efforts of Messrs. Brian J. Read and John A. Schmit in carefully checking and criticising this text. Sincere thanks must go to Mrs. Linda Sandy, Mrs. Susan Maher and my wife, June, for their time taken in typing this manuscript.

FUNDAMENTALS AND INSTRUMENTATION

Chapter 1

Introduction and historical background

The earliest reported account of a separation that can be considered as an example of liquid chromatography has been attributed to Tswett, born in Asti, Italy, in 1872. In 1903, while working as a chemist in Russia, he described[1] the separation of green plant pigments in a column filled with powdered chalk. From that time little appears to have been reported until, in the 1930's, Reichstein adopted the method for the isolation of natural products[2].

The next significant advance in the technique was the work on liquid partition chromatography which, in 1941, led to Martin and Synge[3] being awarded the Nobel Prize. In 1948, Moore and Stein reported the use of ion-exchange chromatography for the separation of amino acids[4]. This application alone must have been largely responsible for the very considerable interest which was later shown in liquid chromatography by those working in the biomedical field.

The technique, as practiced up until the mid-1960's, generally involved using a fairly large column containing a packed bed of adsorbent, most commonly silica gel or alumina, coated with a stationary liquid for partition applications. The separation was carried out by percolating liquid through the bed under the force of gravity. The scale of the operation was large by modern standards in that a single separation consumed considerable quantities of solvent and adsorbent. The progress of the separation was most often monitored by collecting fractions of the column effluent and subsequently performing some independent method of quantitation. This usually involved evaporating the fractions to dryness so that the residue could be weighed. More specific monitoring of a component could be achieved by redissolving the residue and carrying out a spectrophotometric assay. It should be apparent that a method involving so many steps and that is wasteful of reagents, operator time and sample material tends to be unpopular in an era when great demands are made for rapid and precise data on, all too often, minute quantities of sample. Because of the limitations of the existing technique, a number of closely related methods have been developed for separating mixtures of chemical substances in the liquid phase. The most widely practised of these chromatographic methods include paper chromatography (PC) and thin-layer chromatography (TLC), which may be considered as semi-micro techniques involving partition or adsorption mechanisms, respectively, capable of producing fairly good resolution of small quantities of sample but lacking, except in specialised instances, an easy method of obtaining quantitative results. Although separations performed by both of these methods may often take less than 1 h, particularly in the case of TLC, measurement of the area or density of the spots must be performed after the completion of the separation. This step is time consuming and even then a precision better than 5% is seldom achieved. Neither PC nor TLC are strictly suitable for large-scale separations, as increasing the thickness of the paper or adsorbent layer to increase capacity leads to a progressive deterioration of the separating power of the system. Larger-scale samples can be handled by dry-column chromatography. This is a form of column chromatography where the sample is applied to the head of the dry adsorbent bed and then washed down the column with the appropriate solvent. In certain preparative applications of this method, the sample has been recovered by dissecting the column and extracting the sample components from the adsorbent bed – clearly this approach does not lend itself to repetitive analysis as the column packing must be renewed for each sample.

The most recent developments in column chromatography have been concerned with the transition of the technique from these fairly slow, laborious methods to a refined instrumental method. It is the practical aspects of this more modern form of column chromatography with which this book is concerned. Terms used to describe this latest approach to column chromatography include high-speed..., high-performance..., modern... – all attempting to convey the significance of these developments. For the sake of avoiding any unwanted inferences in this text the most recent ramification of column chromatography will be simply described as "liquid chromatography" (abbreviated as LC).

The evolution of LC as a highly sophisticated analytical technique results from the need to have a separation system in the liquid phase which is complementary to gas chromatography (GC), *i.e.* a method which is capable of rapidly separating complex chemical mixtures and providing simultaneously a continuous record of the separation from which the quality of the separation and, when suitably calibrated, a quantitative assessment of the composition of the original sample may be deduced.

LC in its most modern form is able to achieve separations in a matter of a few minutes which by previous techniques may have taken hours or days or may not have been possible. This achievement has come from the results of much intensive research and development work associated with improving our theoretical understanding of the factors involved in separations in the liquid phase and in the consequent design and construction of suitable apparatus with which to perform the separations. Of particular importance to this development has been the availability of specialised chromatographic column packings and sensitive in-line detection systems for continuously monitoring the separation being carried out. These developments have led to systems which, in favourable instances, can on the one hand detect part per billion (1 in 10^9) levels of impurities in samples and on the other hand be used for collecting gram quantities of pure chemicals. In this latter application, *i.e.*, preparative chromatography, LC has much to offer relative to GC in that the sample does not have to be vaporised when introduced into the column and conversely does not have to be condensed from the vapour phase in order to collect the sample after separation. If pure, relatively volatile, carrier solvents are employed, recovery of a component of a sample collected from effluent from a LC column can be simply a matter of removing the solvent from the collected fraction by evaporation, if necessary, under reduced pressure.

Quantitation of analytical results generated in modern LC systems is achieved in much the same manner as in GC, where digital integrators or dedicated computing systems have been established as the most time-saving methods. A precision of better than 1% has been reported by many independent workers in the field of modern LC, suggesting that the technique is directly suitable for many assays of commercial importance. Unlike GC, the precision of the method does not normally vary a great deal from sample to sample, presumably since vaporisation of the sample, with attendant possibilities of decomposition or variations in the rate of evaporation, is absent in the liquid phase.

Developments in the technique have now reached a particularly exciting stage, as many of the apparently conflicting views that were held a few years ago are becoming rationalised and clarifying those aspects of the method that need greatest attention to detail.

REFERENCES

1 M. Tswett, *Proc. Warsaw Soc. Nat. Sci., Biol. Sect.*, 14 (1903) No.6.
2 T. Reichstein and J. van Euw, *Helv. Chim. Acta*, 21 (1938) 1197.
3 A.J.P. Martin and R.L.M. Synge, *J. Biochem.*, 35 (1941) 1358.
4 S. Moore and W.H. Stein, *Ann. N.Y. Acad. Sci.*, 49 (1948) 265.

Chapter 2

Basic principles and terminology

Although strictly a misnomer, the term "chromatography" has been adopted universally to cover the "science of separations". More accurately, the term embraces techniques which enable samples of chemical mixtures to be separated by exploiting differences in their physical or chemical properties. These differences govern the rate of migration of the components of a mixture passing under the influence of a moving fluid through a "bed" of stationary phase. The stationary phase may be a finely ground solid or a liquid coating thereon and the form of the "bed" may be a thin layer or packing within a glass or metal tube which, as such, is referred to as a chromatographic column.

Column chromatography is concerned with the separation of components of a mixture by establishing conditions under which the individual components flow at different rates through a packed column, under the influence of a moving liquid phase, referred to as the mobile phase or carrier. This action is known as the elution of the sample from the column.

The total liquid issuing from the column is referred to as the column effluent. The portion of the effluent originating from the mobile phase is termed the eluent and the part originating from the sample is termed the eluate.

The differential rates of elution arise from interactions between the components of the sample and the material used to pack the column or a coating thereon. There are four principal mechanisms in LC by which components of samples are selectively retained. These are the exploitation of differences in partition coefficients (liquid–liquid chromatography), adsorption effects on surfaces such as silica gel (liquid–solid chromatography), dissociation of weak or strong electrolytes (ion-exchange chromatography), or in molecular size or shape (steric exclusion chromatography).

The interaction of sample with the column packing is referred to as retention. For any given chromatographic system the degree of retention of a compound is a characteristic of that sample, since it is dependent on the solubility, adsorption, size and ionisation characteristics of that compound in that specific environment of the chromatographic system employed. This retardation of a sample in a column system is expressed quantitatively as the retention volume, which is defined as the volume, usually in millilitres, of mobile phase which flows through the column system from the moment of sample introduction to the appearance of the maximum concentration of the eluting peak at the detector. When a sample does not experience any interaction with the material packed in the column it passes through without retention and is said to elute in the void volume (or dead volume) of the column. The void volume is usually represented by the symbol V_0. Physically this volume represents the interstitial spaces between the particles packed in the column and any readily accessible pores within the packing material itself which are occupied by the mobile phase. It will be seen later that in practice it is an ideal to minimise this particular parameter of a column since it represents time lost while waiting for samples to pass through the column. It follows that a sample which is retained on a column will elute in a volume larger than the void volume. Its retention volume, V_R, will be the sum of the void volume and the volume of mobile phase necessary to overcome the interactions between the sample and the column packing.

Retention volume is a characteristic of a given sample–chromatographic system combination expressed in absolute terms. In certain circumstances it is preferable to express retention of a sample relative to the elution of a non-retained sample. This is commonly referred to as the relative partition coefficient or the capacity factor, k', and is defined by the expression:

$$k' = \frac{V_R - V_0}{V_R}$$

When no change in the mobile phase flow-rate occurs during the elution of the sample, the expression may be considered as

$$k' = \frac{t_R - t_0}{t_0}$$

where t_R and t_0 are the retention times of a retained and non-retained sample, respectively. Fig. 2.1 shows how, at constant flow-rate, capacity factors, retention and void times are directly measurable from a chromatographic trace since a recorder chart invariably moves at constant speed.

In partition chromatography*, the capacity factor is related to the distribution coefficient, K, *i.e.*, the ratio of the concentrations of the sample component in the two liquid layers.

The capacity factor is also related to the mass of component in the mobile and the stationary phase within the column. The two terms are related as follows:

$$k' = \frac{\text{Mass in stationary phase}}{\text{Mass in mobile phase}}$$

$$= \frac{\text{Concentration in stationary phase}}{\text{Concentration in mobile phase}} \cdot \frac{\text{Volume of stationary phase}}{\text{Volume of mobile phase}}$$

$$= K \cdot \frac{\text{Volume of stationary phase}}{\text{Volume of mobile phase}}$$

This expression indicates that the retention of a component in a given column will only be increased by either a change in the distribution coefficient or an increase in the volume of stationary phase relative to mobile phase in the column. The distribution coefficient will be dependent on the chemical nature and temperature of the liquid phases forming the system, whereas the volume of stationary phase is governed largely by the surface area of the chromatographic support.

If the mass of component in the mobile phase and the stationary phase is expressed graphically, curves like those in Fig. 2.2 may be obtained. The slope of the graph is the capacity factor, k'. The point marked D indicates the limit of linear behaviour, *i.e.*, the

*Although partition chromatography is described here, the same treatment applies to other modes of separation, except in place of stationary phase one uses surface area (adsorption), ion-exchange capacity (ion-exchange), or total pore volume (steric exclusion).

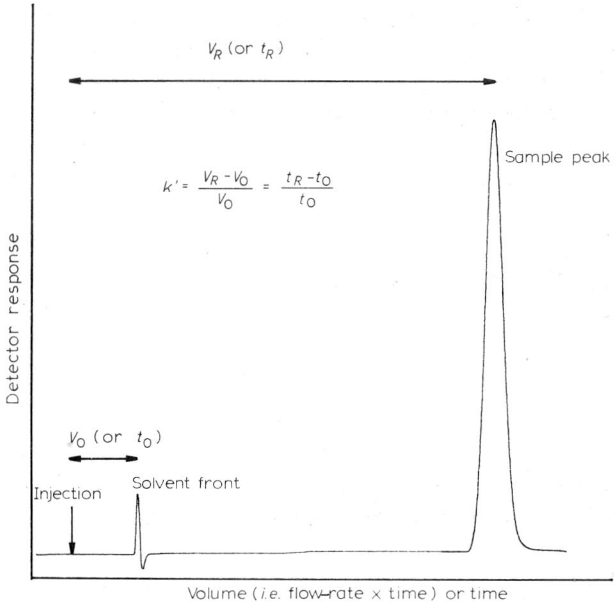

Fig. 2.1. Measurement of capacity factor, k'.

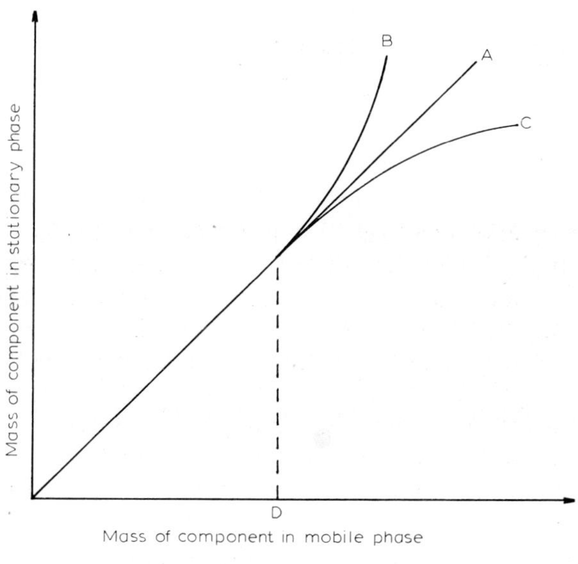

Fig. 2.2. General characteristics of sorption isotherms. (A) Linear curve; (B) concave curve; (C) convex curve.

sample (or linear) capacity, and its value is dependent on the chromatographic system being used. It is of importance to be aware of any deviation from linearity since deviations of type B (concave) tend to give rise to peaks having a pronounced leading edge and retention times which increase as the sample size is increased. The convex form (type C) leads to peaks with a trailing edge and retention times which decrease as the sample size is increased.

Since chromatographic analysis is a separation method, one is always concerned with more than one component and it is important to be able to define the separating power of a column for the samples being studied. It was shown earlier that the capacity factor, k', and retention volumes are characteristics of individual chemical species in a given chromatographic system. When dealing with chemical mixtures it is easier to give these values suffixes, thus

$$k'_{\mathrm{a}}, k'_{\mathrm{b}} \ldots k'_{\mathrm{i}} \text{ and } V_{R_{\mathrm{a}}}, V_{R_{\mathrm{b}}} \ldots V_{R_{\mathrm{i}}}$$

which relate to components a, b, and i in a mixture. For any separation to be possible it is essential that each component has a different value for the capacity factor, *i.e.*, each component must be retained to a different extent. In these circumstances the system is said to be selective towards the compounds being analysed. In chromatographic terms selectivity, α, is expressed as the ratio of the capacity factors of the two components of interest. Thus:

$$\alpha = \frac{k'_{\mathrm{b}}}{k'_{\mathrm{a}}} = \frac{t_{R_{\mathrm{b}}} - t_{R_0}}{t_{R_{\mathrm{a}}} - t_{R_0}}$$

A separation between components A and B in a mixture will only be possible in a chromatographic system if the selectivity factor, α, has a value other than unity. Perhaps one of the most important uses of this term is in the reporting and recording of chromatographic data and correlation of the same with the type of sample and the experimental parameters.

A favourable selectivity factor does not, however, indicate whether or not a separation will be achieved on the chromatographic system used. For a separation to occur the individual component bands must occupy a sufficiently small volume of the mobile phase in the column so that the bands do not overlap. A selectivity factor with a value other than unity merely indicates that the points of maximum concentration of the two components are not coincident.

In practice, a sample is introduced rapidly at one end of the column as a concentrated "band". This band moves through the column bed under the influence of the mobile phase. If diffusion, or mixing, phenomena do not occur, it is reasonable to expect that the same discrete band of sample will eventually be eluted from the column in its original volume, *i.e.*, with the concentration unchanged, and the recorded profile will be rectangular.

This situation is not the case in practice since diffusion phenomena lead to dilution of the sample. The dispersion of sample bands, which results in a chromatographic peak, creates a distribution of sample concentration rather than a sharp line or rectangular distribution. Although inevitable, diffusion of this kind must be minimised if many components are to be separated in a column. The spreading of the sample bands during their passage through the column tends to produce (on the strip-chart recorder) a distribution curve of sample concentration which approximates a Gaussian curve.

Each sample band, although contained in a discrete volume, can be considered as occupying a certain length of the chromatographic column. For a separation to be possible not only must the selectivity be favourable (*i.e.*, the α value $\neq 1$) but the lengths of column occupied by each consecutive band must not overlap. This depends on the extent of spreading of the band and how much length of column is available in which to achieve the separation.

In practical chromatographic work, the band spreading is always discussed in terms of the shape of the eluted peak as produced on the resultant chromatogram. This is most easily appreciated by reference to Fig.2.3.

Clearly, the narrower the width of an eluting peak (*i.e.*, the lower the volume containing the eluting component), the greater the chance of separating a multi-component mixture in a column. The ability of a column to minimise peak spreading is referred to as the efficiency of a column. A column which minimizes the peak spreading of a component as it passes through the column is referred to as being highly efficient and is one of the prime objectives in the development of modern LC. One of the features of chromatographic columns is that their efficiency is dependent on the velocity of the carrier liquid passing through the column. The reasons for this effect are discussed in Chapter 3, but it is mentioned here to emphasize that the efficiency value assigned to a column depends a great deal on the manner in which it is used.

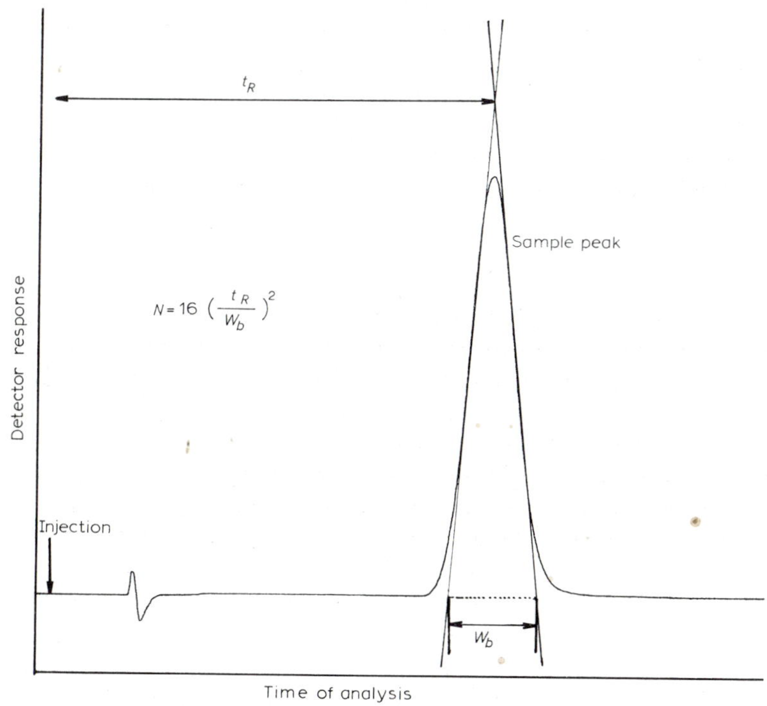

Fig. 2.3. Measurement of column efficiency, *N*.

The efficiency in all chromatographic techniques is expressed quantitatively as the number of theoretical plates, N, of the column. This value is calculated from the following expression

$$N = 16\left(\frac{t_R}{W_b}\right)^2$$

where W_b is the base width of the peak or, more strictly, the base width of the triangle constructed on the peak. The generally accepted assumption is that the eluted chromatographic peak approximates a Gaussian distribution. In these circumstances W_b, being the base width of the constructed equilateral triangle, as shown in Fig. 2.3, represents 4 σ (standard deviation units of the Gaussian peak). Several expressions have been proposed for calculation of column efficiency. These generally differ in the point where the peak width is measured, *i.e.*, at the base of the triangle (as above), at half the height of the eluted peak, at half the height of the constructed triangle, or at the height of the peak where the deviation from the mean is exactly 1 σ (*i.e.*, the peak width is then equal to 2 σ; this occurs at approximately 60% of the height of the peak). These different expressions all tend to give similar values for the overall column efficiency since the value of the proportionality constant used in the calculation is given a different value depending on where the peak height is measured. The proportionality constant becomes unity when the peak width becomes equal to 2 σ (at 60% height). The use of the base width of the constructed triangle tends to be the method most often used.

The theoretical plate concept is a very useful and almost universally accepted method of assessing the performance of chromatographic systems. The concept has its origins in the theoretical treatment of fractional distillation columns. A detailed understanding of the fundamentals of this theory is not needed for practical interpretation of chromatographic performance and it will not be dealt with here.

Calculation of column efficiency using the last equation gives simply a number of theoretical plates and as such gives no indication of the dimensions of the system employed. For instance, based on this calculation one could test two columns and find that both exhibited an efficiency of 1,000 theoretical plates. One column could be 10 m in length, *i.e.*, have an efficiency of 100 plates per metre, whereas the other column, being 10 cm long, is exhibiting the equivalent of 10,000 plates per metre. Both of the values are quite possible in modern LC; however, it is a matter of deciding rationally which is the better of the two columns. The choice in this instance would be quite apparent to anyone with any idea of chromatography, but there are instances when the decision is not so clear-cut. The choice can be made by considering the parameter defined earlier, *i.e.*, the void volume of the column. Since good separations at high speed are the ultimate objective, the column with the minimum void volume would be the one giving the best overall performance.

The characteristics of a column may be defined more precisely by a number of other related terms. The ambiguity demonstrated above may be avoided by using the term height equivalent to a theoretical plate, HETP, which is more commonly being referred to as the plate height, H. This is calculated by dividing the column length by the number of theoretical plates, thus

$$H\,(\text{mm}) = \frac{\text{length (mm)}}{N}$$

The lower the value of H, the better is the column performance. The examples given earlier yield H values of 10 mm and 0.1 mm, respectively, indicating the superiority of the 10-cm-long column.

Having defined the peak width it is now possible to describe the resolving power of a chromatographic column. It was shown earlier that for complete separation of two chromatographic peaks, the eluting bands must not be coincident or overlap. The selectivity factor, α, defines the former. The latter characteristic of a column is defined by the resolving power, which relates the width of the eluted peaks to the distance between the peak maxima. More strictly, this treatment applies to the positions and widths of the constructed triangles rather than the peaks.

The resolving power, sometimes referred to as the resolution factor, R, of a column is calculated as follows

$$R = 2\left(\frac{t_{R_b} - t_{R_a}}{W_b + W_a}\right)$$

where t_{R_a}, t_{R_b}, W_a and W_b are the retention times and base widths, respectively, of peaks A and B shown in Fig. 2.4.

Unity resolution is achieved when the difference in retention time (or volume) between the maxima of peaks A and B, $(t_{R_b} - t_{R_a})$, is equal to the sum of the half widths of the bases of the constructed triangles, *i.e.*, the adjacent triangles just touch at the baseline.

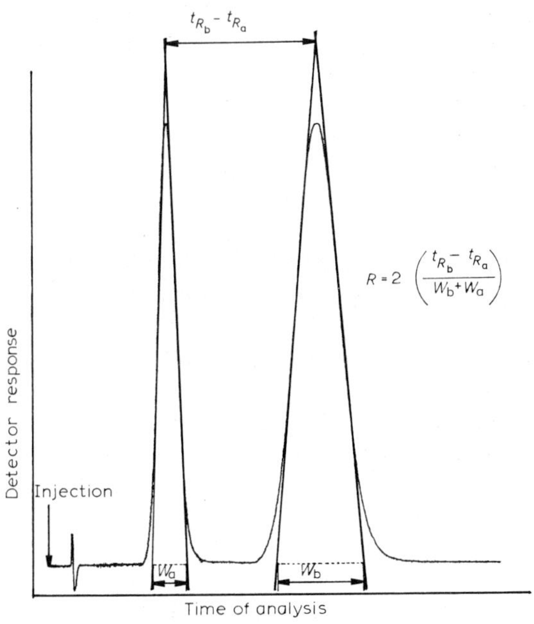

Fig. 2.4. Measurement of resolution, R.

The resolution factor thus calculated defines the separation achieved in a chromatographic analysis. Since in practice the peak shapes approximate to a Gaussian distribution rather than an isosceles triangle, then when $R = 1$ there is still a slight overlap of the peaks (approximately 2%) and when the resolution is improved to $R = 1.5$ the contribution of the area of one peak to the area of the next one is reduced to approximately 0.03%, *i.e.*, essentially complete separation. When considering peaks of equal size, this small amount of peak overlap is unimportant. However, if one eluting peak is present only in minor proportions, the contribution in height and area from the overlap of the larger peak may become significant.

It is important not to ignore the column performance parameters when involved in day-to-day chromatographic separations. The significance of the measurements becomes apparent when wishing to reproduce chromatographic conditions to a high degree of precision.

The column efficiency and selectivity characteristics of a freshly packed, or recently received, column should always be tested with a suitable sample mixture under carefully standardised conditions. A record of such a test is invaluable if, at some later date, the performance of the column is in doubt. One simply can repeat the test and compare the results. It is good practice to establish a test procedure for each column type and check them as a matter of routine. It is also advisable to keep a record of the indicated inlet pressure necessary for the flow of a given solvent through the column at a given rate of, say, 1 ml/min. Any marked change in the resistance to flow of the column indicates that material is being built up in the column (either particulate matter or completely retained components of the sample) in which the resistance to flow will increase. Conversely, a marked decrease in the resistance to flow is usually indicative of packing material being lost from the column. These procedures will often pin-point problems before they become sufficiently serious for the column to be no longer serviceable.

GENERAL RESOLUTION EQUATION

Expressions for the resolution, column efficiency and capacity factors are calculated from the widths and retention characteristics of the eluting peaks, *i.e.*, all derived from easily measured parameters taken from the chromatographic trace. These individual expressions of chromatographic performance can be integrated into a single expression which describes resolution in terms of column efficiency (number of theoretical plates), selectivity (nature of chemical interactions related to the phases used), and capacity factors (giving the extent of phase interaction).

The form of this integrated expression is as follows,

$$R = \frac{1}{4}\sqrt{N}\left(\frac{\alpha - 1}{\alpha}\right) \cdot \left(\frac{k'_b}{k'_b + 1}\right)$$

and is referred to as the general resolution equation. Examination of this equation indicates that the resolution is a function of the square root of the column efficiency, thus to improve the resolution between two peaks by efficiency will require a considerable increase.

On the other hand, resolution is directly dependent on the selectivity and capacity of the chromatographic system. (An outline of the derivation of this equation is given in Appendix 1.)

CALCULATION OF OPTIMUM COLUMN LENGTH

Having obtained a general equation for the resolution in terms which are readily measurable, the equation may be used to derive an expression which enables one to calculate the optimum length of a column necessary to obtain a certain (selected) resolution based on one chromatographic analysis carried out previously under non-idealised conditions.

The above equation may be rearranged and squared to give an expression for column efficiency, N, *i.e.*

$$N = 16R^2 \left(\frac{\alpha}{\alpha - 1}\right)^2 \cdot \left(\frac{k'_b + 1}{k'_b}\right)^2$$

Substitution of experimental results in this equation enables the minimum number of plates required for a given separation, hence the optimum column length, to be calculated. This procedure is best illustrated with a worked example.

Referring to Fig. 2.5, a preliminary analysis using a 10-cm-long column gave incomplete resolution. What length of column is required to obtain baseline separation between the two peaks?

On the basis of the result shown in Fig. 2.5, the column efficiency, N, may be calculated.

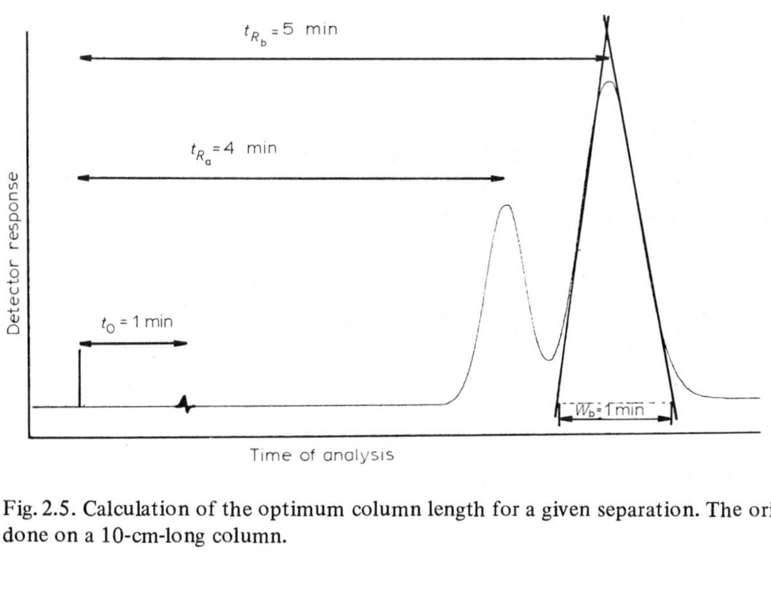

Fig. 2.5. Calculation of the optimum column length for a given separation. The original separation was done on a 10-cm-long column.

Thus

$$N = 16\left(\frac{t_R}{W_b}\right)^2 = 16\left(\frac{5}{1}\right)^2 = 400 \text{ theoretical plates}$$

Since HETP = L/N, the plate height is 0.25 mm.

Similarly from Fig. 2.5, the selectivity factor, α, may be calculated

$$\alpha = \left(\frac{t_{R_b} - t_0}{t_{R_a} - t_0}\right) = \left(\frac{5-1}{4-1}\right) = 4/3 = 1.33$$

and the capacity factor, k'_b, for the last peak

$$k'_b = \left(\frac{t_{R_b} - t_0}{t_0}\right) = \left(\frac{5-1}{1}\right) = 4$$

To obtain just baseline resolution between the two peaks it is necessary for the resolution factor to have a value of 1.5. This value will give approximately 0.03% of overlap for two adjacent Gaussian peaks of similar size.

Substituting these values in the above equation indicates that the minimum number of plates necessary to give this degree of resolution is

$$N_{min} = 16 \circ (1.5)^2 \circ \left(\frac{1.33}{1.33-1}\right)^2 \cdot \frac{4+1}{4} = 915 \text{ theoretical plates}$$

Since the plate height was 0.25 mm, this number of theoretical plates represents a column length of 915 × 0.25 mm = 23 cm.

The use of a 23-cm-long column in place of the 10-cm-long column will give the desired resolution. The most likely choice would be one of 25 cm long, *i.e.*, the nearest standard column dimension.

Where more than adequate resolution is obtained in a separation, it is possible to calculate, in a similar manner, the length of column providing just sufficient resolution. In this way analysis time and inlet pressure requirements can be reduced substantially. This "optimisation" is of most value when designing equipment for quality control applications, since the sample is well defined and there is little chance of much increased resolution being required. In these circumstances the minimum requirements of column materials and, perhaps of greater importance, the pressure capabilities of the instrumentation may be determined. In many instances instrument design dictates a certain unit length of columns, which can be increased by using multiple columns; thus, changing resolution characteristics is more easily achieved by increasing the velocity (flow-rate) of the mobile phase, thus reducing the retention time of the components. This effect is due to the retention volumes having a constant value in a given chromatographic system.

Some chromatographers regard parameters which influence peak shape, such as particle size, column efficiency, velocity of mobile phase, and dead volume, collectively as the "kinetic parameters" of the chromatographic system. Similarly, the features dependent on the chemical nature of the system, *i.e.*, capacity factors, selectivity, partition coefficients, adsorption coefficients, and dissociation constants, can be regarded as the "thermodynamic parameters" of the system.

It is important to appreciate that the simple expression for theoretical plate calculations and column efficiency does not take into account any retention of the sample on the column. Thus it is possible to calculate the efficiency of a column using the width and elution time of a non-retained peak. Such a measurement gives a very good indication of the void spaces and uniformity of column packing and for this reason this quantity is one of the most commonly studied with reference to a change in some property of the system such as column or particle diameter, mobile phase velocity or viscosity, etc. These effects are described in detail in Chapter 3.

In many instances in this text the term liquid or mobile phase velocity will be used rather than the more easily measured mobile phase flow-rate. The reason for this action is that velocity can be directly related to the speed of analysis, whereas the flow-rate depends additionally on the dimensions, particularly the cross-section, of the column and the volume of the column occupied by the packing material and any stationary phase. The linear velocity is determined experimentally by injecting a compound known to be unretained on the column being tested and measuring the time taken for the compound to pass through the column. Knowing the length of the column, the velocity can be calculated.

When seeking to optimize a separation of chemical substances, one has to operate with retained peaks and the selectivity of the phase system becomes important. This was illustrated mathematically earlier. In these instances it is often more interesting to calculate the column performance on the basis of effective theoretical plates. The effective theoretical plate number, N_{eff}, is calculated in a similar manner to the more conventional efficiency except that the retention time of the component is reduced by the void time of the column. Thus

$$\text{whereas } N = 16\left(\frac{t_R}{W_b}\right)^2 \qquad N_{\text{eff}} = 16\left(\frac{t_R - t_0}{W_b}\right)^2$$

In circumstances where the peak(s) being studied have no retention, the number of effective plates will be zero. As the name implies, this term indicates the effectiveness of a column to be able to separate a sample. Since the nature of the phase system in the column also governs the separating ability, *i.e.*, as expressed in terms of the selectivity factor, it is possible to calculate the number of effective plates required to yield a desired resolution between peaks given a certain selectivity factor for the phase–sample system.

Substitution of the expression for effective theoretical plates in the equation describing resolution in terms of selectivity, efficiency and capacity (p. 14) yields the following relationship between resolution of peaks, selectivity and effective plates

$$R = \frac{1}{4}\sqrt{N_{\text{eff}}}\left(\frac{\alpha - 1}{\alpha}\right)$$

Using this equation, the number of effective plates required to achieve a desired resolution between two adjacent Gaussian peaks of approximately the same size may be calculated. The values given in Table 2.1 correspond to the effective plates required to give baseline resolution between Gaussian peaks, *i.e.* a resolution factor equal to 1.5 (as defined earlier).

TABLE 2.1

NUMBER OF EFFECTIVE PLATES NEEDED TO GIVE BASELINE RESOLUTION BETWEEN TWO ADJACENT GAUSSIAN PEAKS AS A FUNCTION OF COLUMN SELECTIVITY

Selectivity, α	No. of effective plates
1.00	∞
1.01	367,236
1.05	15,876
1.10	4,356
1.15	2,116
1.20	1,296
1.50	324
2.00	144

These figures emphasize that to achieve a satisfactory separation both the selectivity (*i.e.*, the thermodynamic factors) and the efficiency must be considered simultaneously. A phase system offering a selectivity of unity will, clearly, be incapable of providing a separation no matter how efficient the column may be. Even when the selectivity between the two peaks is 1.01 the number of effective plates is beyond that offered by any currently available system, particularly when it is remembered that the number of effective plates exhibited by a column is always less than that of actual theoretical plates. In these circumstances it would almost certainly be a simpler matter to change the phase system to improve the selectivity of the system, *i.e.*, altering the chemical nature of the mobile or the stationary phase.

It can clearly be seen from the table that if a highly selective phase system is employed, columns with low or modest efficiencies will still give good results. Perhaps the extreme case can be considered as a separation achieved by simply distributing the sample between two immiscible liquids in a separatory funnel.

Before leaving the subject of ways of expressing column performance, there is one further method by which the column efficiencies may be calculated. This is in terms of the reduced plate height, which is obtained by dividing the actual plate height by the mean particle diameter of the column packing material. This produces a dimensionless number for the reduced plate height.

This treatment has been developed by Knox and coworkers to describe and compare the efficiency characteristics of columns differing in overall size and also in the nature of the packing material.

Column performance is very dependent on the velocity of the mobile phase passing through the column – a feature which is dealt with in detail in the next chapter. By plotting graphs of the reduced plate heights against the reduced mobile phase velocity (calculated by taking diffusion and viscosity into account) the performance of columns of different design may be compared.

The theoretical treatment and reasoning behind this method is beyond the scope of this book. Interested readers are recommended to refer to the publications and work of J.H. Knox (*e.g.*, ref. 1).

REFERENCE

1 G.J. Kennedy and J.H. Knox, *J. Chromatogr. Sci.*, 10 (1972) 549.

Chapter 3

The chromatographic support and column

INTRODUCTION

Of all the factors contributing to the advances in the practice of LC in recent years, the characterization of the influence of the chromatographic support and the subsequent development of specialised materials must be regarded as the most important.

LC has traditionally been a slow technique, offering only a limited separating power. Attempts to increase the speed of analysis by increasing the velocity of liquid passing through a column proved unsatisfactory as the efficiency and hence resolving power were found to decrease rapidly as the liquid velocity increased. Following an increased understanding of the factors responsible for this phenomenon, modern support materials have been designed to provide, in ideal circumstances, high column efficiencies and their performance is much less dependent on mobile phase velocity. This can lead to a realization of high-speed liquid phase separations which compete with GC in terms of analysis times and resolving power.

In this chapter it will be seen that no one design of chromatographic support offers all the advantages without any disadvantages, so that selection of a support depends a great deal on the application of the technique.

Classical column chromatography invariably relies on a flow-rate of mobile phase, generated by the influence of gravity, through a column bed which contains a chromatographic packing having particles in the size range of 60–120 U.S. mesh (250–125 μm). A table for converting either A.S.T.M. or B.S.S. sieve sizes to micrometres is given in Appendix 2.

The separating power of columns operated in this mode has traditionally been limited since to ensure a liquid flow under gravity the diameter of the particles has to be relatively large. As efficiencies per unit length of these columns were low, *i.e.*, they had large HETP values, it was often necessary to employ long columns. Under these conditions the overall time taken to complete a separation was frequently measured in hours, with a consumption of considerable quantities of solvents and sample material.

Attempts to improve the speed of a separation by increasing the head pressure and thus accelerating the liquid flow resulted in a rapid decrease in the already low column efficiency. Not surprisingly, under these circumstances LC did not rate as an attractive technique and was often neglected in favour of TLC and GC, which offer higher speed, higher resolution, and whose sample requirements are low.

The dependence of the efficiency of a typical classical column, expressed as HETP, on the mean linear velocity of the mobile phase is shown in Fig.3.1.

Much of the understanding of LC has been illucidated using the reasoning previously developed for the theoretical treatment of GC. It has been found that both systems can be described by qualitatively similar processes, but the quantitative influence of each of these terms varies considerably in the gas and liquid phases.

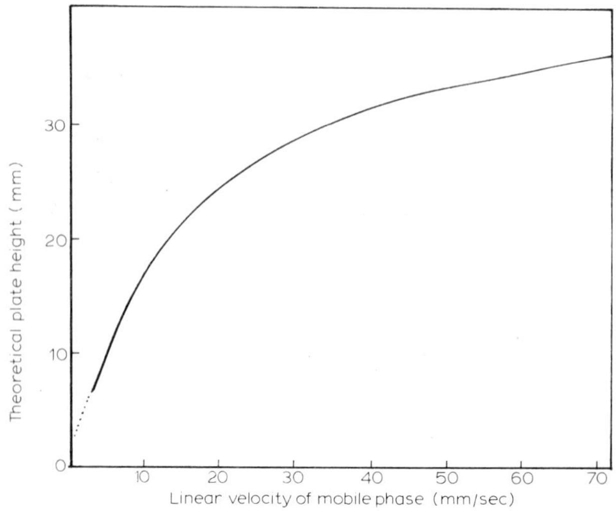

Fig. 3.1. Typical curve of efficiency *vs.* carrier velocity for a classical LC column. The data are for a porous packing having a mean diameter of 150 μm.

SOURCES OF BAND BROADENING

The general effect of a sample band spreading to occupy a larger volume during its passage through the chromatographic system was indicated in the last chapter. This spreading of the sample will result in a widening of the peak observed on a chromatographic trace. The recorded peak, however, indicates the total dispersion of a sample during its passage through the apparatus. It is important to distinguish between dispersion of the peak which takes place within the column, due largely to the nature of the column packing material, and dispersion or mixing which can occur before or after the column, in places such as the injector, the interconnecting tubing, and the detector. This extra-column band broadening becomes progressively more important as high efficiency is demanded from the equipment and when high-performance columns are used it can become the limiting feature if insufficient attention has been paid to the design of these parts. These latter aspects are discussed in detail in the chapters describing the instrumental requirements of HPLC. It suffices at this stage to point out that not all band broadening occurs within the column.

It is generally accepted that there are four principal sources of band broadening which may occur in a chromatographic system. These are known as: (1) Eddy diffusion; (2) longitudinal diffusion; (3) mass transfer of sample between the phases; (4) extra-column diffusion. Each of these terms contribute to the band broadening, thus the overall HETP can be considered as the sum of the individual "inefficiencies", thus

$$\mathrm{HETP}_{\mathrm{total}} = H_{\mathrm{eddy\ diffusion}} + H_{\mathrm{longitudinal\ diffusion}} + H_{\mathrm{mass\ transfer}} + H_{\mathrm{extra\ column}}$$

Depending on the operating conditions one or several of these factors will dominate.

Eddy diffusion

This term relates to the flow paths of unequal length that must exist through any, less than perfect, packed column. Some sample molecules will find themselves swept through the column close to the column wall where the density of packing is comparatively low, while others will pass through the more tightly packed centre of the column bed at a correspondingly lower velocity. In consequence, molecules following an easy path will elute ahead of those following a more difficult route, leading to a broadening of the eluting sample band (Fig.3.2).

This effect in a packed column is in direct contrast to the flow profile that would be expected in an unpacked tube. In this latter situation, there would be a streamlined flow profile across the column such that the liquid furthest from the walls would travel at the highest velocity.

A state of laminar flow exists in the chromatographic column under normal operating conditions. Turbulent flow, which would greatly improve lateral mixing in the column, has been calculated to require a liquid velocity in the order of a thousand times faster than those currently employed[1]. It is conceivable that this approach may be investigated at some future date.

These flow path inequalities are dependent largely on the uniformity of column packing and the diameter of the packing material used. To minimize this effect the mean particle diameter of the packing should be as small as possible consistent with obtaining a uniformly packed bed. This contribution to band broadening is essentially independent of mobile phase velocity and hence is a constant contribution to the overall plate height of a column.

The magnitude of eddy diffusion is controllable to some extent by the method used to

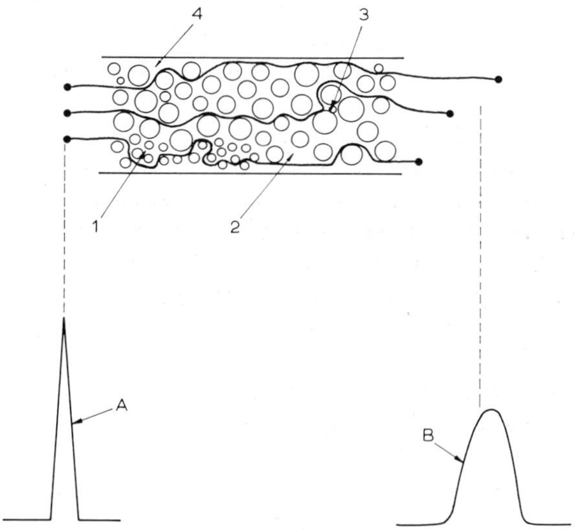

Fig. 3.2. Sample band broadening due to eddy diffusion. (A) Initial concentration profile; (B) final concentration profile. (1) Fine particles; (2) coarse particles; (3) agglomerated particles; (4) low density of packing near column wall.

pack the column. A novice will often experience difficulty in obtaining a homogeneous column bed. With experience or the use of a well designed packing machine, a more uniform column may be obtained.

Longitudinal diffusion

In GC this term has proved to be of considerable significance, and relates to the dispersion of a sample band under the influence of molecular diffusion (*i.e.*, random molecular motion, very much like Brownian movement). The high diffusion rates in the gas phase cause sample bands to disperse longitudinally along the column, particularly at low mobile phase (gas) velocities, leading to peak broadening, hence inefficiencies. In principle, the same effect is possible in the liquid phase and this would become important at very low mobile phase velocities, leading to a decrease in column efficiency. In practice, due to the fact that diffusion in the liquid phase is about 10^5 slower than in the gas phase, this effect is rarely observed as the magnitude of the mobile phase velocity where this occurs is far below the practical working range. Analysis carried out at velocities where this term is important would take an excessive time unless very short columns, *i.e.*, 1–5 cm long, were being employed. For most practical purposes the longitudinal diffusion term may be ignored in all work except where very low flow velocities are being employed.

Mass transfer

If a sample is to be retained on a column packing material, then while the sample is passing through the column there must be some interaction between the packing material and the sample. This interaction may be an adsorption of the sample on or a partition into the column packing, followed at the next moment, when fresh mobile phase is in contact with the packing, by desorption (or repartition) of the sample molecules, after which they once again return to the mobile phase. Such exchange interactions occur repeatedly with all sample molecules during their passage through the column. As the liquid (mobile) phase is moving relative to the column packing material, molecules of sample which at one instant happen to be in the stationary phase "see" fresh mobile phase and vice versa. If one assumes that equilibration of this transfer of sample is not instantaneous, then that portion of the sample in the mobile phase is always ahead of that portion in the stationary phase at any one instant. The faster the mobile phase is moving through the column and the slower the rate of equilibration of sample molecules between the stationary and the mobile phase, the wider will be the sample band which eventually elutes from the column. As one might imagine, the contribution of the mass transfer term to the overall plate height increases with the velocity of the mobile phase. It is also dependent on the thickness and the viscosity of the stationary phase layer. A thin layer of stationary phase of fairly low viscosity will allow the most rapid transfer of the sample.

The chromatographer has some control over liquid phase mass transfer by the choice of the solvent used as mobile phase, *i.e.*, he should use one with a low viscosity. It is also possible in some cases to reduce stationary phase mass transfer by operating at elevated temperature.

Fig. 3.3 illustrates the contribution to the overall plate height by the eddy diffusion,

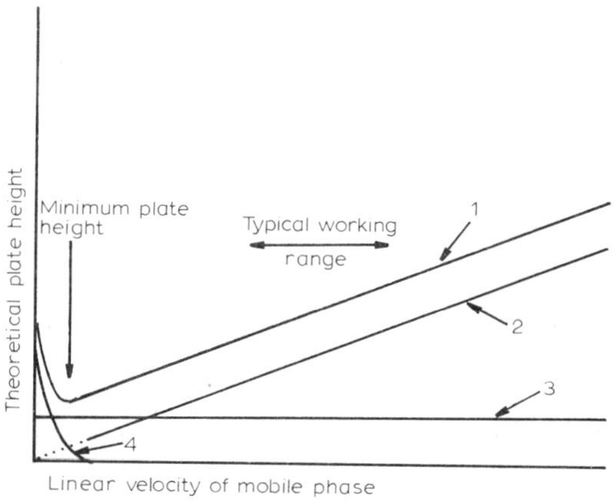

Fig. 3.3. Contributions to the overall plate height. (1) H_{total}; (2) $H_{mass\ transfer}$; (3) $H_{eddy\ diffusion}$; (4) $H_{longitudinal\ diffusion}$.

longitudinal diffusion and mass transfer terms individually and when combined. In the latter case a curve is produced of similar outline to that obtained experimentally. In practice, however, the complex flow characteristics of the mobile phase at high velocity tend, if anything, to reduce the slope of the HETP *versus* velocity curve. It is considered that this phenomenon is due to an interaction of the eddy diffusion and mass transfer effects.

In a packed LC column there is another phenomenon which may be regarded as a mass transfer characteristic originating from the slow diffusion rates in the mobile phase. In most column packing materials there exists some form of internal pore structure, traditional column packings being almost exclusively totally porous in their nature. When mobile phase is pumped through the column, these pores within the packing become filled with mobile phase. Due to the slow rate of diffusion this mobile phase tends to stagnate in the pores. When subsequently a sample is passed through the column, some molecules diffuse into these pores and their exit from the pores is retarded by their very slow movement in the mobile phase. The net result is that the molecules are held back relative to the main band of sample thus giving rise to peak broadening. In this instance the slow rate of mass transfer responsible for the broadening is "partition" between "moving" mobile phase and "stationary" mobile phase.

The concept of "stagnant pools" of mobile phase being trapped within chromatographic packings is one of the most useful when attempting to explain the characteristics and developments in LC column technology. To overcome inefficiencies produced by the mobile phase mass transfer phenomenon it is necessary to minimize the pores or sites where mobile phase is able to stagnate. In the following sections, it will become apparent that this effect can be minimized by either making the internal pore structure impervious, reducing the overall diameter of the column packing material or preparing supports with very wide pores so that liquid can flow easily in and out or even through the particles. The ultimate aim in the development is to achieve a high inherent efficiency, *i.e.*, low HETP value, which remains essentially unchanged by the mobile phase velocity. In Fig. 3.3 such a performance might be indicated by a straight-line plot of HETP *versus* velocity parallel with, and close to, the horizontal axis. Having achieved such a performance it would be reasonable to suppose the velocity of the mobile phase could be increased indefinitely to achieve faster and faster analyses. Understandably there is a limit to this supposition, usually measured in terms of the capabilities of the chromatograph being used. These limitations will become apparent in forthcoming paragraphs and in the chapters dealing with chromatographic instrumentation.

In practice, various phenomena are responsible for band broadening and a combination of these factors indicates that the minimum plate height, *i.e.*, maximum efficiency, will be found at very low mobile phase velocity. This velocity is, unfortunately, too low for most practical purposes, except when using very short columns, *i.e.*, less than 5 cm long, and it is common practice to make use of the decrease in the slope of the HETP *versus* velocity curve that occurs at higher mobile phase velocities and to accept some decrease in column efficiency in return for a substantially reduced analysis time.

Let us return now to the design of chromatographic support materials necessary to minimise band broadening. The effects described earlier indicated that:

(1) Eddy diffusion can be minimised by reducing the diameter of the support consistent with maintaining a uniform packing structure.

(2) Longitudinal diffusion is essentially eliminated at high mobile phase velocity, thus is of little consequence in high-speed LC.

(3) Mass transfer, although made worse by increasing the mobile phase velocity, can be minimised in the mobile phase by reducing the diameter of the support and/or eliminating long, narrow pores within the particles. In the stationary phase, the mass transfer is minimised by using, where possible, phases of low viscosity, thinly coated on the support material.

From these conclusions it is easy to understand why in recent years so much effort has been applied to the study of columns packed with very small particles. These developments are summarised in the following paragraphs.

ROLE OF PARTICLE SIZE IN LC COLUMNS

It was noted earlier and shown in Fig. 3.1 how the efficiency of a classical LC column, *i.e.*, diameter of support particles in the size range 125–177 μm, deteriorated as the velocity of the mobile phase was increased. Based on the conclusions on the nature of the effects giving rise to band broadening much effort has been devoted to the study of the chromatographic characteristics of columns packed with smaller particles of support. Results of many independent studies have confirmed that in general more efficient columns, the performance of which is less dependent on mobile phase velocity, could be achieved with finer packings. An illustration of this improvement in performance is given in Fig. 3.4.

This figure can be considered representative of the improvement in performance typically achieved with irregular-shaped, totally porous materials such as diatomaceous earths and silica gels, *i.e.*, simply by using finer grades of the classical support materials.

Although in the early 1970's many independent studies have confirmed this trend,

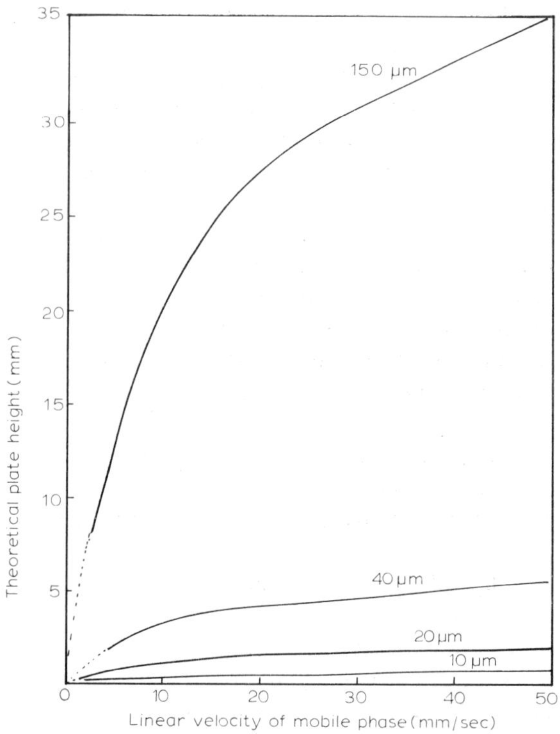

Fig. 3.4. Influence of particle diameter of column packings on efficiency.

when the diameter of the support particles used was decreased to a value in the region of 50 μm and below there appeared to be a disparity in the results, some confirming a continued improvement of performance with decreasing particle size, while others reported an optimum below which efficiency started to decrease. This apparent inconsistency of results has subsequently been rationalised in that the dry packing methods for preparing columns which were acceptable for coarse particles were not adequate for the efficient packing of columns with fine-grained particles. It is now generally accepted that as the particle size is reduced, the chances of agglomeration of the particles by static charges are increased, leading to a less dense packing structure, which gives rise to voids or dead volume within the column bed. This results in a lower than expected column performance.

The point where any particular packing method no longer produces acceptable columns depends considerably on the nature of the material being loaded into the column for use as the chromatographic support. The literature contains numerous accounts of methods for packing columns with various types of chromatographic supports. Some methods work best with spherical particles and others with irregularly shaped particles. Unfortunately many appear to give poor reproducibility, particularly from operator to operator.

A very definite improvement in the performance of columns packed with very small particles was achieved by the development of "wet" methods of packing columns. Although wet (slurrying) methods have been used for organic support materials for a long time, *i. e.*, ion-exchange resins and porous polymers for steric exclusion work, the method was generally found unacceptable for packing columns with coarse inorganic materials, such as silica gel. However, re-examination of the method showed that it held advantage over dry methods for the packing of very fine material, *i.e.*, particles less than 20 μm. Although, based on reduced plate height studies of Kirkland[2], there is reason to believe that the methods are still not perfect, they are the best available at the present time. The broken lines added to Fig. 3.4 close to the horizontal axis represent the typical efficiency characteristics reported for supports of approximately 13 and 6 μm diameter. The very significant improvement in column performance with small particles reflects the improvements in the technique of packing and additionally in the methods currently available for classifying heterogeneous materials into fractions having in themselves a very narrow particle size distribution. Methods of packing columns are detailed in later sections of this chapter.

It will be appreciated that the gain in performance possible by using finer support particles has to be paid for in terms of the pressure required to achieve a certain liquid velocity through a column of given length. The resistance to flow increases exponentially as the particle size decreases. Putting this statement into practical terms, if a column is to be operated at very low velocity, for example, at a velocity of 1 mm/sec, then the pressure required to achieve this liquid flow is minimal, *i.e.*, less than 1 bar (15 p.s.i.g.) for a column packed with large particles (100 μm) even for a column of 500 mm in length.

This combination is actually the arrangement used in classical column chromatography. For a reduction of the diameter of the support materials in such a column to 10 μm an inlet pressure of approximately 11 bars (160 p.s.i.g.) would be required for the low velocity of 1 mm/sec. With a column packed with 5-μm particles the pressure requirement for the same mobile phase velocity would be approximately 110 bars (1600 p.s.i.g.). Precise values are dependent on the viscosity of the mobile phase and on the porosity of the support. The values quoted are derived from data reported by Majors[3] and are presented to give an indication of the magnitude of the pressure requirements as the particle size is decreased. The figures given above relate to the inlet pressure required to achieve a low flow velocity through the column, *i.e.*, 1 mm/sec. This value means that the void time of a 500-mm-long column will be 500 sec. Therefore, the earliest peak to elute, a non-retained peak, would take over 8 min to reach the detector.

Earlier in this chapter it was mentioned that in practice the speed of analysis was often increased by raising the mobile phase velocity and sacrificing some column efficiency. Currently, a practical velocity which may be considered typical is 10 mm/sec, although, as indicated in Fig. 3.4, higher velocities could be employed without significant loss of efficiency. Even so the pressure requirements to yield a velocity of 10 mm/sec through the columns mentioned earlier would be in the region of 110 and 1100 bars (1,600 and 16,000 p.s.i.g.) for the 10- and 5-μm-diameter supports, respectively. From these values it can readily be appreciated that if high-speed analyses are to be attempted with 500-mm-long columns packed with 5-μm-diameter support material of this type, then exceedingly high operating pressures, *i.e.*, greater than 1030 bars (1.5×10^4 p.s.i.g.) would be necessary. Currently, it is the practice to use much shorter columns, *i.e.*, 50–250 mm in length packed with these fine materials. This choice reduces the inlet pressure requirements for a given velocity and the overall void time, essentially in proportion to the reduction in column length. At the same time, of course, the overall number of theoretical plates available from the column drops similarly. However, the high efficiency per unit length (low HETP value) of columns packed with 5-μm support particles can be high enough for a short column to still provide adequate effective plates for the separation of many sample mixtures.

POROUS LAYER SUPPORTS

So far the effect of particle size has been described for columns filled with supports differing from the classical types only in the diameter of the supports and in the method of packing the column. Following the realisation of the deleterious influence of slow mass transfer on column performance, notably at high mobile phase velocities, there have been many attempts to minimise the problem by designing synthetic supports for optimum mass transfer. These studies have led to a number of very successful chromatographic supports which offer practical improvements such as ease of column packing and low inlet pressures yet still offering high-speed analyses.

Perhaps the most significant improvement in support design was the introduction of the material known by such names as porous layer, pellicular or controlled surface porosity supports. Although differing technically in their design and method of manufacture, these materials share the common feature that the chromatographic support is based on an impervious sphere, usually glass, on the surface of which is the active chromatographic layer formed as a crust of approximately 1–2 μm thickness. The aim with this design of materials is to restrict the depth of pores into which the mobile phase and the sample molecules flow, thereby reducing the stagnant pools of mobile phase described earlier, which leads to a very significant reduction in the inefficiencies originating from the mobile phase mass transfer limitations. Their HETP *versus* velocity profiles accordingly compare well with those of totally porous material of much smaller particle size. Depending on the manufacturer, these supports are prepared with an overall bead diameter in the size range 20–50 μm. Done and Knox[4] and Kirkland[5] have reported in-depth studies on the performance of Zipax, a commercially available controlled surface porosity support (DuPont), using fractions of various mean particle diameters, within the range of 20–106 μm. This type of chromatographic support possesses free flowing, quicksand-like, properties enabling a very dense bed of packing to be built up by straightforward dry packing techniques; the resultant columns offer high efficiencies.

The larger diameter of the bead also leads to less resistance to flow in the column; hence, a lower inlet pressure is needed to achieve a given mobile phase velocity compared with that required when using very fine supports.

The sustained efficiency at high liquid velocity and the ease of use of these materials was probably largely responsible for the revival of interest in LC in recent years. These porous layer types of support suffer from a common limitation in that the surface available for interaction with sample, or on which to apply stationary phase, is low, hence the sample capacity of the support is limited. This restriction is of little consequence when dealing with analytical-scale separations using very sensitive detection systems, but can produce problems if large samples are required for subsequent collection, to offset detector sensitivity limitations, or where trace impurities are to be determined. For this latter application it is necessary to introduce a large quantity of sample in order to obtain a detectable amount of the impurity component.

The HETP *versus* mobile phase velocity profile of these materials varies considerably with the nature of the surface layer. It would appear that the most rapid mass transfer occurs when the surface layer contains wide pores rather than narrow pores. Pictorially it can be imagined that the surface needs to have an open texture allowing free access and exit of molecules to and from all the regions of the layered surface. In this respect the work of Kennedy and Knox[6] has shown that the performance of controlled surface porosity supports, where the surface is built up of multilayers of even finer beads of say 200-nm diameter, offers a mass transfer superior to that of materials where the surface layer is formed of silica gel. This latter material contains a range of pore sizes including some which are quite narrow. These narrow pores tend to trap mobile phase, leading to "stagnant pools of mobile phase".

Superficially porous supports of the general type described are typified by chromatographic packing materials available under the trade names Corasil (Waters), Perisorb (Merck), and Zipax (DuPont). Specific details of commercially available packings are given in chapters devoted to separation methods, *i.e.*, adsorption, ion-exchange, etc.

TOTALLY POROUS (MICROPARTICULATE) SUPPORTS

A second type of support which has been designed for optimum performance is the totally porous, spherical packing, where the dimensions of the internal pores are controlled during the manufacture. A packing with pores of very large diameter will allow mobile phase to permeate freely through the column and, in the case of packings with a diameter less than 10 μm, this can lead to a significant reduction in the inlet pressure required to produce a desired velocity of mobile phase through a column. Depending on the size of the pores within the support material, it is possible to achieve a situation where only molecules below a certain size can enter the support, whereas other, larger, molecules cannot enter the pores and are said to be excluded. Such large molecules are only able to move through the column via the inter-particle spaces. It should be apparent that the exclusion phenomenon depends on the combination of the diameter of the pores and the "size" of the molecules passing through the column. By tailoring the support material to give a range of pore sizes it is possible to achieve an exclusion range, the largest pores allowing both large and small molecules to enter the support whereas the smaller pores allow only the small molecules to enter. The difference in permeability of a column packing towards molecules of different sizes forms the basic concept of separations performed by steric exclusion chromatography (SEC), an important LC method for characterising samples of high molecular weight or those in which the molecular weights of the individual components differ widely. The method is described in detail in a later chapter. At this stage it suffices to be aware of the phenomenon, remembering that the chances of a molecule entering a pore depend on its "size" as "seen" by the chromatographic support. This "size" will be a function of the molecular weight of the sample, its shape, and the degree of solvation occurring in the mobile phase.

In producing supports with rapid mass transfer characteristics for techniques other than

SEC, it is important that the pore sizes are large enough not to impede the passage of molecules of mobile phase or sample through the column. Although, as mentioned earlier, this will depend on the molecular size of the compounds being studied, assuming these are generally less than 2000 amu (this is the range in which LC methods are most successful, excepting SEC, which is the method of greatest value above 2000 amu) then it is considered that only pores smaller than approximately 40 Å will restrict the movement of these molecules.

In addition to the diameter of the internal pores, it was described earlier that for best mass transfer the depth of pore should be as shallow as possible. Since only totally porous supports are being considered here, the pore depth can only be reduced by diminishing the overall particle diameter. Practical approaches to the achievement of this goal have been to make a series of porous silica or glass supports offering different mean pore diameters. Products of this type are available commercially under such trade names as: Controlled Porosity Glass (CPG) (Electronucleonics), Porasil (Waters), and Spherosil (Rhone-Progil). Full details of the available products are given in chapters dealing with the separation methods. These products are generally of spherical form for the particles of larger diameter, but smaller size ranges, when offered, are produced as irregularly shaped materials, which might prove more difficult to pack into a homogeneous bed. Specific methods of preparation of these materials tend to be proprietry information, however, it is believed they are produced by the selective leaching of heterogeneous glasses – the pores are created when a more easily attacked region of the bead is dissolved.

More recently a different method of preparing small-diameter, totally porous supports has been described. This method relies on the agglomeration of extremely small (50 Å) particles in a controlled manner which yields spherical particles of very narrow size distribution. The range of pore dimensions may be controlled during the preparation. These porous microspheres may be produced in the 5-μm size range and offer very high efficiencies in a manner analogous to that of the 5-μm materials described earlier, but with the advantage that larger pores can be incorporated leading to even better mass transfer and a higher column permeability, *i.e.*, a lower resistance to liquid flow through the column bed, which enables high velocity of mobile phase to be achieved with a significantly less inlet pressure. These materials have been developed and described by Kirkland[7,8]. From his data it is possible to derive an idea of the pressure requirements of these porous microspheres compared with the finely ground silica gel types of support given earlier (p.26). A 500-mm-long column packed with porous silica microspheres is estimated to require an inlet pressure in the order of 40 bars (580 p.s.i.g.) for a linear velocity of 1.0 mm/sec. The pressure required for 10 mm/sec mobile phase velocity would be in the order of ten times higher than this value. Microspheres of silica, similar to those described by Kirkland, are available commercially under the trade name Zorbax (DuPont). Support materials which, from the limited data available, might be expected to perform in a similar manner have been developed by the United Kingdom Atomic Energy Authority and are available under the trade name Spherisorb (Phase Separations).

Apart from the gain in efficiency which is achieved when using a column packed with very fine, totally porous supports, the most significant advance is the increase in sample capacity, which is in the order of 1 mg of sample per gram of support. This value is an approximately tenfold increase over that when using the superficially porous packings, permitting larger sample sizes to be separated, leading to improved detection of minor components, and giving the possibility of using less sensitive detection methods and a chance to collect separated components in worthwhile quantities for examination by alternative techniques.

DEPENDENCE OF COLUMN EFFICIENCY ON OPERATIONAL CONDITIONS

When calculating HETP values derived from a chromatographic trace containing a number of peaks having different capacity factors, it is sometimes observed that the efficiency is dependent on the capacity factor and yet another column may give an efficiency value which is relatively constant and thus independent of the capacity factors of the peaks. Whichever situation arises depends largely on which of the effects contributing to the mass transfer term is dominant, *i.e.*, whether the rate determining step is diffusion in the stationary phase or in the mobile phase, or mass transfer to and from stagnant pools of mobile phase[8]. An apparent low efficiency of a chromatographic column as measured on peaks with low capacity factors, *e.g.*, k' less than unity, is often indicative of extra-column band broadening due principally to dead volume in the injection and detection systems.

The efficiency of all chromatographic columns is dependent on the mobile phase velocity, thus to place these various columns into some relative order of merit it is useful to extend some of the definitions described in the previous chapter so that the time or speed element can be included. One of the most widely accepted methods of achieving this is to compare columns by the maximum number of effective plates that are generated per second, N_{eff}/sec. Since the resolving power of a chromatographic system is directly related to the number of effective plates and the selectivity of the phase system (see p.17), the term N_{eff}/sec gives a positive indication of the high-speed separating capabilities of the system. It is often observed that the numerical value of N_{eff}/sec differs with the capacity factor, k', of the peak used for the calculation. The in-depth theoretical reasoning behind this effect is considered beyond the scope of this book, but the overall conclusion from the theory and practice is that the maximum value of N_{eff}/sec for a particular system is given by a peak having a capacity factor in the range 2–3. Although, of course, it is not possible to achieve a separation where all the components being analysed have the same capacity factor, optimum performance in the terms described will be obtained when the component peaks elute in the region of $k' = 1–10$ (ref.9). The stationary phase/mobile phase combination should be adjusted so that the maximum number of components of the sample elute in this region.

On this basis, it is of interest to compare the various types of materials that have been proposed for use as supports in modern LC in terms of their maximum observed value of N_{eff}/sec. These values, given in Table 3.1, are taken from the scientific literature and serve as an indication of the relative performance of the materials. Because it has not been possible to obtain all data taken at one value of the capacity factor, *i.e.*, $k' = 2.0$, little significance can be attached to small differences in the value of N_{eff}/sec.

From these data the reason for the current practice to use either superficially porous supports or particles of less than 10 μm diameter is quite apparent. It is also of interest to

TABLE 3.1

COMPARISON OF THE PERFORMANCE OF DIFFERENT LC PACKINGS

Column type	Mean particle diameter (μm)	Max. N_{eff}/sec	Reference
Classically packed	150	0.02	10
Closely sized silica gel	20	2	3
Superficially porous beads (Zipax)	27	10	5
As above – infinite diameter*	27	16	11
High-performance silica gel	5–10	10	3
High-performance silica gel	5	23	3
High-performance silica gel	5 (in drilled tubes)	100	16
Porous silica microspheres	4.6–5.6	36	8

*The term "infinite diameter column" is described later in this chapter.

compare these values with those obtained by other related techniques, notably TLC and GC. Snyder has estimated that for a TLC separation, a value of 0.05 effective plates per second could be considered realistic, which when compared with a value of 0.02 for classical column chromatography explains the earlier held view that TLC was faster than LC. The data given in the above table clearly show how the development in column packing technology has considerably changed this situation. In GC, classically packed columns offer typically ten effective plates per second and this value can be improved by using capillary columns packed with particles of 10 μm diameter to give approximately forty effective plates per second. It can be seen that the most recent developments in LC supports and column packing techniques have overcome the earlier criticisms that LC was a very slow technique relative to GC.

Column dimensions and geometry have a pronounced effect on the performance which is achieved with any given support material as also has quality of the surface on the inner wall of the column.

Many papers have been published which attempted to correlate good chromatographic efficiency with column size and also with the ratio of the particle diameter to the internal diameter of the column. Many apparent contradictions occur in the literature which are difficult to rationalize. For simplicity, this text will outline results and conclusions taken from a series of independent papers which appear to complement each other so as to present a reasonably consistent picture of the situation.

COLUMNS FOR HIGH-PRESSURE LC

Currently, column sizes employed in LC range in length from about 50 mm to 1.2 m and in diameter from 1 to 25 mm. Perhaps a notable exception is the Varian LCS-1000 nucleotide analyser, which uses a 3-m × 1.0-mm-I.D. coiled column. When lengths of columns greater than these are required, it is common practice to couple two or more columns in series, using low-volume capillary connectors. Various designs have been proposed for column connectors. The one illustrated in Fig.3.5 can readily be formed from two precision reducing union tube fittings and a short length of 0.25-mm-I.D.

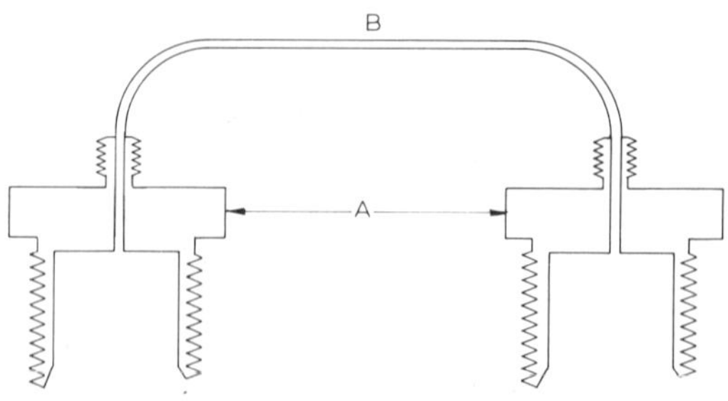

Fig. 3.5. Construction of a low-dead-volume coupling for connecting two columns. Nuts, ferrules and columns have been omitted for clarity sake. (A) Reducing unions (drilled out); (B) capillary tubing (0.25 mm I.D.).

capillary. For the lowest dead volume it is necessary to machine away the inner shoulders of the reducing union as described in Chapter 4 (see Fig. 4.1).

Columns having internal diameters in the range 1–5 mm are used for analytical separations, whereas the larger sizes tend to be used for either steric exclusion chromatography or preparative separations. The development of packing techniques for supports of very small diameter (5–10 μm) has resulted in columns of such high efficiency that short lengths, *i.e.* 100–250 mm, of column are adequate for many separations.

The use of straight columns is almost universally accepted as the best method of attaining the highest column efficiency. Reports of the use of columns which are coiled or formed into other configurations[12] without significant loss of efficiency tend to be restricted to the examination of columns which are not of high performance by today's standard. In other words, if the chromatographic support and packing technique are not capable of giving a high-performance system, then the shape of the column is of little consequence. The same may be said about the nature or quality of the inner wall of the column. The best results which have been reported to date have been obtained using precision bore tubing of stainless steel[13], glass[14], or tantalum[15]. An alternative method of producing a pore-free inner surface has been demonstrated by Asshauer and Halász[16], who employed a drilled tube as a chromatographic column. Tubing used for making columns should be free from roughness and any microporous surface structure on the inner wall. Pores in the column wall will create inefficiencies due to slow mass transfer in the mobile phase in much the same way as fine pores will do in a support material. Fine longitudinal scratches can also lead to poor performance by providing an easy flow path for the mobile phase.

COLUMN EFFICIENCY AND INTERNAL DIAMETER

Following the development of reliable methods of packing columns with particles of small diameter, it has become apparent that the efficiency of a column does vary with the column diameter, higher efficiencies being obtained with the wider-bore columns.

Wolf[17] has reported that columns of 2.1, 7.7 and 23.6 mm I.D. packed with identical chromatographic materials gave efficiencies of 600, 1325 and 2350 theoretical plates per

50 cm length, respectively, when tested under comparable conditions of mobile phase velocity. These data indicate an almost fourfold improvement in efficiency by using the largest diameter column. In these columns the packing material was retained in the column by porous metal frits fitted at either end and the sample was introduced immediately upstream of the column inlet. As well as retaining the packing material in the column, this frit also had the effect of dispersing the plug of sample uniformly across the head of the column. Although perhaps an over-simplification, the gain in efficiency in large-diameter columns in this case can be considered to be due to the decreased deleterious influence of the non-uniform column packing in the vicinity of the column wall. Adverse wall effects are well established in all branches of LC; these arise from the non-uniformity of the packing, as mentioned above, or in some instances where there is an interaction between the sample and the column wall, *i.e.*, adsorption.

An alternative technique of sample introduction to the one described above is to inject the sample directly into the column packing at the inlet of the column. Based on experience gained in GC, many feel this technique should be the most satisfactory for LC. Ideally, if the sample is injected centrally on to the packing material, it will immediately begin to move through the column under the influence of the mobile phase. Trans-column sample mobility (*i.e.*, from the centre to the wall of the column) will be governed by diffusion in the liquid phase, which as mentioned above is very slow, approximately 10^5 times slower than in the gas phase. In this situation as the sample band passes through the column it expands laterally until it reaches the column wall, thereafter continuing through the column in much the same way as if the sample was initially diffused across the top of the column by means of a porous metal frit, as described above, or by packing the first few millimetres of the column with inert beads such as ballotini beads[18].

In some situations with an appropriate geometry of column it is possible to achieve a situation where the sample will travel to the detector end of the column before it reaches the column wall. Under these circumstances the sample never experiences the less uniform region of the packed bed close to the column wall. Under ideal conditions a high column efficiency can be obtained. This method of performing LC has been described as the "infinite diameter method", since the sample should never reach the wall of the column. It should be apparent that this effect depends on the mean particle diameter of the column packing material and on the geometry of the column, a short, wide column being the obvious choice. However, if small-diameter supports are employed, the infinite diameter effect can be achieved in quite narrow columns. Knox and Parcher[19], for instance, have calculated that a column of 5 mm I.D. and less than 330 mm in length, packed with particles of 30-μm diameter, should exhibit an infinite diameter effect and the sample should never reach the non-uniform region of packing near the column wall.

If the column and packing geometry are such that the sample does reach the region of the column wall, then the diameter has a definite influence on the overall efficiency. It has been reported by De Stefano and Beachell[20] that when using columns of 500 mm length infinite diameter characteristics were observed if the internal diameter of the column was 7.9 mm or greater leading to the highest efficiency characteristics for the less than 37-μm-diameter, superficially porous beads used in their study. Narrower columns, having internal diameters in the range 4.76–6.3 mm, yielded a significantly poorer performance. However, reducing the internal diameter still further to the region of 2–3 mm resulted in an improvement in efficiency relative to columns of 4.7–6.3 mm I.D. Decreasing the column width to 1.6 mm led to a decreased efficiency compared with the 2–3 mm columns, presumably due to the increased difficulty of packing the column uniformly and also to the greater influence of dead volume in the detector and interconnecting lines on such a low-volume column.

The decreased efficiency of a column of intermediate diameter has been attributed to wall effects. With large-diameter columns wall effects can be ignored, as the sample never reaches the wall (infinite diameter effect). At the other end of the scale, with columns of 2–3 mm I.D., the diffusion distance is sufficiently short that, despite slow diffusion rates, sample molecules have time to enter and leave the non-uniform region of the column packing many times, maintaining a kind of trans-column equilibrium. With columns of intermediate diameter, the trans-column diffusion distance is greater and since diffusion rates are unchanged, the movement of sample molecules near to the wall will be at a faster rate than that of those travelling through the more uniformly packed centre part of the column bed.

There are, unfortunately, several practical difficulties associated with attempting to carry out on-column injection in pressurised LC systems, perhaps the most important being that if infinite diameter performance is to be achieved the sample must be injected centrally into the packed bed otherwise the sample will tend to travel down one side of the packing, close to the column wall. This situation would lead to a deterioration in performance since the sample would be passing through the less well packed region of the column bed. Other problems which can arise from this approach are that repetitive penetration of the syringe needle into the packing material can disturb the uniformity of the top layers of the packing leading to a deterioration of performance and blocking of the syringe needle with the fine particles of support material. These latter problems can be reduced by inserting PTFE fibre or a porous PTFE plug into the head of the column, although porous PTFE has been known to collapse after prolonged use. The alternative methods of inserting a porous metal frit or ballotini beads into the column, as described earlier, minimise these problems, but also rule out the possibility of obtaining an infinite diameter effect as the sample would be diffused across the entire width of the column immediately following injection. Porous frits have an additional advantage in that they prevent particulate matter, such as fragments of septum material, from entering the column. In practice it is generally easier to clean or replace a porous frit rather than to extricate foreign particulate matter from the top layers of a packed column.

METHODS OF PACKING CHROMATOGRAPHIC COLUMNS

A brief survey of the literature dealing with LC soon reveals that many methods have been proposed enabling one to pack efficient chromatographic columns. If the field of GC can be taken as a guide, many more are likely to be proposed in the future. Unfortunately, this situation can be very confusing, particularly to a beginner, since many methods work well for one type of packing, *e.g.*, dense spheres, yet are totally unsatisfactory for other materials. In this text two methods will be described. One seems to work well with the superficially porous type of beads having diameters in the region of 30 μm. The other is a slurry technique, which is most suitable for packing columns with particles of less than 10-μm diameter. Restriction to these two types of support has been made as these materials have contributed most to the realisation of high-speed high-resolution liquid phase separations.

Dry-packing method for superficially porous beads of approximately 30-μm diameter

Materials of this type are very dense and free flowing. These features permit such supports to be dry packed in very much the same manner as columns filled with much coarser material as in GC. The commonest procedure is to place small quantities of support (say 30 mg) in the column, which is being held in an upright position and bounced on a hard surface.

Although the procedure outlined appears very straightforward, attention should be given to the following points which have been known to cause difficulties:

(1) The tubing selected for the column must be free from internal scale and longitudinal scratches.

(2) The tubing must be scrupulously clean. If a column is to be re-used, it may be cleaned using a pipe cleaner or a small piece of cloth, soaked in solvent, and drawn through the column on a fine cord or nylon thread.

(3) Carefully insert a retaining frit at the column outlet and for the duration of the packing procedure cover with a protective cap so that the frit does not become blocked, distorted or damaged with the bouncing action.

(4) Ensure that during the packing procedure the support is added at a constant rate and the column is bounced with a constant amplitude.

(5) When the column appears to be full, bounce for at least 5 min to ensure that no further settling occurs.

(6) If a frit is to be inserted at the inlet, ensure that it is not forced down hard on to the packing. This will simply block the frit, reducing its porosity.

If done with care this technique will work well for superficially porous supports.

Variations in packing structure have been known to occur if the support material is not closely sized. During the packing procedure segregation of the relatively coarse and fine particles can give rise to regions with different density and mass transfer characteristics. For many years the procedure of separating support materials into very narrow ranges of particle size, *i.e.*, where the ratio of the diameter of the largest to the smallest particle is minimal, has been adopted as the only way to achieve high performance[21]. However, recent work reported by Halász and Naefe[22] and by Done *et al.*[23] suggests that for particles greater than 20 μm, a maximum to minimum diameter ratio of 2.0 does not adversely affect performance. If this proves to be general, the methods of separating fractions of support for packing columns will be greatly simplified.

To overcome the variation of support being added to the column and changes in the packing method mentioned above many prefer to employ a mechanical procedure. Machine-packed columns offer two distinct advantages in that they minimise column-to-column variation and remove the tedium which is associated with methodically packing a column by hand, thus ensuring that the technique of addition or bouncing does not vary during the course of packing the column. The commonest mechanical method of packing

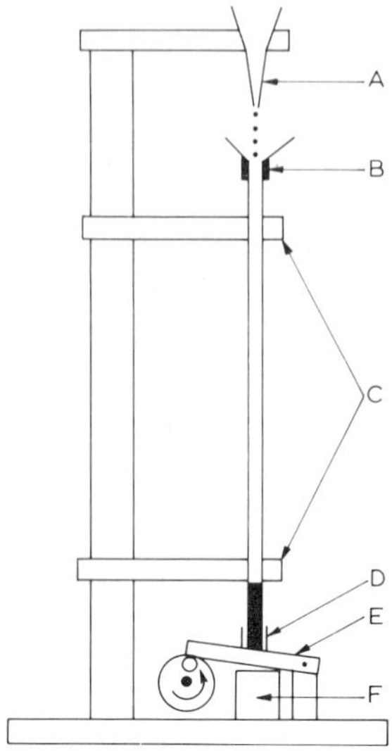

Fig. 3.6. Machine for the dry packing of chromatographic columns. (A) Feed funnel for packing with restricted orifice; (B) detachable funnel; (C) supports allowing column to be held vertical, but move in an up-and-down manner; (D) protective end cap; (E) cam-driven arm, raising column on each revolution; (F) hard metal block.

columns with dry support is to use a machine which simulates the hand-packing method, *i.e.*, the column is held vertically over a motor-driven cam which bounces the column continuously with constant frequency and amplitude. The packing material is fed into the column as a continuous fine stream from some delivery device. Opinions vary widely on the magnitude of the bouncing action and whether or not tapping or vibrating the column walls is beneficial. The drawing shown in Fig. 3.6 conveys the general lay-out of such a machine.

Several workers have observed that rotating the column can also improve the packing characteristics. Done *et al.*[23] have found that rotating the column at a speed of 180 rpm while simultaneously bouncing the column at a rate of about 100 times per minute with a vertical displacement of 10 mm has given consistently superior results in their experience compared with other dry packing methods. They also found that lightly tapping the column at the position of the top of the packed bed was beneficial. The values reported can probably be taken as guide lines rather than critical characteristics if machines for this purpose are being constructed. By following such a procedure columns of 1 m in length can be packed in less than 1 h. A detailed drawing together with construction information of a similar column-packing machine has been reported in the literature by Hazelton[24].

High-pressure slurry method for packing columns with materials of less than 20-μm diameter

Support materials of less than 20-μm diameter have failed to be packed satisfactorily by dry methods of the type described above, due in part to their slow settling characteristics and static charges, which tend to cause the particles to aggregate, giving rise to a non-uniform packing structure. In these circumstances a better packing structure can be obtained by employing the slurry methods. These rely on initially dispersing the support in a liquid medium of such a density that the particles neither float nor settle. A balanced density slurry of this type enables the support to be pumped into a column with minimal risk of sedimentation occurring during the packing procedure. If sedimentation does occur, regions of different packing density will be created within the column which lead to poor column performance. By using a high-pressure method the column bed is formed very quickly, reducing still further the risk of sedimentation.

Balanced slurries of most inorganic support materials are achieved by blending liquids of high specific gravity, *i.e.*, tetrabromoethane and tetrachloroethylene either together or with the addition of a solvent of lower specific gravity such as acetone, by trial and error, until the support is suspended in the liquid medium. As a guide, silica gel particles can be suspended in this manner in a liquid mixture containing approximately 60% tetra-bromoethane and 40% tetrachloroethylene by weight. An alternative procedure for suspending silica microspheres has been reported[7] where the liquid medium is a very dilute ammonia solution (0.001 *M*) and the suspension is created by ultrasonic action. This method apparently works because the very uniformly sized spheres become charged, which causes the individual particles to repel each other. With materials having particle diameters in the region of 5 μm a stable suspension may be obtained in this manner.

During this procedure it is important to eliminate air bubbles in the packing since initially these will keep the particles buoyant. However, when pressure is applied in order to pack the slurry into the column, the air bubbles will either dissolve or be compressed, thus upsetting the stability of the balanced slurry.

A schematic outline of the apparatus for slurry packing columns is given in Fig. 3.7.

The system comprises a solvent reservoir, a high-pressure pump – ideally of a design which will deliver high liquid flow-rates and operate up to at least 300 bars (4500 p.s.i.g.), some form of pressure-indicating device, and a slurry reservoir connecting with a wide-bore union to the chromatographic column. Additionally, for convenience in operation it is useful to have some provision to drain solvent from the pump and reservoir system, so

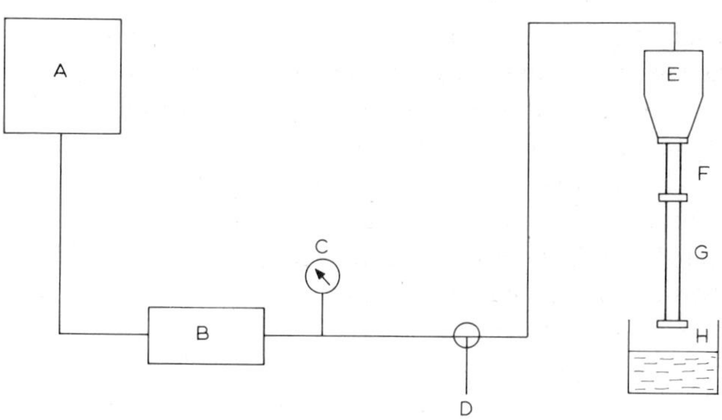

Fig. 3.7. Apparatus for slurry packing chromatographic columns. (A) Solvent reservoir; (B) pump; (C) pressure gauge; (D) drain valve; (E) slurry reservoir; (F) extension; (G) column; (H) beaker. (Reproduced from *Basic Liquid Chromatography*, Varian-Aerograph, with permission.)

that the solvent may be quickly changed without having to pump the entire volume of the previous liquid through the system.

The first step in the packing procedure is to take a clean column and fit a porous stainless-steel frit at the outlet end to retain the support material. The porosity of the frit depends largely on the particles of the smallest diameter likely to be present in the support materials; a 2-μm porosity frit is suitable for most applications. However, for the finest materials (less than 5 μm, nominal) a frit of 0.5-μm porosity is to be preferred. The porous frits are fitted either directly into a small recess in the end of the column or in the coupling which holds the column to the detector. The former position retains support material in the column, whether the column is in use or not, preventing packing from coming loose when storing or transporting the column. The latter method facilitates unpacking of the column or changing of the porous frit should it become blocked in service.

The column is initially filled with solvent of the same composition as the balanced slurry held in the feed reservoir. It is important that the connection between the reservoir and the column does not restrict the flow, *i.e.*, the internal diameter should be at least as wide as the bore of the chromatographic column. To ensure the most rapid filling of the column it is useful to estimate the quantity of support material required to fill the column and to employ a slight excess, say 20%, in the reservoir, as this will avoid unnecessary wastage of material and excess resistance to liquid flow during the packing process. Above the space occupied by the balanced slurry, a layer of an immiscible liquid of lower density – such as water – is carefully placed. The remaining volume of the reservoir and the rest of the apparatus are filled with an even less dense solvent, such as hexane, taking care to eliminate air pockets in the system. The operation of packing varies slightly, depending on the type of pump used in the apparatus.

If the pump employed is a constant-pressure pump, *i.e.*, commonly one driven by pneumatic pressure, it can be adjusted to give maximum pressure almost as soon as it is actuated. This action results in a very rapid flow initially, followed by a progressive decrease in flow-rate as the column bed is being packed into place. The pressure applied should be in excess of that envisaged for subsequent column operation but not so high that the support material is crushed. Most inorganic support materials designed for modern LC will withstand pressures up to at least 300 bars (4500 p.s.i.g.). A positive displacement pump, *i.e.*, one which has a mechanical drive, can be used for the column packing procedure by initially setting it to give maximum delivery of liquid. In this case, as the column bed is consolidated, the pressure in the system increases. When the point is reached where the inlet pressure in the system approaches the desired pressure, or the maximum permissible for the equipment used, the output of the pump is progressively reduced in order to maintain a constant pressure in the system.

Whichever approach is employed, the pumping is continued until water starts to elute from the column. The pump is then switched off and the pressure in the system allowed to fall to atmospheric pressure. The reservoir and column are removed from the rest of the apparatus, which is then flushed with a water-miscible solvent such as alcohol. The column is then carefully separated from the reservoir, avoiding any disturbance of the column packing.

Some workers recommend that a short pre-column be used which protects the real column from being disturbed during these manipulations. The pre-column is removed only when the column is ready to be used for a chromatographic analysis. At this stage the column is packed with the desired support, but in a hydrated form, since water was the last liquid pumped through the column.

The last stage of column preparation is to flush the column to remove water and any residual traces of balanced slurry solvent and to activate the support material for chromatographic analysis. Inorganic types of supports, *e.g.*, silica gel and alumina, can be activated by pumping a series of dry solvents of decreasing polarity through the column. The solvents used are selected from the eluotropic series which is discussed in Chapter 6. As an example, Scott and Kucera have reported that a silica gel packing can be conditioned by flushing with the following solvents in turn: ethyl alcohol, acetone, ethyl acetate, trichloroethane and heptane[25]. The quantity of each of these solvents required to completely remove the previous solvent is the subject which causes some controversy. However, Snyder[26] has suggested that several hundred column volumes of solvent may need to be pumped through the column before equilibration with the new solvent is achieved.

To complete the packed column for use in the liquid chromatograph it is usually advisable to fit some form of packing retainer in the column inlet. This may be in the form of a metal or PTFE frit or, alternatively, woven stainless-steel mesh or PTFE fibre. This latter type is the most suitable when an on-column injection technique is practised, since the syringe needle will easily pass through the fibres. Many organic types of column packing such as the styrene–divinylbenzene beads used for steric exclusion chromatography and the support matrix of some ion-exchange resins, cannot be handled by the above-mentioned techniques, since a change of solvent can lead to swelling or shrinking of the packing material. Methods for these more specialised materials will be discussed in the chapters dealing with their use.

Having packed or purchased a chromatographic column, it is very advisable to test its performance by injecting a test mixture under carefully controlled conditions. Similarly, a performance check can be repeated from time to time if deterioration is suspected. The choice of a mobile phase and test samples depends on the column being studied, but the test mixture should contain at least two components: one which elutes with a low capacity factor, *i.e.*, $k' < 1$, and one which is more strongly retained, having a capacity factor of at least 4. The theoretical plate height calculated from the early eluting peak will give an indication of how well the column is packed since, when k' is low, there is very little mass transfer contribution to the overall plate height. The efficiency of the column derived from the more strongly retained peak will give in addition a measure of the quality of the packing material since slow stationary phase mass transfer characteristics will lead to a marked decrease in plate heights. It is important to note, while on the subject of testing columns, that a reversal of the direction of liquid flow will in most cases lead to disruption of the packing and is therefore not recommended.

In practice, one occasionally experiences difficulties in emptying a column prior to re-use. After removal of the end fittings, some very fine packings show remarkable reluctance to be loosened from a well packed bed. The use of stiff wire and tapping the column to dislodge the material are not recommended because of the risk of damage to the internal wall of the tubing, which for the highest performance must be free of the slightest defects. One of the most effective methods is to couple a length of PTFE tubing to the outlet of the LC pump and use the same to deliver as high a flow-rate of water as possible. This produces a miniature hose-pipe, which can be fed into the column. The force of the water jet is usually sufficient to dislodge particles, which are carried away in a dilute slurry. For this approach, a pressure-driven pump usually holds advantage over mechanical pumps as exceedingly high liquid flow-rates can be readily obtained. Once emptied columns should be cleaned with a long pipe cleaner soaked in a solvent the nature of which is dependent on the most likely contaminants, followed by flushing with re-distilled acetone or alcohol and then blowing dry with clean nitrogen.

In the concluding paragraphs of this chapter the characteristics of chromatographic supports may be summarised as follows. A support with a large surface area will accept a higher quantity of "active" surface, *i.e.*, stationary phase, which will lead to columns with a high sample capacity. A support with no internal pores will offer good efficiency since there are no stagnant pools of mobile phase which lead to poor mobile phase mass transfer. Small-diameter supports, if less than 10 μm, enable inter-particle distances to be decreased leading to a more densely packed bed and reducing inefficiencies due to eddy diffusion. Particles having an open pore structure in addition to a small diameter, *i.e.*, in the region of 5 μm, do not suffer from the presence of stagnant pools of mobile phase which can limit the rate of mass transfer in large particles. In the smaller particles the pore depth is insufficient for stagnant pools to form.

For optimum performance in terms of efficiency, sample capacity and speed of analysis, supports which are of small diameter (say 5 μm) having wide internal pores should be used. The high capacity of these supports makes them most suitable for preparative applications and where fairly large samples are required to offset limited detector sensitivity, particularly when minor components are to be monitored.

For maximum operator convenience, columns should be easy to pack and be capable of giving rapid analysis with an acceptable inlet pressure. If these latter criteria are important, the superficially porous supports might be preferred, as these offer good efficiency with ease of manipulation. The limited surface area of these supports can be their greatest limitation, since the sample capacity is comparatively low.

REFERENCES

1 T.W. Smuts, K. DeClark and V. Pretorius, *Separ. Sci.*, 3 (1968) 43.
2 J.J. Kirkland, in S.G. Perry (Editor), *Gas Chromatography 1972*, Applied Science Publishers, London, 1973, p. 39.
3 R.E. Majors, *J. Chromatogr. Sci.*, 11 (1973) 88.
4 J.N. Done and J.H. Knox, *J. Chromatogr. Sci.*, 10 (1972) 606.
5 J.J. Kirkland, *J. Chromatogr. Sci.*, 10 (1972) 129.
6 G.J. Kennedy and J.H. Knox, *J. Chromatogr. Sci.*, 10 (1972) 549.
7 J.J. Kirkland, *J. Chromatogr. Sci.*, 10 (1972) 593.
8 J.J. Kirkland, *J. Chromatogr.*, 83 (1973) 149.
9 L.R. Snyder and J.J. Kirkland, *Introduction to Modern Liquid Chromatography*, Wiley-Interscience, New York, 1974, p. 68.
10 L.R. Snyder, *J. Chromatogr. Sci.*, 7 (1969) 352.
11 H.C. Beachell and J.J. De Stefano, *J. Chromatogr. Sci.*, 10 (1972) 481.
12 L.R. Whitlock and R.S. Porter, *J. Chromatogr. Sci.*, 10 (1972) 437.
13 J.J. Kirkland, *J. Chromatogr. Sci.*, 7 (1969) 361.

14 B. Versino and H. Schlitt, *Chromatographia*, 5 (1972) 332.
15 U. Prenzel, R. Schuster and W. Strubert, *C.Z. Chem.-Tech.*, 3 (1974) 105.
16 J. Asshauer and I. Halász, *J. Chromatogr. Sci.*, 12 (1974) 139.
17 J.P. Wolf, III, *Anal. Chem.*, 45 (1973) 1248.
18 R.P.W. Scott, D.W. Blackburn and T. Wilkins, *J. Gas Chromatogr.*, 5 (1967) 183.
19 J.H. Knox and J.F. Parcher, *Anal. Chem.*, 41 (1969) 1599.
20 J.J. De Stefano and H.C. Beachell, *J. Chromatogr. Sci.*, 8 (1970) 434.
21 C.G. Scott, in J.J. Kirkland (Editor), *Modern Practice of Liquid Chromatography*, Wiley-Interscience, New York, 1971, p. 304.
22 I. Halász and M. Naefe, *Anal. Chem.*, 44 (1972) 76.
23 J.N. Done, G.J. Kennedy and J.H. Knox, in S.G. Perry (Editor), *Gas Chromatography 1972*, Applied Science Publishers, London, 1973, p. 145.
24 H.R. Hazelton, *Lab. Pract.*, 23 (1974) 178.
25 R.P.W. Scott and P. Kucera, *J. Chromatogr. Sci.*, 11 (1973) 83.
26 L.R. Snyder, in J.J. Kirkland (Editor), *Modern Practice of Liquid Chromatography*, Wiley-Interscience, New York, 1971, p. 225.

Chapter 4

Liquid chromatographic instrumentation

INTRODUCTION

Apparatus used for LC analysis differs considerably both in complexity and in the way in which the various functions are performed. Before discussing the individual components in any detail the lay-out of a chromatographic system will be described in general terms.

The various features of a LC system are summarised in Table 4.1. The absolutely essential components from which a very basic instrument can be built are underlined.

It can be seen from Table 4.1 that the number of individual components which make up a comprehensive LC system is quite large. Owing to the diversity of applications which may be studied, *i.e.*, steric exclusion, preparative separations, high-precision quantitative analysis, or high-resolution trace analysis, one must dedicate or optimise small and moderately sized instruments for certain applications or choose a more comprehensive or "research" system by which, with little modification, most types of application can be accomplished. The latter style of equipment, although highly desirable, tends to be costly, particularly since not all equipment features are likely to be used simultaneously, for instance, some detector types are quite unsuitable for monitoring a separation achieved using gradient elution. (The latter procedure is a method whereby the chemical composition of the mobile phase, hence sample retention, is changed systematically during the course of the separation.)

In many instances the selection of a certain design of one component dictates the use of other components which would otherwise not be needed, *e.g.*, a pumping system which

TABLE 4.1

FUNCTIONAL COMPONENTS OF A LIQUID CHROMATOGRAPH

Function	Components
Solvent delivery	Liquid reservoirs (temperature controlled), pump, gradient elution device, flow controller, pressure indicator
Solvent equilibration	Pulse damper (depends on pump design), heat exchanger, pre-column, in-line filter
Sample introduction	Septum-type syringe injector, valve, auto-sampler
Separation	Column(s) – size depends on application, interconnecting couplings, temperature control
Detection	Choice of a number of detector types, which may be linked in series; these are discussed in detail in Chapter 5
Collection	Manual or automatic fraction collector
Data output	Integrator, recorder, computer (possibly controlling auto-sampler and instrument)

produces a pulsating liquid flow must be "damped" to give a smooth flow whereas other pump styles do not need such a device. The following sections describe the options available in the types of units which are employed currently in the various designs of liquid chromatographs.

TUBING AND TUBE FITTINGS

Before discussing the basic units contained in a liquid chromatograph, a few words on the materials of construction of these instruments could be of value. Most commercial chromatographs are fabricated from stainless steel, grade ANSI 316. This grade has a very high degree of corrosion resistance to many organic solvents, oxidising agents, acids, and bases. The achievement of this corrosion resistance is based on the formation of an oxide layer on the metal surface. This surface layer will protect the steel from further corrosion in most situations. Applications where the oxide layer has been known to deteriorate, leading to corrosion of the metal, are those involving mobile phases containing halide ions, mineral acids and certain simple carboxylic acid anions. Often this problem is first recognised by the unexpected coloration of the column effluent and also by severely tailing peaks.

In the event of these compounds having to be used in a liquid chromatograph made from stainless steel, the equipment should be rinsed thoroughly after use. If necessary, steps should be taken to re-form the oxide layer. For this purpose strong nitric acid is usually recommended, but before use reference should be made to the manufacturer's handbook.

The other materials commonly used in the construction of liquid chromatographs are PTFE, silica, and glass. The materials of construction of pumping systems, *i.e.*, the seals and valves, are probably the most common cause for concern, especially if the pump has not been designed specifically for LC. Care should be taken when considering the purchase of an unusual pump from a company who do not manufacture liquid chromatographs.

Most of the tubing used for containing the mobile phase is made from seamless stainless-steel capillary. Up to the point of sample introduction the internal diameter is not critical and tubing of 0.75 mm (0.030 in.) I.D. is to be recommended. Beyond the point of sample introduction, dead volume is critical and here capillary tubing no wider than

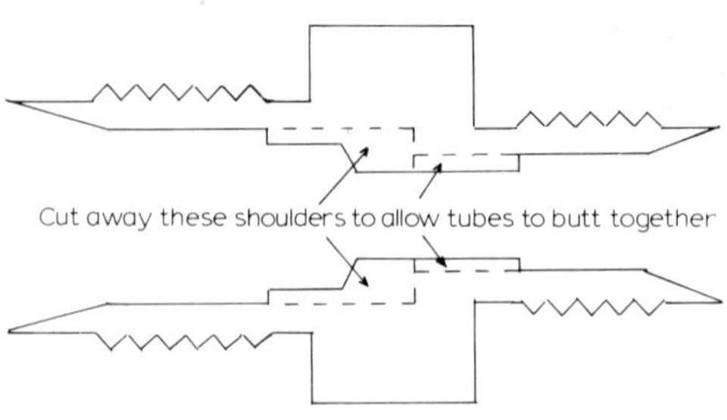

Fig. 4.1. Manufacture of a zero dead volume coupling from a commercially available tube fitting.

0.25 mm (0.010 in.) I.D. should be used for inter-connecting lines, except where the system is being optimised for preparative chromatography, when wider-bore tubing must be used to reduce the resistance to liquid flow.

There are a number of companies who manufacture precision tube fittings of stainless steel which may be used directly for the assembly of the chromatograph. However, in regions where dead volume is critical, *i.e.*, after the sample injector, these tube fittings should be drilled through, as shown in Fig.4.1, so that the tubes butt together.

All seamless stainless-steel tubing up to 6 mm (approx. 1/4 in.) O.D. will withstand the pressures currently encountered in HPLC. Points liable to fail under extreme pressure include: the septum injector (unless specifically designed for high-pressure operation), glass columns*, and detector cell windows and gaskets (depending on the design and then usually only if the outlet of the system has been blocked accidently).

Working with very high pressure in the liquid phase does not represent a serious operator safety hazard as the compressibility of liquids is very low, a rupture of the system creating a leak rather than an explosion. The greatest risk to liquid chromatographers is perhaps the same as that to any who work with highly inflammable and often toxic organic solvents. A well ventilated laboratory and common sense are the most important safety requirements.

SOLVENT DELIVERY SYSTEMS

Systems designed for discontinuous operation

Systems having no true "pump"

In the systems having no true "pump" the supply of a liquid flow at high pressure is obtained by applying a gas pressure, usually nitrogen or helium, of equal magnitude directly to the surface of the mobile phase or through a diaphragm. It is possible to perform this task in several ways, as is illustrated by Fig.4.2.

A common disadvantage of these simple "pumps" is that they can deliver only a finite volume, after which the system must be stopped and refilled. However, assuming a constant gas pressure, each system will deliver a very smooth flow of liquid at constant pressure until the volume of mobile phase is almost exhausted. Models A and C in Fig.4.2 have liquid directly in contact with the high-pressure gas which will tend to dissolve in the liquid. Restricting the area of gas–liquid interface reduces the rate of dissolution considerably, enabling a greater proportion of the liquid held in the "pump" to be used without problems due to gas bubble formation. Model B can be produced by either using metal bellows or collapsible plastic bottles as the reservoir. A commercial pump operating on this principle is available (from Pye-Unicam) but only for operating pressures up to approximately 6.5 bars (100 p.s.i.g.). In principle types B and D could be constructed in such a way that little if any gas comes in contact with the liquid phase.

The overwhelming advantage of all these "pumps" is that within the limits of pressure and gas solubility there is very little that can go wrong. Commercial variants of C are

*Steel columns lined with glass are available commercially and overcome this problem.

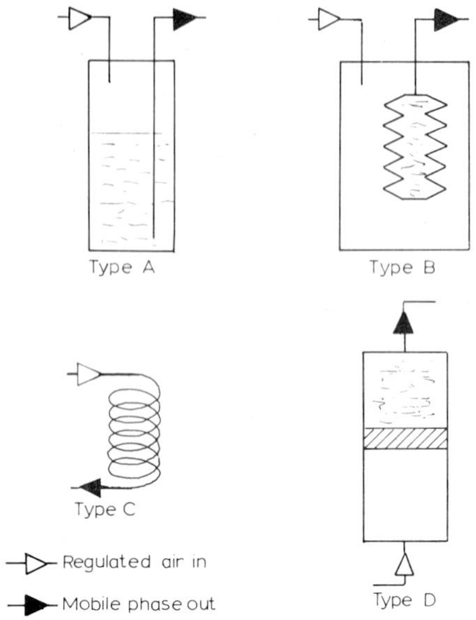

Fig.4.2. Designs of simple pumps using gas pressure as the driving force. In type B use is made of a collapsible plastic bottle or metal bellows. In type D a sliding piston is used.

capable of operating at pressures up to 200 bars (3000 p.s.i.g.), although most gas cylinders are not pressurised to this extent; thus, more realistic working pressures are probably in the range of 100–150 bars (1500–2000 p.s.i.g.).

Although these "pumps" are of very simple construction, safe operation is an important consideration. Each of these designs relies on an appreciable volume of gas and liquid compressed under high pressure and in this respect any large leak in the system or inadvertent sudden release to atmospheric pressure can pose a significant safety hazard. Most commercial pumping systems utilise some form of interlock device on the control valves so as to avoid accidental release of high pressure to the atmosphere. Similarly, components used in the construction, *i.e.*, valves, tubing, etc., are used well within their pressure capabilities.

It is probably apparent that these pumping systems tend to be employed in simple or low-cost apparatus and in home-built equipment which is likely to be used for educational or quality control work. Although lacking some of the capabilities of the more sophisticated units, when properly designed such systems are capable of producing remarkably good results and their utility should not be ignored.

Mechanically driven syringe-type pumps

A very significant improvement over the former type of "pumps" is given by the mechanically driven syringe pumps, although they share the common feature that a finite volume is put into the pump and the system must be stopped for refilling when the liquid has been exhausted. In a well designed pump this need not be a great problem in that the volume contained within the pump initially may be adequate for very many analyses.

This style of pump comprises a large cylinder in which the mobile phase is contained and a tight-fitting piston which is driven into the cylinder by some mechanical means, displacing the liquid at a rate equal to the rate of advance of the piston. The liquid output of such pumps is controlled by electro-mechanical means, thus, within the limits of the materials of construction and the power of the drive system, a pump of this type will displace a constant volume of liquid irrespective of the resistance to flow in the chromatographic system. Clearly, no pumping system has infinite power and thus each pump will have some finite limiting pressure. Some of the more recently developed pumps of this type can deliver liquids against pressures in excess of 400 bars (6000 p.s.i.g.). The advantages of pumps of this type are that they will deliver liquids at a constant pre-selected flow-rate and the liquid flow is free of pulsations. Perhaps their most serious drawbacks are that they have a finite volume and analysis must be stopped while the pump is refilled. One of the consequences of having a mechanical drive for the piston has made some models rather slow to refill. Similarly, changes from one solvent to another, when flushing with fresh solvent is necessary, can be time consuming. Some of the present models are not designed to work at high flow-rates, *i.e.*, in the order of 50 ml/min and above, as required for high-speed preparative work and thus could limit their overall versatility. It should be appreciated that the design and specification of LC pumping systems change rapidly and it is quite probable that the availability of good pumps of this type with few, if any, of the drawbacks mentioned will tend to increase as LC develops. Perhaps then the only criticism will be the cost, which, based on current prices, will tend to be high.

Syringe-type pumps are commonly regarded as producing a constant flow of mobile phase. Although this is a reasonably accurate description, it should be realised that they really operate with a constant displacement of the piston. Under the extreme pressures that may be generated in a modern liquid chromatograph some compressibility can occur to the extent of a few per cent. Additionally a similar volume change sometimes takes place when certain liquids are mixed during a solvent gradient, the decrease in volume which occurs when water and alcohol are mixed being perhaps the most widely known example. Unless special provision is made to compensate for these "deviations", the flow will not be strictly constant. However, like many aspects of LC, the change is of little consequence if the same effect occurs each time the instrument is required to perform the same task. It is the reproducibility of the system which is of paramount importance.

Pumping systems capable of continuous operation

Pneumatic amplifier pumps

It was mentioned above that one of the drawbacks of the mechanically driven syringe pump was the relatively slow refilling action. Pneumatic amplifier syringe pumps overcome this problem by utilising air pressure to drive the piston. Fig.4.3 indicates the delivery and refill strokes of this type of pump.

The pneumatic section contains a piston which is typically 23 or 46 times the area of cross-section of the piston in the liquid section. This difference in piston area gives the pump a built-in compression ratio so that, for example, one bar of gas applied will yield

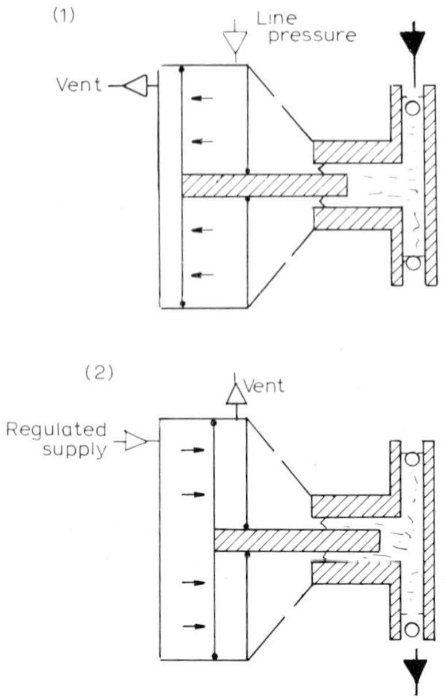

Fig. 4.3. Operation of a pneumatic amplifier pump. (1) Filling stroke; (2) delivery stroke.

a pressure in the liquid section of twenty-three (or forty-six) bars. Application of the air during the delivery stroke generates a compressed liquid; the flow-rate with which the liquid leaves the pump depends entirely on the solvent viscosity and the resistance to flow at the pump outlet. In modern LC the greatest resistance is provided by the column packing. The volume of liquid present in the pump body varies with the individual model but is usually in the range of 2–70 ml. In use, the piston advances smoothly under the constantly applied gas pressure, displacing liquid from the pump. When this piston has reached its limit of travel a microswitch is activated by the movement of the air piston which operates a shuttle valve reversing the gas pressure applied to the air piston and exhausting the "drive" side of the air section. This results in the piston moving rapidly backwards and liquid is sucked into the hydraulic section from an adjacent reservoir. Another microswitch is positioned at the rear of the pump which, when actuated by the air piston, reverses the gas flow to recommence liquid flow in the normal way. The refilling action of large-volume pumps of this type is normally accomplished in less than 2 sec, the models of smaller volume taking a fraction of a second. This rapid filling stroke enables these pumps to deliver liquid at rates up to 100 ml/min. Thus, although the action of these pumps is strictly discontinuous, the very rapid refilling does not interfere with the chromatographic analysis and thus they may be considered as continuously operating systems. Like the mechanically driven syringe pumps their delivery of liquid is pulse free except during the refill step, this action being very rapid only in the case of air-driven pumps. Some models may also be operated up to pressures in excess of 600 bars, but these are not currently offered in commercial liquid chromatographs. Since the motive force of the pumps is provided by compressed gas, the liquid output is controlled by the applied pressure and the resistance to flow in the system. This feature has considerable merit in that, if the outlet of the pump is opened to drain the liquid remaining in the reservoir and pump, it may be rapidly discarded, making the change-over from one solvent system to another a very rapid process. On the other hand, the reproducibility of the liquid flow through the column and detector relies very much on the constancy of the applied pressure, resistance to flow in the column, and the mobile phase viscosity. Where analyses are performed with a mobile phase having a constant composition, the flow-rate is essentially constant, provided no particulate matter is introduced which could change the resistance to flow through the column. During gradient elution work, where the liquids being mixed differ in viscosity or where a swelling or shrinking of the column packing may occur as in ion-exchange chromatography, the flow-rate can vary. In many instances, the actual value of the flow-rate is less important than the system producing the same effect each time the same operations are carried out. Thus, as with the mechanically driven syringe pumps, it is the reproducibility of the system that is of major importance. Some sophisticated units are equipped with flow controllers so that, by choice, one may operate under constant-flow conditions for the most predictable retention times (which is particularly important in steric exclusion work) or operate with constant pressure enabling very rapid solvent change-over in the pump and the solvent delivery lines. Flow controllers usually operate by using a differential pressure transducer fitted across a known resistor, for example a fine-bore capillary. Ohm's Law applies equally to liquid streams as it does to electricity, thus a change in flow will be sensed as a change in the differential pressure and this signal is fed back to the air pressure regulator in a closed servo loop reducing or increasing the gas pressure as required.

Reciprocating (or metering) pumps

The metering pump was one of the earliest types of pump used for LC. A typical outline of a metering pump is shown in Fig.4.4.

In most models the piston is in direct contact with the liquid mobile phase being pumped, however, in some models, known generally as diaphragm or membrane pumps, the piston action is transmitted to a flexible stainless-steel membrane via a hydraulic system. In a diaphragm pump the mobile phase does not come into direct contact with the piston and its seals, hence allowing selection of the materials of construction for their wear resistance alone rather than having to consider possible corrosive aspects of the mobile phase. The liquid throughput of reciprocating pumps is a function of the pumping frequency of the piston and its displacement volume. Until very recently it has been the practice to operate with a constant piston frequency, usually in the order of a hundred strokes per minute. In this case changes in liquid flow-rate are achieved by adjusting the displacement volume of the piston either directly, by reducing the stroke or, in the case of membrane pumps, by transmitting part of the pumping energy to a second "dummy" piston, which is an adjustable, spring-loaded shock absorber. The maximum displacement volume, hence flow-rate range, is governed primarily by the cross-sectional area of the piston.

The reciprocating action of these pumps results in the liquids being delivered in a rapid series of pulses, rather than at a smooth flow-rate. For maximum stability of the column

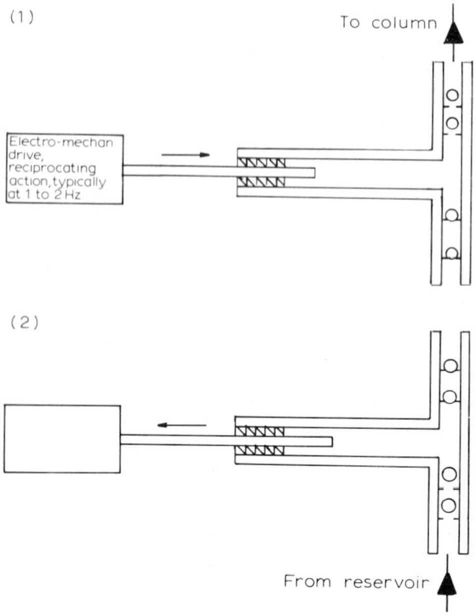

Fig. 4.4. Action of a reciprocating (metering) pump. (1) Delivery stroke; (2) refill stroke.

packings and minimum detector noise, the mobile phase flow must be free from pulsations. When using reciprocating pumps of the types discussed it is common practice to install a pulse damper using a T-piece fitted between the pump and the column in order to smooth the liquid flow. This is usually a capacitance-resistance network comprising a Bourdon tube or pressure gauge which provides an expansion volume, coupled to a capillary restriction. Unfortunately, in order to effectively remove pulsations generated by most reciprocating pumps a considerable resistance is necessary; this in turn means that in applications where a high flow-rate of mobile phase is needed a considerable build-up in pressure occurs in the pulse damper. Many liquid chromatographs employing this type of pump use a pressure gauge as the "capacitor" in the pulse damper so that the operator can ensure the pump is not made to operate beyond its recommended pressure range. One limitation of the simple pressure gauge is that the inner tube is sealed at one end yet liquid is free to enter the tube. When it becomes necessary to change the mobile phase for another analysis, unless precautions are taken, the small amount of original mobile phase held up in the pressure gauge will slowly bleed into the new mobile phase, causing contamination; this may be particularly serious if the former phase is immiscible with the new phase. This situation may be overcome by employing either a flow-through Bourdon tube in place of the simple pressure gauge or, alternatively, one with the Bourdon tube filled with liquid and sealed by a diaphragm so that the mobile phase does not become trapped in the gauge.

A particular disadvantage of using any capacity-resistance pulse damper is that much of the performance of the pump is sacrificed in the pulse damping system. In practice, if an essentially pulse-free liquid flow is to be achieved, as much as half of the total pressure drop in the system can occur in the damping system, thus limiting the maximum pressure available at the injection port quite significantly. For this reason it is often useful to use a high-pressure metering valve as a variable restrictor rather than using a simple capillary restriction in the pulse damper. The valve may be adjusted to give either minimum pulsation or minimum pressure drop, depending on which is more critical for the application in hand. This arrangement is quite an acceptable compromise, as maximum liquid throughput, which would cause most pressure build-up in the pulse damper, is most likely to be needed for preparative applications where it is unnecessary to operate the detectors at maximum sensitivity.

Pulsations in the liquid stream may be reduced without loss of pumping capability by using two or more reciprocating pumps which are linked in parallel but operate out of phase. Most manufacturers of reciprocating pumps offer models where two pump heads may be mounted 180° out of phase on the same drive shaft so that one pump head is filling whilst the other is delivering liquid to the chromatograph. The type of smoothing of the liquid flow achieved is shown in Fig.4.5; the most effective damping is obtained when the volumetric outputs of the individual pump heads are identical. Contamination or wear of the check valves in the liquid inlet and outlet ports can make this latter requirement a challenge. With this arrangement the pulses in the liquid flow are very much reduced, allowing a less restrictive pulse damper to be employed.

A noteworthy development in reciprocating pump design has been the introduction of twin piston pumps where the piston displacement is held constant but the pumping frequency is adjusted electronically, depending on the flow requirements. This system incorporates an eccentric piston drive mechanism, which results in a considerable reduction of the pulsations associated with the more conventional twin piston head design shown in Fig.4.5, obviating the need for a complex pulse damper. In certain critical applications a "high-sensitivity noise filter" is employed to eliminate minor residual fluctuations in the

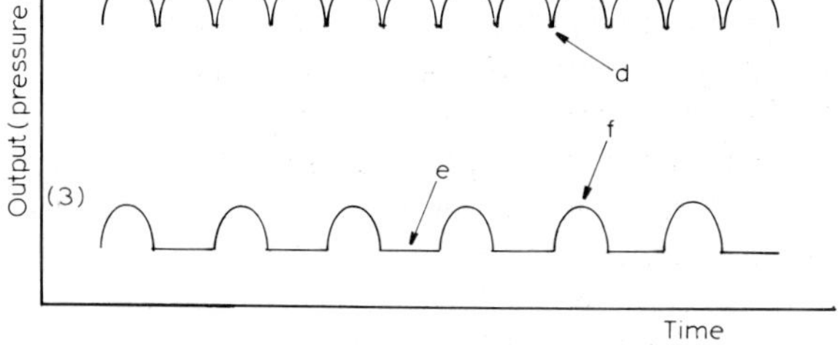

Fig.4.5. Output from a twin-headed reciprocating pump. (3) Single-headed pump. (e) delivery stroke; (f) refill stroke. (2) Twin-headed pump (180° out of phase). (c) Delivery stroke; (d) end of refill stroke of head 1; start of fill stroke of head 2. (1) Resultant flow pattern in chromatograph (after some resistance, *i.e.*, pulse damping). (a) Delivery rate of solvent; (b) static liquid condition, *i.e.*, no flow.

flow-rate. The "filter", like other pulse dampers, restricts the maximum liquid flow and pressure capabilities of the unit.

GRADIENT ELUTION DEVICES

The very pronounced dependence of sample retention on the composition of the mobile phase has already been indicated and is described fully in Chapter 6. In many applications of LC to the separation of complex mixtures or samples containing widely dissimilar components, it is frequently necessary to modify the chemical composition of the mobile phase in order that all of the components present in the sample may be satisfactorily eluted from the column. Snyder has described this situation as the "general elution problem"[1] and wider aspects of this are discussed in Chapter 6. At this stage it is only necessary to indicate that to overcome this elution problem it is very often desirable to change the chemical composition of the mobile phase being supplied to the column during the course of the separation. This technique is known as gradient elution or solvent programming.

A number of devices have been described that allow gradient elution to be achieved and these vary considerably in design. For any device to be of any practical value it must be flexible in its operation, easy to use, and, above all, reproducible. The types of gradient elution device employed in commercial liquid chromatographs are largely dictated by the characteristics of the pumping system used. All systems may be subdivided into two categories, those which mix the solvents prior to their entry into the pump that provides the liquid flow in the chromatographic system – the so-called "low-pressure gradient systems" – and those which mix the solvents in the high-pressure region of the chromatograph immediately before the solvents enter the separating column. Understandably these systems are generally referred to as "high-pressure gradient systems".

Low-pressure gradient systems

This type is most often employed in chromatographs which use a reciprocating piston or diaphragm pump.

The greatest attraction of these pumps, apart from low cost in the simpler models, is that they possess a relatively low internal volume, usually only several millilitres. In this case it is possible to vary the composition of the solvent feeding the high-pressure pump without causing too great a lag in the time for the various compositions of solvent to reach the column. However, the design of the internal parts of the pump head needs particular attention, *i.e.*, low volume and the elimination of poorly swept regions, if serious distortion of the slope of the gradient profile is to be avoided. The sweeping characteristics of a pump can often be improved by making PTFE liners for the pump head to reduce its internal volume[2].

The simplest arrangement of the low-pressure gradient system is to add the modifying liquid to the reservoir feeding the pump from a separating funnel whilst ensuring the contents of the reservoir are well mixed. Alternatively, a second, low-pressure pump can be used to transfer the modifying solvent to the reservoir holding the mobile phase for

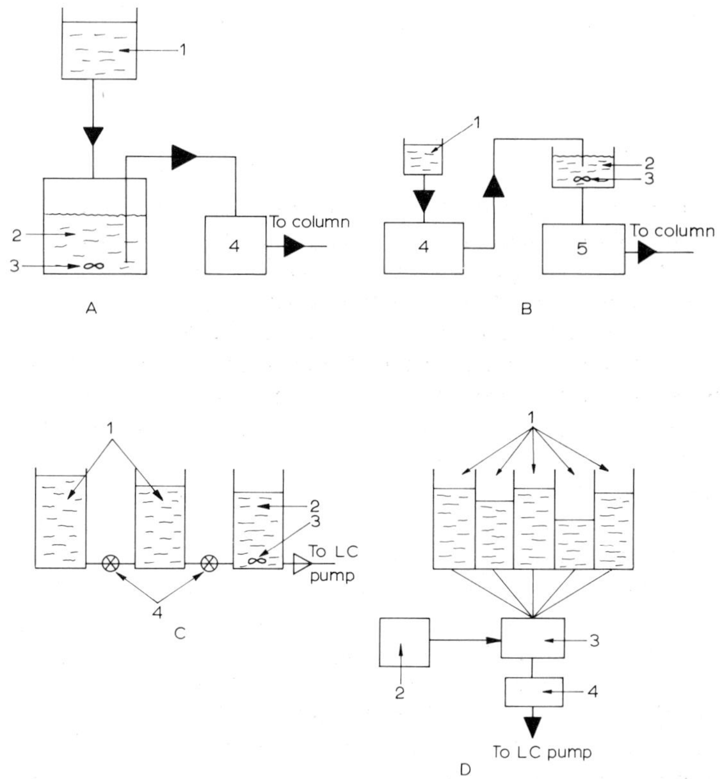

Fig.4.6. Some types of low-pressure gradient system. (A) As liquid is drawn into the pump, an equal volume of modifying solvent enters the reservoir holding the mobile phase. (1) Modifying solvent; (2) starting solvent; (3) stirrer; (4) pump (high pressure). (B) Modifying solvent is transferred to mobile phase with a second pump. (1) Modifying solvent; (2) starting solvent; (3) stirrer; (4) transfer pump; (5) pump (high-pressure). (C) Multiple reservoirs containing different solvents permit complex gradient profiles to be produced. (1) Modifying solvents (many possible); (2) starting solvent; (3) stirrer; (4) valves. (D) Apparatus for incremental gradient elution[2]. (1) Reservoirs of different solvents; (2) programmer; (3) multiport valve; (4) dilution and mixing volume.

delivering to the high-pressure pump. The various possible arrangements for low-pressure gradient systems are illustrated in Fig.4.6. In all cases the volume of liquid originally contained in the mixing chamber or reservoir feeding the pressurising pump and the rate of adding modifying solvents significantly affect the shape of the gradient profile and consequently the elution characteristics of compounds from the chromatographic column. These low-pressure solvent gradient systems potentially offer greater versatility than high-pressure gradient systems as they are capable of handling a series of different modifying solvents while no more than two solvents are handled by the high-pressure systems. Similarly, with low-pressure systems it is possible for the practically minded chromatographer to custom design his own gradient system with little difficulty and cost. The disadvantages, however, are often measured in terms of ease of operation, reproducibility, and speed to respond to a change in desired solvent composition – particularly if the pulse-damping system contains a significant volume of mobile phase. Another inconvenience with these systems is that if one decides to use an alternative gradient composition, after one has already been initiated, a considerable volume of solvent must be discarded as it contains solvents mixed in some intermediate proportions.

High-pressure gradient systems

Systems of this type are incorporated into the more sophisticated and, necessarily, more expensive commercial liquid chromatographs.

When syringe-type pumps are used in equipment offering gradient elution capability two pumps are generally employed, each containing a different liquid. Any proportion of the two liquids can be supplied to the analyser by each pump operating at a fraction of the desired flow-rate. Gradient elution is achieved by progressively increasing the displacement rate of one piston while retarding the other piston by the same amount. The liquids issuing from the pumps are then mixed in a low-volume mixing chamber which relies on either diffusion mixing or mechanical stirring; in the latter case a magnetic follower is often used. The mixed liquids then pass into the analysing system.

The reciprocating or diaphragm pumps may be used in parallel in a manner similar to the system described for mechanically driven syringe pumps. In this case each pump will have its own individual reservoir, which can have any desired volume, hence operate continuously. The high-pressure liquid streams are mixed immediately prior to entering the separating column. In practice this system is, unfortunately, quite difficult to accomplish when using simple reciprocating pumps, as the delivery of liquid from one pump must be reduced as the other is increased. This is achieved with a mechanical vernier control, intended for hand operation. A fairly complex controller is needed if this system is to operate in the highly reproducible and finely adjustable manner that can be obtained from the previously mentioned gradient systems. The use of two twin-piston pumps where the frequency of the piston action is controlled electronically is much more feasible, since the frequency of the pistons may be altered by means of the electronic programmer. This approach is the basis of a successful commercial gradient elution system.

The reproducibility and accuracy of most gradient systems using a pair of small-volume pumps tend to be rather poor when the output from the two pumps is greatly dissimilar, for instance when delivering a 98%:2% mixture.

When employing a pump which has any form of reciprocating action, particular attention must be given to the design of the valve operation between the outlets of the two pumps, or to carefully synchronise their refilling action. If this is not done, then as one pump is refilling, the other pump may force some of the second solvent back into the tubing normally associated with the first solvent. This can lead to a discontinuity in the solvent programme reaching the chromatographic column.

The pneumatic amplifier pump may also be employed in a similar manner, but since the flow characteristics normally rely on the applied pressure and the resistance in the chromatographic system, there is an even greater risk of solvent being backflushed from one solvent delivery line to another during refill. This problem may be overcome by driving both pumps from the same air line and arranging their operation so that they refill at the same instant even though only one pump may be completely empty.

An alternative approach that overcomes the problems associated with the synchronisation of pumps is to employ a single pump. A system based on the use of a single pneumatic amplifier pump is offered commercially and its operation is outlined diagrammatically in Fig.4.7. One solvent is passed through the pump, while the other solvent is being contained in a holding coil.

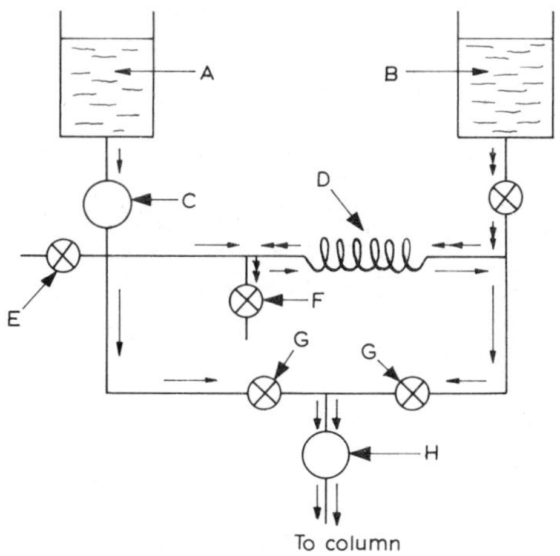

Fig.4.7. Commercial single-pump gradient system. (A) Primary liquid; (B) secondary liquid (C) pump; (D) holding coil; (E) purge valve; (F) drain valve; (G) proportioning valves; (H) mixing chamber. The direction of flow of secondary liquid during the coil-filling step is indicated by a double arrow; the direction of flow of liquids during operation is indicated by a single arrow. (Reproduced by courtesy of DuPont.)

The primary liquid flowing from the pump at high pressure can reach the column by either of two routes, depending on which one of the two proportionating valves is open. These valves are activated by solenoids controlled electronically in such a manner that only one valve is open at any given instant. The valves function on a pre-set switching frequency, *i.e.*, within a given time the period each valve is open depends on the composition of mobile phase selected by the operator or gradient programmer. In common with gradient systems employing two mechanically driven pumps, instruments incorporating such gradient systems can be easily programmed to deliver a mobile phase of constant composition formed by mixing the two liquids in any desired proportion or to produce a gradient change of mobile phase composition in any form, *i.e.*, a linear, non-linear or stepwise change with respect to time.

High-pressure gradient elution systems offer perhaps the greatest operator convenience and most rapid response to a change in operating conditions. Their most serious limitation is that they are normally designed to deliver gradient mixtures of only two solvents, although for very many applications this presents no sacrifice in versatility. There are, however, areas of work where multi-solvent gradients could be useful. In these circumstances the low-pressure gradient systems offer some advantage.

OTHER COMPONENTS OF THE SOLVENT DELIVERY SYSTEM

It should be apparent from the preceding pages that the choice of one component, such as a pump, often dictates the design characteristics of other parts of the liquid chromatograph. For instance, only a reciprocating pump will need a pulse damper and a mechanically driven syringe pump will not need a solvent reservoir in the generally accepted meaning of the word. There are other features which are more universal in their use and these will now be discussed.

Solvent degassing

In all forms of modern LC the mobile phase is pressurised and then passes through the chromatographic column reaching the detector at the column outlet at essentially atmospheric pressure. In all systems there is always a risk that small gas bubbles may be formed as the liquid issues from the column. If this occurs in the region of a detector which employs a flow cell, *e.g.*, an ultraviolet photometer or refractive index detector, then the gas bubbles will cause severe baseline stability problems on the recorded trace (chromatogram). Deterioration of the separating power of a column can also occur if gas bubbles are formed within certain chromatographic columns, particularly those designed for steric exclusion chromatography. The necessity of the removal of any dissolved gas from a liquid immediately before its use as a mobile phase is almost universally accepted. How and where this is carried out varies with the design of the solvent delivery system of the instrument. There are two very effective ways of removing dissolved gas, the first being simply to heat the liquid(s) to boiling point under reflux conditions for 5–10 min. This method is very straightforward and may be carried out away from the chromatograph or in a built-in reservoir if one is provided with a suitable heater and a water-cooled condenser. The only disadvantage of this method is that a change in temperature cannot be accepted with mobile phases which have been equilibrated with a stationary phase for certain types of liquid–liquid partition chromatography or partially saturated with water for liquid adsorption chromatography. In these last-mentioned cases, the second method of degassing liquids is more acceptable. This involves agitating the mobile phase by rapid stirring, ultrasonic vibration or rapid recycling from the reservoir through the pump and back to the reservoir whilst the atmosphere in the reservoir is partially evacuated by a low-pressure vacuum line, *i.e.*, a pressure reduction of about one half of a bar. The rapid recycling capability of the pneumatic amplifier type of pump is most suitable for this "*in situ*" degassing method. The success of vacuum degassing depends a good deal on the agitation of the mobile phase*.

Pressure indication

With a technique such as LC where high pressures are encountered, it is important to have a continuous indication of the maximum pressure within the apparatus for the

*It has been known that the stainless-steel membranes in the pump heads of certain membrane pumps will not withstand a vacuum being drawn on the mobile phase, thus care should be exercised when considering the design of custom-built liquid chromatographs.

benefit of operator safety, the avoidance of damaging the equipment or column packing by overpressure, and as an indicator of the operating conditions. Two pressure indicating devices are available: the simple pressure gauge and a flow-through pressure transducer.

The simple pressure gauge is attractive in that it is of low cost and readily available in models covering a wide pressure range. Pressure gauges are usually installed using a T-piece in the tubing leading to the injector. Gauges do suffer from one quite serious drawback in that the tube within the gauge has a significant hold-up volume which can lead to contamination of one mobile phase with the previous one unless the gauge is carefully emptied during each solvent change or isolated from the rest of the chromatographic system. A gauge may be effectively isolated from a system to prevent contamination by separating the gauge and the solvent feed line by a length, say 1 m, of capillary tubing and having a drain valve situated near to the pressure gauge end of the tubing. During normal operation the drain valve is closed and the capillary and gauge are filled with the mobile phase. When it is necessary to change mobile phases, the drain valve is opened and the fresh solvent is allowed to flow along the capillary flushing out the previous mobile phase. Although this does not change the liquid within the gauge itself, the length of capillary minimises back-diffusion of this liquid into the chromatographic system.

In liquid chromatographs which employ a gas-driven pump – either the simple gas displacement type or the pneumatic amplifier type – it is sometimes more convenient to measure the applied gas pressure with the pressure gauge. This will be virtually identical to the liquid pressure in the case of the simple gas displacement pumps or a constant fraction of the liquid pressure for the amplifier type of pumps. In the latter instance, special pressure gauges are invariably available which, although measuring low-pressure gas, are calibrated in terms of high-pressure liquid, the compression ratio of the pump being built in. This approach is quite attractive in that the liquid flow path from pump to injector may be made with a low volume and designed to be swept efficiently.

Pressure transducers on the other hand are attractive as they generally have a lower internal volume and the pressure-sensing element (strain gauge) may be designed as a flow-through unit, allowing it to be installed directly in the mobile phase line from the pump to the injector. This arrangement overcomes the cross-contamination problems associated with the hold-up volumes within the simpler pressure gauge. Since a pressure transducer gives a change in electrical characteristics for a change in pressure, it is a relatively simple matter to provide the pump with a safety cut-out in the event of the pressure rising higher than any desired value. Thus with any mechanical pumping system, where a blockage in the pipework could lead to an extremely rapid rise in pressure, a sensitive cut-out should always be employed to prevent damage to the pumping system.

Filters

It has already been indicated in Chapter 3 that current high-performance chromatographic columns may be packed with support particles having diameters in the region of 5 μm and there is no reason to believe that some workers may wish to develop columns packed with even finer material. Thus, it should be appreciated that the packed chromatographic column is capable of acting as an extremely efficient solvent filter removing any particulate matter from the mobile phase. This situation, if it were allowed to occur, would be extremely deleterious to the chromatographic column, which would be open to the risk of becoming blocked. To avoid this problem and to offer some safeguard to other parts of the equipment, *e.g.*, the pump, some chromatographers filter all solvents prior to use. Although this procedure goes a long way to minimise the problem, there is always a possibility of particulate matter being produced within the equipment and this should be removed using an in-line cartridge filter fitted immediately ahead of the sample injection system. There are several sources of particulate matter within instruments, the most common being: wear in the mechanical parts of the solvent delivery system, dust in the reservoirs, and precipitation of salts if an organic mobile phase is used in an instrument which has previously contained inorganic buffer solutions which have not been completely washed out in the change over sequence. Another quite common occurrence is bacterial growth in solvents, particularly in buffer solutions, which have been allowed to stand in the apparatus. A porous metal filter having a porosity of 2 μm will effectively remove all of these contaminants, reducing the risk of blocking the column. It is possible that even finer porosity filters, *e.g.*, 0.5-μm pore size, will be necessary if columns packed with particles less than 5 μm become accepted practice. In either case it is important that the filters are checked regularly, for they may easily become blocked. The same care should be taken with samples injected into the apparatus, ideally filtering them before analysis. This aspect will be discussed in more detail in later sections.

Heat exchanger

Most forms of LC are temperature dependent to some extent, liquid partition and ion-exchange being the most sensitive to temperature change. If all analyses are performed at ambient temperature in a laboratory with good temperature stability, then for all but the most critical work no further temperature control of the column and solvent supply is required. In some applications it is found desirable to operate the chromatographic column at an elevated temperature so as to improve sample solubility and the mass transfer characteristics of the system. In these circumstances it is important that the mobile phase entering the column is pre-heated to the same temperature as the column, in order to avoid a temperature gradient in the first few centimetres of column packing. If the mobile phase flows to the injection port and column via metal capillary tubing, which typically has an outside diameter of 1.59 mm (1/16 in.), the heat transfer from the tubing to the mobile phase is quite rapid. As a guide one estimate has suggested that the length of tubing required to equilibrate the temperature of a mobile phase flowing at 10 ml/min is in the region of 4.5 m; lower flow-rates would need a correspondingly shorter length of heat-exchange tubing. When gradient elution is to be employed, a compromise must be made between thermal equilibrium and the delay in the solvent gradient reaching the chromatographic column due to the volume of the heat exchanger. The method of thermostatting the heat exchanger is usually governed by the overall temperature control system of the chromatographic columns; some apparatuses employ a forced air oven in a similar manner to that used in most gas chromatographs, whereas in other systems jackets are fitted round the columns and water circulated through them from a precision thermostatic bath. The relative merits of these two methods of temperature control are discussed later in this chapter, in the section dealing with the chromatographic column. It suffices at this point to mention that the capillary tubing forming the heat exchanger must be as efficiently heated as the other parts of the chromatographic system. The lay-out of the components within the apparatus should be in close proximity, so that a uniform temperature is maintained. This situation is similar to that in GC, however in LC the effect is by no means as critical because of the high thermal capacity of liquids.

Pre-columns

For maximum life of columns and highest reproducibility of results, especially when performing true liquid–liquid (partition) chromatography*, it is necessary to ensure that the chemical composition of mobile phase entering the separation column remains absolutely constant. In columns having a physically held stationary phase, its useful life depends almost entirely on the care taken to preserve the coated layer. If a mobile phase is used which is not saturated with respect to the stationary phase, then the latter will gradually dissolve in the mobile phase, leading to a steady decrease in capacity factors for the samples being examined. The normal practice is to ensure saturation of the mobile phase as closely as possible by equilibration with stationary phase before the separation is attempted. This is achieved by shaking and stirring the mobile phase with excess stationary phase. As an additional precaution, the mobile phase is pumped through a pre-column held at exactly the same temperature as the separating column and filled with a coarse support coated with a high percentage of the same stationary phase as used in the separating column. The pre-column allows intimate mixing of the mobile phase and the stationary phase, ensuring, within the limits of experiment, that the mobile phase is truly saturated. Subsequent passage of this mobile phase through the column should not lead to any depletion of the level of stationary phase on the chromatographic support.

In situations where the main emphasis of chromatography is based on liquid partition systems needing carefully saturated mobile phases, it could be worthwhile considering some form of temperature control system for the mobile phase reservoir. It must be borne in mind, however, that the pump will almost invariably not be temperature controlled and its large thermal mass makes the idea impracticable.

SAMPLE INTRODUCTION

Most of the sample introduction devices employed in LC are very similar in their mode of action to those used in GC. Detail differences in design are necessary to minimise dead volume and particularly to avoid badly flushed regions where sample molecules could be trapped and held back relative to the main sample plug. In a number of instances sampling systems which proved poor for GC are good for LC and vice versa. This is due to the great difference in diffusion rates in the two phases (gas phase being approximately 10^5 faster)

*It will be seen later that liquid partition chromatography may be accomplished in several slightly different ways. The use of the term "true" is taken to mean columns where the stationary phase is physically coated on the surface of the chromatographic support as distinct from the more recently developed packings where the stationary phase is chemically bonded to the support.

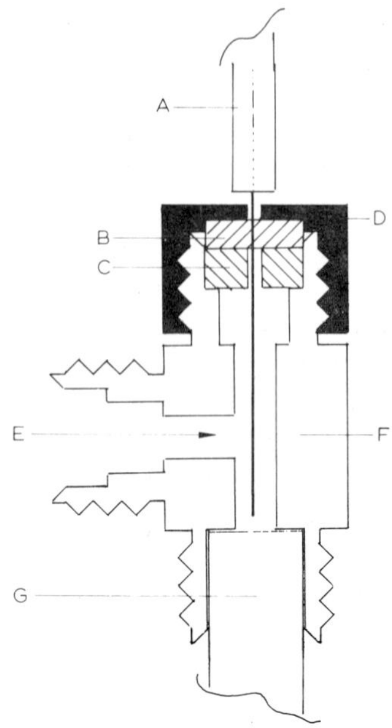

Fig. 4.8. Simple syringe-through-a-septum injection system. (A) Syringe; (B) silicone septum; (C) PTFE support; (D) nut with hole reduced in size; (E) mobile phase inlet; (F) T-piece tube fitting (drilled out); (G) column.

and due to the fact that in LC the sample does not expand immediately after introduction. There are essentially five different modes of sample introduction in LC. These may be summarised as follows:

(i) Injection with a microsyringe, *viz.* (A) through a septum into the head of the column while the mobile phase is flowing (on column injection), (B) as above, but with the mobile phase stopped, *i.e.*, at atmospheric pressure (stop-flow injection), or (C) through a septum into the moving liquid stream immediately ahead of the column.

(ii) Using a microsampling valve, *viz.* (A) small fixed-volume (four-port) valves or (B) external-loop (six-port) valves.

Each sample introduction method possesses some advantages and some limitations. These are described in the following sections.

Septum injector

This sample introduction device is probably the simplest and most widely used in LC. A very basic septum injector can be easily constructed in a manner as shown in Fig. 4.8 from a standard T-piece as supplied by any of the manufacturers of precision tube fittings.

The arm of the T-piece taking the column should be machined out in the manner described earlier to reduce dead volume. The other arm of the T-piece in line with the column connection should be machined flat to improve the sealing of the septum. This very simple device is capable of giving quite good results for injections made into the packing material (on-column injection) and for stop-flow injections. The major problems likely to be encountered are more associated with the method of injection rather than with the design of the injection port. The practical difficulties with on-column injection were discussed earlier in relation to the attainment of highly efficient columns (p.33). They are: the difficulty of placing the sample centrally on the column packing, disturbing the first few millimetres of the column packing leading to a deterioration of column performance and the serious risks of blocking the injection microsyringe with particles of column packing. An alternative approach which gives much more acceptable results is to inject the sample into the mobile phase immediately before it enters the chromatographic column. This procedure enables the column to be fitted with a porous plug which prevents any disturbance of the packing material and any particulate matter from entering the column. The life of microsyringes is also greatly extended. The only sacrifice is the small amount of dead volume at the column inlet; however, in a well designed system this effect is only significant when working with weakly retained components on columns of the highest efficiency. Fig.4.9 shows a cross-section diagram of a commercial injector of this type. Note how the sample is contained within an efficiently swept capillary until it reaches the column.

Should microsyringes become blocked during an injection it is important not to attempt to force the offending particles of packing or septa from the syringe with the action of the plunger of the syringe as this can lead to the barrel splitting. The preferred approach is to remove the plunger and simulate an injection into the liquid chromatograph. Having pierced the septum the high-pressure liquid will force the material blocking the needle further back into the wider part of the syringe body, where it will be flushed away rapidly.

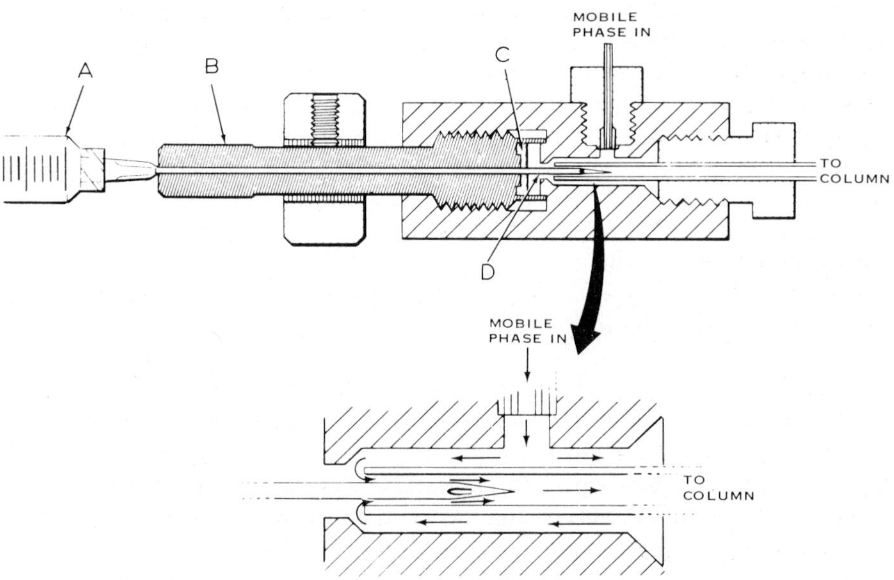

Fig.4.9. Commercial syringe-type injector. (A) Syringe; (B) needle guide; (C) septum; (D) syringe needle. (Reproduced by courtesy of DuPont.)

This action should be carried out using a high liquid flow or pressure setting, but in the case of instruments using on-column injection, care should be taken not to push the tip of the syringe needle into the column packing. If the use of syringes with replaceable needles is considered as an alternative approach, considerable care should be taken to ensure that the seal between the needle and the barrel will withstand the high pressures that are employed in LC since many syringes of this type are intended primarily for GC where the inlet pressures are considerably lower.

Failure to effectively fill a microsyringe with a sample and failure to clean it thoroughly between injections are the most elementary, yet most common errors made by the inexperienced chromatographer. As a guide, syringes should be flushed, by drawing up and discharging the sample solution, at least ten times prior to injection. Equally important, the syringe should be rinsed a similar number of times with pure solvent after use. This feature can be easily demonstrated by filling a syringe with a highly coloured liquid, *e.g.*, blue ink, and then observing the rate of disappearance of the colour in the syringe barrel with successive rinses with water.

A wide range of different materials has been proposed for making injection port septa. However, most are based on silicone rubber and these, unfortunately, tend to deteriorate very rapidly in the presence of certain organic solvents, notably the chlorinated hydrocarbons such as chloroform. This problem has been partially overcome by the introduction of speciality materials such as PTFE-faced septa or those having a layered or "sandwich" structure. More recently fluorinated elastomeric materials have become available which are not affected by the chlorinated and other solvents which are responsible for the deterioration of the more conventional materials.

Septum injection techniques are attractive in that the volume of sample injected may be easily changed, and they are particularly useful when handling small samples. Depending on design, the upper pressure limit where injections may be made is in the region of 100–150 bars (1500–2200 p.s.i.g.). Above this pressure, stop-flow techniques are to be preferred. For routine quantitive analysis, valve injection devices hold advantage over septum injectors as they tend to be more reproducible, particularly by minimising the contribution made by the operator, and because the sometimes troublesome septum can be eliminated. Valves, however, are generally much more expensive.

Valve sample introduction systems

Small fixed-volume (four-port) valves

Valves in this category are capable of introducing very precise volumes of the sample liquid into the chromatographic system. Two somewhat different designs are available currently. They are depicted in Fig.4.10.

The former, Type A, is a hand-operated valve whereby the cavity cut through the centre shaft is first filled with sample solution. When the shaft is turned, this cavity is introduced into the mobile phase stream immediately ahead of the separating column. These valves are generally available with interchangeable shafts so that different sample sizes, ranging from about 0.1–5.0 μl, may be accommodated.

A change in sample volume is thus achieved only after dismantling the valve and

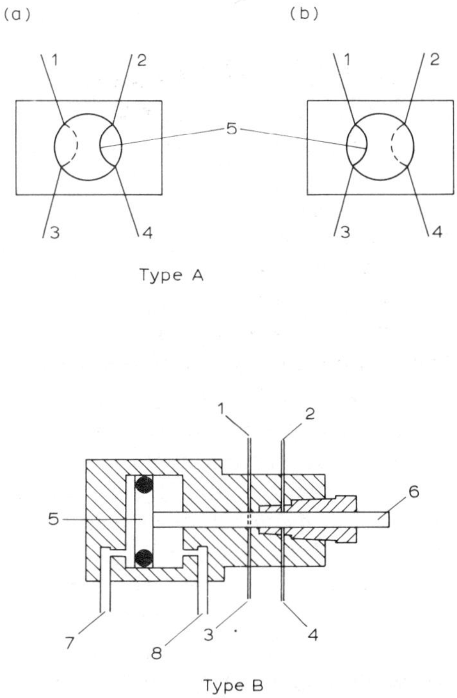

Fig.4.10. Fixed-volume (four-port) valves. Type A: (a) Fill valve position. (1) Mobile phase in; (2) excess sample out; (3) to column; (4) sample in; (5) calibrated groove in rotatable shaft. (b) Inject sample position. (1) mobile phase in; (2) to drain; (3) to column; (4) flushing solvent; (5) calibrated groove in rotatable shaft. Type B: (1) Excess sample out; (2) mobile phase in; (3) sample in; (4) to column; (5) pneumatic piston; (6) shaft (groove shown in sample line); (7) air input – advance shaft; (8) air input – return shaft. (Type B valve redrawn by courtesy of Hamilton.)

changing the shaft. This procedure is time-consuming and, since the shaft is high precision fit in the valve, could easily result in damage if not carried out correctly. Current models of this type are capable of being used in systems operated at pressures up to 330 bars (5000 p.s.i.g.).

It is often found that valves of this type and the external loop valves require considerable torque to operate the valve and there is a risk of blocking the liquid flow path if the change from sample load to inject is not effected quickly. This can result in a disturbance on the resultant chromatograms or, even worse, if it is not realised immediately that the system is blocked, could lead to overpressure in systems employing positive displacement pumps. The chance of this situation arising can be reduced by easing the tension applied to the valve seat until the valve just starts to leak and then re-tightening slightly. This action will then allow minimum effort to be applied when operating the valve and also reduce internal friction as much as possible, consistent with a leak-free system.

The second style of four-port valve, Type B in Fig.4.10, is pneumatically operated in much the same way as the pneumatic amplifier pumps. The sampling and mobile phase connectors are in line. Sample transfer is achieved by filling a groove machined in the centre shaft and activating the air piston which pulls the shaft bringing the groove into the second liquid stream. A reversal of the air pressure returns the shaft to its original position. The advantage of this design of valve is that the push–pull action of sampling can easily be automated, making the device ideal for automatic sampling systems. This valve does suffer from the same limitation as the other design, *i.e.*, it has a restricted range of injection volumes, which are varied only by dismantling the valve and changing the shaft, an operation which must be carried out with surgical care and cleanliness.

External loop (six-port) valves

A small change in the design of the hand-operated four-port valve described earlier makes it possible to locate the volume of sample to be injected outside the valve in a length of capillary tubing rather than in a cavity within the shaft. A valve of such a design is commonly known as an external loop valve and is shown diagrammatically in Fig.4.11.

In valves of this type, the external loop is detachable and a series of loops can be made from capillary tubing each having different volumes. It is a simple matter to change these sampling loops as no high-precision part of the valve need be disturbed. For efficient flushing with mobile phase sampling loops should ideally be long and narrow; however, if large volumes, say 1–5 ml of sample solution, must be injected as in some preparative applications, the loop can be of a large coil of tubing and some compromise with internal diameter must be made. The data given in Table 4.2 may be useful as a guide when preparing sample loops; however, as most tubing is supplied in "nominal" dimensions, calibration will be necessary if accurate volumes are required.

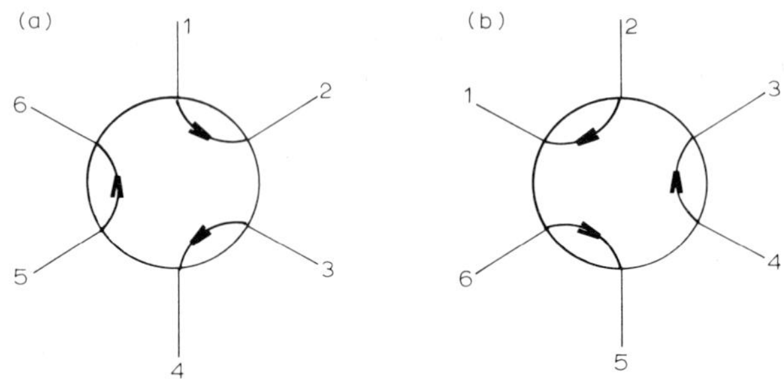

Fig.4.11. External loop (six-port) valve. (a) Loading operation. (1) Mobile phase in; (2) to column; (3) from loop; (4) to waste; (5) sample solution in; (6) to loop. (b) Sample introduction. (1) to loop; (2) mobile phase in; (3) to column; (4) from loop; (5) to waste; (6) flushing solvent.

It is not necessary to have a separate loop for each desired injection volume, since if a loop contains a volume larger than required it is possible to activate the valve for a short time interval, *e.g.*, 10–60 sec, so that only a proportion of the sample is introduced into the chromatographic column. This is achieved by measuring the flow-rate of mobile phase through the chromatographic system, which gives the time taken to displace the sample solution from the loop. Thus opening a valve for a known fraction of this time will result in the introduction of a corresponding fraction of the volume of sample held in the loop.

TABLE 4.2

APPROXIMATE VOLUME-TO-LENGTH CONVERSION FOR THE PREPARATION OF EXTERNAL SAMPLE LOOPS

Internal diameter of capillary (mm)	Approximate volume (μl/cm)
0.25	0.49
0.50	1.96
0.75	4.40
1.00	7.85

This practice holds some advantage when injecting large volumes, as taking a fraction of the loop volume will give a plug injection of sample solution whereas in the complete flushing of a large loop some dilution of the sample solution with mobile phase can occur leading to the sample being introduced into the column over a significant time period, resulting in poor peak shape. It should be appreciated that the precision of this method relies very much on the ability to actuate the valve for very precise time intervals; for this reason, inexperienced hands may be unable to obtain the desired reproducibility. Automatic operation of the valve, with the aid of electronic timers, greatly improves the precision of injection.

When seeking to achieve the highest reproducibility of sample introduction from any valve, particular attention should be given to the following points:

(a) The valve and associated tubing must be kept free from contamination by thorough flushing with pure solvent between each injection.

(b) Air bubbles have been known to form in the cavity of valves leading to variations in the volume of sample solution held in the valve. A check valve giving approximately two bars (30 p.s.i.g.) back pressure fitted to the return line from the sampling stream will minimise this effect.

(c) Valves in which the sample is held in internal cavities, *i.e.*, a four-port valve, have been known to suffer from internal leakage across the seals. In some examples of valve this fault is particularly difficult to detect.

Combination injection devices

It will be apparent that the syringe-through-septum and valve methods of sample introduction both have some merit. The former, syringe injection, is attractive as the requirements in terms of sample volume are low and the volume introduced can be easily varied. Valve injection is more precise and reliable since the problems associated with septa are eliminated.

A combination of these advantages is attempted in the so-called septumless injectors (currently available from Waters and Rheodyne). In effect, provision is made to inject any volume of sample into the loop of a six-port valve by means of a microsyringe. This is carried out when the loop is switched out of the main solvent stream from the pump to the chromatographic column; at this point the loop is at atmospheric pressure and initially contains only mobile phase. After the sample has been loaded, the valve is actuated, allowing the entire contents of the loop, *i.e.*, sample solution and mobile phase, to be swept into the column. This system owes its success to the slow rate of diffusion mixing of the liquids held in the capillary tubing of the loop.

Automatic sample injectors

As the field of LC develops, more emphasis is being placed on the routine or quality control applications of the technique with a subsequent need for apparatus capable of unattended operation.

Automatic sampling systems used in GC are not generally suited to LC applications as most rely on the use of a mechanised microsyringe. The safety problems attendant with repeated piercing of a septum which is being subjected to a high pressure of liquid mobile phase are too great to consider the approach suitable for LC. It is possible, however, that a microsyringe sampling technique could be devised if the injection is performed at atmospheric pressure by the use of a combination device as described in the previous section.

One particularly versatile automatic sampling system specifically designed for modern LC (from DuPont) uses a pneumatically actuated six-port valve to introduce the sample to the chromatographic column. On command from the control system, solutions are transferred to the valve by air pressure via a needle assembly which pierces septum-capped sample vials. Ancillary features available on the control system include automatic actuation of the mobile phase pump, gradient elution, recorder, and data handling system.

CHROMATOGRAPHIC COLUMN AND COUPLINGS

Much of the detail of the design of chromatographic columns has been described in Chapter 3. In this section it is necessary only to expand on the all important matter of dead volume within the system and to discuss methods of controlling the temperature of the separation.

Dead volume in the chromatographic system

The design of every part of the chromatographic system from the injector, through the column and the detectors must aim to reduce to an absolute minimum the void space within the components. Equally, if not more important, is the need to avoid regions where the mobile phase can stagnate, for in these regions part of the sample will inevitably be swept leading to a considerable broadening of the sample bands with an associated loss of resolution. Much attention should be given to the design and assembly of connections within this region of the chromatograph. Although, at present, there are few companies who offer a complete range of zero dead-volume tube fittings suitable for modern LC, it is a fairly straightforward matter to modify the more conventional precision tube fittings which are available from many suppliers. Fig. 4.1 indicates how standard tube fittings may be modified to yield suitable components.

Care should also be taken to ensure the ends of tubing are cut "square" so that sections of tubing may be butted together without creating any dead space.

Column connectors

Two lengths of column may be linked together in series by using two drilled out reducing union tube fittings, as shown in Fig.4.1, which are joined with a short (50-mm) length of narrow-bore (0.25 mm-I.D.) capillary tubing. In the less frequent situation where there is sufficient space within the instrument a simple drilled-through union may be used to permit the columns to be butted together. The procedure of linking columns together is universally accepted in the field of steric exclusion chromatography, where the selectivity of different columns is due largely to the pore structure of the column packing and the nature of mobile phase has only a secondary influence on the separation. In other forms of LC, the nature of the mobile phase is more critical, each column type most often requiring a different mobile phase in order to chromatograph the same sample. Considerable care is needed to select columns if it is considered necessary to have columns of different selectivity connected in series, otherwise, as one frequently finds, the separation may be achieved almost exclusively on one column and the other simply contributes unwanted and unnecessary dead volume. Some applications where coupled columns have achieved some degree of success have been in the area of ion-exchange chromatography[3].

Guard columns

Some workers, when studying complex samples, prefer to use short guard columns or pre-columns fitted after the injector to retain any unwanted components of the sample or particulate matter which might otherwise be retained very strongly on the high-performance separating column. These guard columns are replaced when they have become seriously contaminated. In areas of work where column contamination is likely to be a regular problem, for instance those where samples from biological origin are handled, it is perhaps of more use to consider installing a guard column into the system where it is possible to back-flush the guard column as the analysis of the less strongly retained components is being performed in the main separating column. Back-flushing of columns can only be used where the column packing material will not be disturbed by the reversal of flow. Columns containing microparticulate materials cannot normally tolerate such an action. Filling the guard column with a comparable support of larger particle size often reduces the practical difficulties.

Temperature control of the separating column

During the early years in the development of modern LC, there were many conflicting reports regarding the importance of controlling the temperature of a LC system. These differences in opinion almost certainly arose because some forms of LC, *e.g.*, adsorption chromatography, were rather slow to respond to small changes in temperature, while workers using liquid–liquid partition were requiring very strict control of temperature to maintain the stability of the mobile–stationary phase system.

A study of the literature reveals that a very similar degree of temperature control is recommended for the various forms of LC. Details of some recommendations are summarised in Table 4.3.

TABLE 4.3

TEMPERATURE CONTROL REQUIREMENTS FOR LC COLUMNS

Separation method	Temperature control (± °C)	Reference
Adsorption	0.26	4
Partition	0.30	5
Ion exchange	0.50	6

Based on these and similar data, Maggs[7] has estimated that, as a general guide, it is necessary to control the column temperature to within 0.2°C if the repeatability of retention volume measurements is to be better than 1%.

In general, operation of the chromatographic column at temperatures above ambient holds several distinct advantages. Raising the temperature increases the solubility of a sample in the liquid phases and also improves the rate of mass transfer. These effects lead to higher column efficiency and, as viscosity decreases with increase in temperature, to lower inlet pressures for a given liquid flow-rate. Elevated temperature is to be recommended in any application where such a rise in temperature would not lead to decomposition of the sample or the column packing material. Temperatures used in typical ion-exchange and partition chromatography (using bonded phases) are in the range 25–75°C. In steric exclusion chromatography, temperatures as high as 130°C are sometimes employed in order to enhance sample solubility, particularly when dealing with polymer samples such as polyolefins.

It is now generally agreed that more reproducible results are possible if the temperature at which the separation is performed is held constant. In many instances this is conveniently accomplished simply by working at ambient temperature in a modern efficiently air conditioned laboratory where the air temperature seldom fluctuates more than a degree. When it is desirable to operate at an elevated temperature or to operate near room temperature under carefully controlled conditions some form of thermostat must be provided. This is achieved using either a circulating liquid thermostat or a forced-air circulating oven.

Circulating liquid thermostat

In this method a liquid, usually water, is pumped from an external thermostatic bath through tubing to "jackets" which are fitted around the columns. These jackets may be readily assembled from two T-piece tube fittings in which the two in-line arms can accommodate tubing of different diameters. These are normally described as "heat exchanger" or "thermocouple" T-pieces by the suppliers of tube fittings. Fig.4.12 shows the construction of one end of such a jacket.

To ensure that other important areas in the chromatograph are temperature controlled, the liquid should be circulated to the pre-column, heat-exchange tubing through which the fresh mobile phase is brought to the column system and, ideally, the injector and detector. Although it is possible to circulate liquid to all these parts or alternatively to immerse all these components in a liquid thermostatic bath, the arrangement can be rather inconvenient when changing columns or if a leak of mobile phase occurs. Circulating

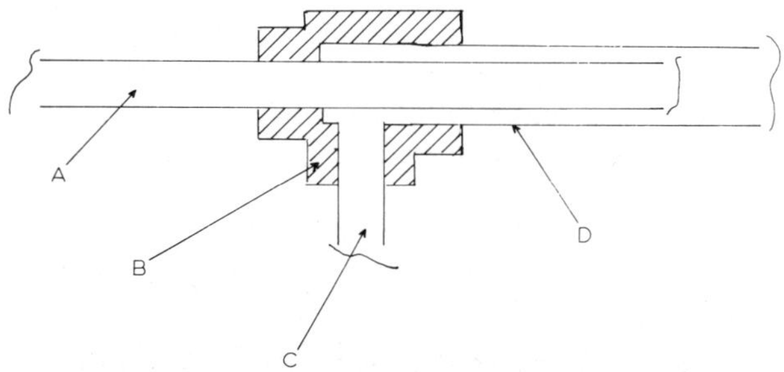

Fig. 4.12. Construction of a jacket for control of column temperature using circulating liquids. (A) Chromatographic column; (B) tube fitting, T-piece, with unequally sized ports (typically, 6, 6, and 9 mm); (C) thermostating liquid circulated through this line; (D) outer jacket.

liquid thermostats can often provide control of the liquid temperature to within 0.01°C of a pre-set temperature, which is certainly more than adequate for most LC separations.

Forced air thermostatically controlled ovens

This approach reflects the influence that GC has had on the development of LC. Temperature-controlled ovens containing all the components which are temperature sensitive, *i.e.*, heat exchanger, pre-column, injector (or valve), chromatographic column and, ideally, the detector, are swept with air driven from efficient fans. Although most air ovens are only able to control the air temperature to within a degree of a pre-set value, the temperature stability within the chromatographic column system is generally within 0.1 or 0.2°C due to the ballasting influence of the high thermal mass of the chromatographic components. This precision of temperature control is quite acceptable for LC separations but is attainable only when the air within the oven is circulated rapidly. An air thermostat is very convenient when operations such as changing columns and detecting leaks in the chromatographic system have to be carried out. Most commercial systems have provision for the fitting of a purge line to the oven so that an inert gas may be flushed through the heated compartment if hazardous, *i.e.*, toxic or inflammable solvents are being used. One slight drawback with these forced air ovens is that without external cooling they cannot control at room temperature due to the energy of the circulating fan(s) ultimately being dissipated as heat. However, a coil of metal tubing fitted in the oven through which is flushed cold tap water or a supply of chilled liquid may be used as a cold spot against which the thermostatic oven will control. This situation parallels the use of cooling water which is necessary for the operation of liquid thermostatic baths at room temperature.

DETECTORS

The details of the various types of detection systems available for LC are dealt with, in depth, in Chapter 5. In this section, detectors are described only in as much as how they fit into the overall chromatographic system.

The importance of very low or near zero dead volume systems has been mentioned earlier. This is never more so than in the detection system. Felton[8] has concluded that dead volume immediately prior to the detector is probably the most critical parameter in the entire liquid chromatograph. Utilising very short and particularly very narrow-bore tubing can lead to some practical problems. For instance, it is imperative to prevent any solid material from entering the fine capillaries otherwise the particles will accumulate and the tubing may become blocked. The use of a 2-μm-porosity plug at the column outlet will normally prevent problems developing from this source.

If a chromatograph is to cover a wide range of applications, *i.e.*, sometimes analytical, sometimes preparative separations, it is useful if the post-column tubing and detector flow cell, where appropriate, can be easily replaced with ones having a slightly larger internal diameter. This will lower the resistance to liquid flow through these components, which is desirable when working with large-bore preparative columns where liquid flows in excess of 50 ml/min can be employed. The associated increase in dead volume in the system is insignificant in preparative applications but would be unacceptable for narrow-bore, high-efficiency analytical columns.

The performance of all detection systems which utilise a flow-through cell is adversely affected by gas bubbles issuing from the column which either pass through or are held up in the detector flow cell. This problem is best eliminated at the source by thoroughly degassing the mobile phase before use in the liquid chromatograph. However, if the liquid has remained in the instrument's reservoir for some time or degassing was not efficient, gas bubbles can be a problem. These may be minimised considerably by applying a small (1–2 bars) back pressure on the outlet of the detector flow cell by either a capillary restrictor or a micrometering valve. In both of these instances the back pressure will also be dependent on the mobile phase flow-rate and thus for the most trouble-free operation a small pressure gauge installed using a T-piece tube fitting at this point is a good investment to protect the detector in much the same way as a pressure gauge in a pulse damper will reduce the risk of pump damage. An alternative method of applying back pressure to the detector is to use a spring-loaded check valve which will maintain a pre-set back pressure irrespective of the mobile phase flow-rate.

One of the commonest sources of gas bubbles in detectors is when a column is replaced by one which is free of mobile phase either because it is new or because it has dried out on storage. If this column is installed into the liquid chromatograph, the air contained in the column will be swept into the detector. This situation may be avoided by initially connecting only the column inlet to the chromatograph, actuating the mobile phase pump and purging the column until free of visible air bubbles before connecting to the detector. If an air bubble becomes trapped within a flow cell, very poor stability of the recorded baseline is observed. These bubbles can normally be removed by a momentary change of back pressure, *i.e.*, releasing the detector outlet to atmosphere or blocking the flow completely for a fraction of a second while the mobile phase pump is still operating. Alternatively, the detector flow cell must be disconnected from the column and back-flushed with a solvent, such as alcohol, using a conventional syringe. A 2-ml glass syringe is ideal for this purpose.

Should a blockage occur within one of the narrow-bore tubes in a liquid chromatograph it is always better to release the mobile phase pressure, to disconnect the offending part, if known, and to back-flush using either a glass syringe filled with liquid or a length of PTFE tubing coupled to the outlet of the mobile phase pump. The other end of the PTFE tubing can be connected to capillaries or flow cells in the instrument and the mobile phase used to back-flush the components. Most PTFE tubing will withstand pressures of 25 bars (350 p.s.i.g.), which is adequate for the purpose. The temptation to use the pump pressure to displace offending particles or swarf by forward flushing usually results in a blockage which is even more difficult to remove than the original one.

FRACTION COLLECTORS

A convenient feature of most commercial liquid chromatographs is the provision of a manually operated fraction collector valve located in the liquid flow line immediately after the detector(s). This valve, as its name implies, enables individual, separated components to be collected for further examination. Unlike in GC, where – in simple systems – the efficiency of sample collection is often quite poor, the procedure in the liquid phase is very straightforward and essentially quantitative.

Provided that the collection valve is of as low internal volume as other parts of the liquid flow path and is located immediately after the detector, before any back-pressure device, as mentioned earlier, the separated component will emerge from the collection valve within a second or so of passing the detector. In many instances the response time of the electronics of the detector and recorder are in the order of 1 sec, thus collection can be made as and when peaks appear on the recorded chromatogram. The characteristics of any particular instrument may be checked by injecting a coloured compound, such as a food dye, into the chromatograph and measuring the time delay between the moment the detector responds to the substance and the moment one sees the colour emerge at the collection point. Such is the simplicity of sample collection that for many high-speed separations the method is quite adequate. Only when the number of components to be collected is quite large and when they elute over a fairly long period of time is it worth considering the use of the automatic fraction collectors of the type which have been used for many years with conventional column chromatography. These fraction collectors are so well established that it is unnecessary to discuss them in any detail in this text. Automatic collectors are normally actuated by a definite increment of time or liquid volume flowing from the column. In this latter case the "sensor" is a drop-counter for low liquid flow-rates or a siphon-counter of 1- to 10-ml capacity for higher flow-rates. An alternative method is to fit a microswitch to the pen recorder which is "triggered" as the pen responds to an eluting peak. Some modern electronic integration systems have an external command facility permitting a similar control of a fraction collector without the need to modify a pen recorder. When considering using any automatic fraction collector, particular care should be taken to avoid any sample carry-over, or loss of resolution, due to dead space in the collecting device.

MEASUREMENT OF MOBILE PHASE FLOW-RATE

Accurate measurement of the mobile phase flow-rate during an analysis is important since the records – the chromatogram or integrator print-out – normally yield only data in terms of time, not volume. However, some recently introduced pumping systems are provided with an electrical output which enables the recorder chart speed to be related directly to the liquid flow-rate. In applications such as steric exclusion work, where the molecular size is related to the elution volume, an accurate conversion from time to volume is essential.

Similarly, the sensitivity of some detection systems and the efficiency of chromatographic columns are dependent on the mobile phase flow-rate (the latter strictly the linear velocity), thus in quantitive work it is necessary to reproduce precisely the operating conditions each time the analysis is performed.

With many of the modern, positive displacement pumps and those equipped with flow controllers, the desired mobile phase flow-rate is selected by adjusting the controls of the pump drive system. In these circumstances the actual flow-rate through the chromatographic column will be essentially as set on the pump controls, unless one suspects a pump malfunction, a leak in the system, compressibility of the liquids (which is seldom more than a few per cent over the pressure range used in LC), or a change in volume occurring when mixing two liquids, *e.g.*, equal volumes of alcohol and water when mixed yield less than the combined volumes.

The simpler mechanical pumps and those driven by pneumatic pressure (but without flow control) give flow-rates dependent on the resistance to flow in the column, mobile phase viscosity, and temperature. In these instances the flow-rate should be measured as a matter of routine.

Methods of flow measurement include: (A) volumetric measurement, (B) gravimetric measurement, and (C) flow meters.

Volumetric measurement

Simply collecting the column effluent in a measuring cylinder for a given period of time is the most widely used method of flow measurement. In steric exclusion work it has been the practice to automate this procedure by using a "siphon counter". With this device each time a certain volume, commonly 1, 5 or 10 ml, has issued from the column, the siphon empties. This is sensed by photocells, which give rise to an event mark "spike" on the chromatographic trace, thus a semi-continuous record of flow is obtained.

Gravimetric measurement

Gravimetric measurement involves collecting the effluent in a pre-weighed container for a given time interval followed by weighing. Although more precise than the volumetric method, it is tedious to perform and is normally only used when wishing to carefully check for one of the faults mentioned above.

Flow meters

Flow meters usually comprise a calibrated tapered glass tube in which a float or ball of known mass is suspended by the upward flow of the moving liquid. An increase in the liquid flow-rate will result in the float being raised within the calibrated tube. This method of flow measurement is of marginal value in LC since the position taken up by the float and hence the indicated flow-rate is dependent on the specific gravity of the mobile phase, which in the case of analyses involving gradient elution is changing continuously. Another disadvantage in practice is that the smallest of air bubbles can become attached to the float and this provides additional buoyancy to the float, leading to a very inaccurate measurement.

An alternative flow-measuring system has been developed commercially in which a small air bubble is injected into a tube through which the mobile phase is flowing. Two photocells are positioned a known distance (volume) apart on the tube. As the air bubble, swept by the mobile phase, passes the first photocell, a digital timer is started, which stops as the bubble passes the second photocell. Using this method very precise flow-rate measurements may be obtained.

PRESENTATION OF RESULTS

It was mentioned earlier that the goal in the development of LC is to achieve a complementary analytical technique to GC particularly in regard to speed of analysis and presentation of results. On the latter point, there is now no difference in the two techniques.

Chromatographic data are presented almost universally in the form of a chromatogram using a strip chart recorder. For quantitive analysis and greater precision in retention time measurements, digital integrators, computing integrators, and computing systems may be employed. Their specification is essentially the same as in the case of GC, *i.e.*, fast response time, wide linear dynamic range, and capable of accepting both narrow (fast-eluting) and wide (slow-eluting) peaks. For maximum convenience, strip chart recorders should be provided with a wide range of chart speeds, as some chromatographic methods take but a few minutes to complete whereas others take hours.

In quantitive work, computing integrators and dedicated computers are becoming popular, for once the detector response data and other basic information have been fed into the system, the analytical results are calculated and printed in report form by the computer. In laboratories where numerous repetitive samples are analysed these systems can offer a significant reduction in time and effort made by the operator. Methods of quantitation are discussed in more detail in Chapter 12.

The features offered on commercial data systems differ in detail from model to model. Specific information is best obtained directly from the manufacturers as specifications and prices on items of this nature tend to change frequently.

AVAILABILITY OF LC EQUIPMENT

Most of the instrumental components that have been described in this chapter are available as commercial products. There have been and probably always will be differences in opinion regarding the decision whether to purchase a complete liquid chromatograph from a commercial source or to construct a home-made liquid chromatograph from the various component parts. Factors in favour of the do-it-yourself approach are most certainly the lower initial capital outlay and to a lesser degree the ability to custom design the apparatus for a specific purpose, provided of course the background know-how concerning the design is available. The drawbacks to this approach arise from the lack of any instrument service back-up and from the fact that building and running repairs can absorb a considerable amount of laboratory time. Another feature which cannot be overlooked is that most instrument manufacturers, while employing readily available components, custom modify them to give a certain performance advantage – details of these modifications and any sophisticated control options may not be available to those who prefer do-it-themselves.

As an aid to those who may wish to obtain details on commercially available LC equipment, Appendix 3 contains the addresses of instrument manufacturers at the time of writing. Details of the products of each company, *i.e.*, type of equipment offered and prices, are not given as these are continually changing as new models are introduced.

REFERENCES

1 L.R. Snyder, *J. Chromatogr. Sci.*, 8 (1970) 692.
2 R.P.W. Scott and P. Kucera, *J. Chromatogr. Sci.*, 11 (1973) 83.
3 C.D. Scott, D.D. Chilcote and N.E. Lee, *Anal. Chem.*, 44 (1972) 85.
4 G. Hesse and H. Engelhardt, *J. Chromatogr.*, 21 (1966) 228.
5 D.C. Locke, *J. Gas Chromatogr.*, 5 (1967) 202.
6 C.G. Horvath, B.A. Priess and S.R. Lipsky, *Anal. Chem.*, 39 (1967) 1422.
7 R.J. Maggs, *J. Chromatogr. Sci.*, 7 (1969) 145.
8 H. Felton, *J. Chromatogr. Sci.*, 7 (1969) 13.

Chapter 5

Liquid chromatographic detection systems

INTRODUCTION

The purpose of a detector in a LC system is to faithfully monitor the composition of the liquid eluting from a chromatographic column and enable, by electronic means, a record of how the composition varies with time to be presented on a strip-chart of a pen recorder. In the ideal situation the detector should be able to monitor a separation but should not influence the extent of the separation. This statement may seem a little strange at first sight, but one of the greatest problems in analytical LC is the deleterious effect on a separation which can be created by dead volume and/or poor flushing characteristics of the parts of the detector through which the mobile phase passes.

Before discussing the operation of detectors in detail it is considered instructive to explain some of the terms that are used to describe the quality of the recorded trace, other than the degree of chromatographic resolution of the peaks, for the benefit of less experienced chromatographers. These terms include "short-term noise", "long-term noise", "drift" and "non-linearity".

High-frequency (or short-term) noise

This symptom is observed on a chromatogram as a fuzzy trace due to high-frequency (usually greater than 50 Hz) oscillations of the recorder pen. This type of noise usually originates from incorrect grounding of the detector and/or the recorder. Alternatively, such oscillations can occur as a result of the gain of the recorder amplifier being set too high or the use of a recorder with too fast a response time. Careful attention to instrument grounding and matching the output impedance of the detector to the input of the recorder are needed to overcome this problem. In many situations high-frequency noise may be reduced using a capacitance-resistance filter, but this should only be considered satisfactory if the resultant decrease in response time does not interfere with the faithful recording of the separation.

Another source of high-frequency or short-term noise is the Schotky effects, *i.e.*, random electron motion, within the electronic components. Since these are fundamental to the nature of the electronics employed, *e.g.*, transistors, the level of this noise can only be improved by selecting components of higher quality. Understandably, there must be a limit on any such improvement, based on existing technology and price.

Long-term noise

This problem covers recorder baselines which are erratic or "lumpy", *i.e.*, have low-frequency random noise. Baseline instability of this type is most often caused by changes in the nature of the mobile phase flowing through the detection system, *i.e.*, impurities. The most common impurities are air and an immiscible liquid, for example, the stationary phase or the previous mobile phase bleeding from the column. If long-term noise is cyclic or regular, *i.e.*, not random, then the source is invariably a heater (thermostat) control or, with some equipment, insufficient mixing of two liquids being delivered to the column system from a two-pump or gradient elution device.

Drift

Characteristically the baseline will continuously move upscale or downscale over a considerable period of time, *i.e.*, 1 h. Such a baseline shift is most often associated with temperature or mobile phase changes or the approach of an equilibrium state of either. A baseline drift is also very common when employing solvent programming techniques, such as gradient elution, to increase the speed of a separation.

Non-linearity

When performing quantitive analysis it is almost essential that the electrical response produced by the detector is directly proportional to the mass or the concentration of the component passing through the detector. If this condition is satisfied, the detection system is said to be linear. Such linearity may be assessed experimentally by plotting a graph, on a logarithmic scale, of the detector response *versus* the mass of sample injected into the chromatographic system. A perfectly linear behaviour will be characterised by a straight-line plot having a slope of unity. Care should be taken, however, to ensure that any observed apparent deviation from linearity of the detector is not caused by other effects within the chromatograph, *i.e.*, limiting sample solubility, column overload or much increased injection volume. No detection system is linear over an infinite mass range although some offer good linearity over three or four orders of magnitude of sample size.

It should be borne in mind that any observed non-linearity may arise from the detector design, the principle on which the detector operation depends or the sample under test. For instance, a compound which is known not to obey Beer's Law cannot be expected to give good linearity when the analysis is monitored with a photometric detector. However, a compound which does obey Beer's Law would not appear to behave in a linear manner if the design of the photometric detector allowed the absorbance of the sample to be measured with polychromatic light.

Beer's Law, one of the fundamental laws of spectrophotometry, states: The intensity of a beam of monochromatic light decreases exponentially as the concentration of the absorbing substance increases arithmetically. Expressed mathematically this becomes

$$\log\left(\frac{I_0}{I}\right) = l \cdot C \cdot \epsilon = A$$

where I_0 is the intensity of the incident light, I the intensity of the transmitted light, l the optical path length of the flow cell, ϵ the molar extinction coefficient, C the concentration of the sample in gram-moles per litre and A is the absorbance of the solution. This relationship is only valid for monochromatic light; the presence of light of other wavelengths, at which the compound of interest does not absorb, leads to a non-linear relationship between absorbance and concentration of sample.

Alternatively, refractive index detectors, for instance, do not exhibit a particularly wide linear range and although one could criticise the design of some detectors, it should be appreciated that the laws of refractometry on which the detectors are based do not suggest that a linear relationship exists between the refractive index and solution composition over a wide range of concentration.

All of the various types of detection systems can be conveniently divided into two categories. Firstly, those systems which by virtue of the principle on which they operate respond to a wide range of substances with much the same order of sensitivity – the so-called non-specific or universal detectors. The second category are those which are unquestionably selective in their response, offering very high sensitivity towards some chemical types, but are of little, or no, use with other substances.

In practice there is no truly universal detector which responds to all species with approximately the same sensitivity. This is perhaps not unexpected since most workers' idea of a universal detector would be one which enabled them to observe the separation of the components in the sample undergoing chromatographic examination yet not to observe minor changes in the composition of the mobile phase or any baseline disturbance during the course of a separation involving the use of gradient elution.

PRINCIPAL REQUIREMENTS OF A LC DETECTOR

The features which need to be considered when assessing a detection system may be summarised as follows. An ideal detector should: (1) be of such a design that the separated components are not re-mixed while passing through the detector, (2) have a low drift and noise level so that small quantities of eluting components may be observed, (3) have a fast response time so as to faithfully record fast eluting peaks, (4) have a wide linear dynamic range so that quantitive analysis may be accomplished in a straightforward manner, (5) be relatively insensitive to changes in mobile phase flow-rate, temperature and composition – within the limits described earlier, (6) respond either to all substances in an equivalent manner, or alternatively, if selective, be readily tunable so that its response to different species may be optimised, and (7) be easy to operate and reliable.

Detectors based on many principles of operation have been proposed for LC, yet only a few have proved sufficiently versatile and robust to be widely used and produced commercially. Indeed, several detectors which were available a few years ago have subsequently been withdrawn from the market. In the following sections only the successful detection systems are discussed in any detail; brief mention only will be made of the other lesser-used and experimental types. This area of liquid chromatography is still one where there is great need of new ideas and necessity for improvement in the design of many of the existing detection systems.

PHOTOMETRIC DETECTORS

Detectors based on the absorbance of light in the visible or UV regions of the spectrum are probably the type most widely used currently in LC. A wide range of detectors are

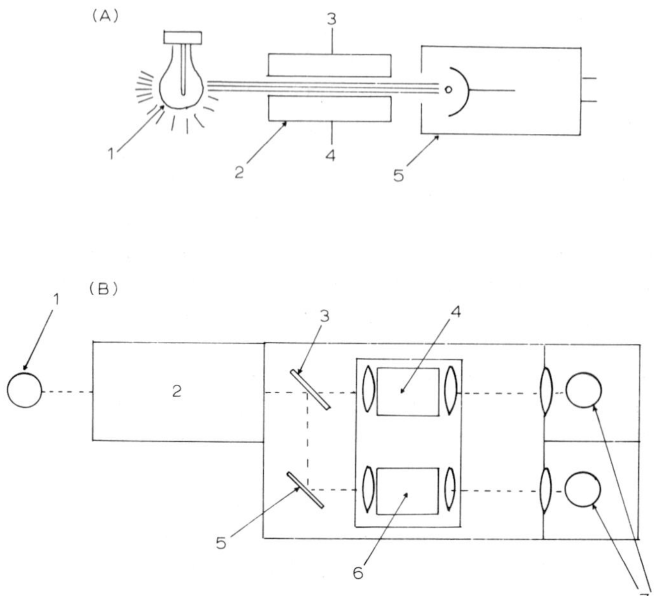

Fig. 5.1. Optical lay-out of single- and double-beam photometric detectors. (A) Single-beam detector, illustrated in the form of a fixed-wavelength photometer. (1) Spectral source, *e.g.*, low-pressure mercury lamp; (2) flow cell; (3) outlet; (4) inlet; (5) phototube. (B) Double-beam detector, illustrated in the form of a variable-wavelength photometer. (1) Spectral source, *e.g.*, deuterium lamp; (2) monochrometer; (3) beam splitter; (4) analytical flow cell; (5) mirror; (6) reference flow cell; (7) photodiodes. (Reproduced by courtesy of DuPont.)

available all based on the principles of photometry. Two basic modes of operation of photometric detectors are possible, depending on whether a reference light path is provided. The two types, depicted in Fig. 5.1, are referred to as either single-beam or double-beam photometers.

In the single-beam mode, the energy from the source lamp passes through the sample flow cell to a photocell via some wavelength selection device. Selection of the operating wavelength may simply rely on the emission characteristics for the light source or, more commonly, on the use of a monochromator or high-quality optical filters. Variations in the intensity of light falling on the photocell due, in favourable cases, only to absorption of light by the liquid occupied in the flow cell are converted electronically to give an output signal suitable for a strip chart recorder. The output signal from the detector may be linear with respect to changes in the transmittance or the absorbance of the liquid in the flow cell. The latter output is to be preferred for most LC applications as absorbance is linearly related to the concentration of an absorbing component in solution. Simple detectors based on a single-beam optical arrangement can suffer from instability problems, as a variation in the light falling on the photocell may be caused by effects other than a change in absorbance in the sample cell, for instance, a fluctuation in the intensity of light emission from the source lamp. However, it should be possible to produce highly stabilised spectral sources which overcome this problem. The second type of photometer, the double-beam system, is generally preferred in most chromatographic work. In this arrangement the reference light beam can be used to monitor all variations in the system other than the change in absorbance in the measuring flow cell. The signal from the two photocells is fed to a differential amplifier. The trace produced on the strip chart recorder represents how the difference between the two signals varies with respect to time. The relative merits of employing a second flow cell in the double-beam photometer system are discussed in a later section of this chapter. The modern photometric monitor is capable of detecting changes in the absorbance as low as 5×10^{-5} absorbance units (twice the short-term noise); typically full-scale deflection of the pen on a strip chart recorder will correspond to 0.01 absorbance units. In terms of sample size, under favourable conditions, this is equivalent to a concentration of approximately 10^{-9} g/ml of a component in a column effluent.

The most established version of a photometric detector is without doubt that employing a low-pressure mercury lamp as a spectral source. This lamp emits light of a very high intensity predominantly at a wavelength of 253.7 nm (usually rounded off to 254 nm). This high-energy output has enabled simple yet high-performance detectors to be constructed with liquid flow cells of quite low internal volume, typically less than 10 μl with an optical path length of 10 mm. A similar detector having a 1-μl volume combined with a 5-mm optical path has also been reported in the literature[1]. It should be appreciated that this low-pressure mercury lamp is not monochromatic although most (approximately 85%) energy is emitted at 254 nm. The other "stray" emission lines must be eliminated if a good linear response over a wide concentration range of a sample is to be obtained. This is achieved by inserting a narrow band-pass interference filter into the optical path. The high-energy output of the simple, yet robust, low-pressure mercury lamp at 254 nm is somewhat fortuitous, as this wavelength is long enough to allow a reasonable choice of organic solvents for use as mobile phases without having an unacceptably high background absorbance while still operating in a region of the spectrum where many organic compounds absorb light quite strongly even though it may not correspond to the wavelength where the maximum absorbance occurs.

As interest in LC as an analytical technique developed, the demand for photometric detectors with variable wavelength capability increased. This need is quite understandable when it is appreciated that the wavelength at which maximum absorption of light occurs varies considerably from one substance to another. Operation of a detector at this particular wavelength will clearly optimise the response of the detector for the compound of interest. Similarly, when dealing with a sample in which a considerable background interference from other components occurs, it is frequently possible to operate the detector at an alternative wavelength where the interferences are less severe. As an example, the detection of the carcinogenic aflatoxins in cereal products is an application of considerable importance in meeting the strict demands set by food regulatory authorities of today. These substances absorb light strongly at both 254 and 365 nm. At the former wavelength, which is offered by most photometric detectors, in this particular instance it is unfortunate that many other compounds also absorb light to a similar extent making detection of the compounds of interest at the sub-part-per-million level impossible. At a wavelength of 365 nm the situation is completely changed, in that most of the sample co-extractives are transparent and no longer interfere with the detection of the aflatoxins. The work of Baker *et al.*[2] has shown that by working at this wavelength the toxins may be detected in samples of peanut butter at concentrations lower than 1 part in 10^8.

Provision to operate simple, single-wavelength photometers at wavelengths other than 254 nm can be made by the use of phosphors which absorb the source radiation and re-emit light at longer wavelengths. Phosphors need to be carefully selected as they can be inefficient in terms of energy output. To ensure good detector linearity interference filters should be used to eliminate all but the desired wavelength. High-quality narrow band-pass filters giving greater than 50% transmission at the desired wavelength are available for the visible region of the spectrum, however, in the UV region filters rarely transmit more than 25% of the incedent radiation.

An alternative approach, while still retaining the comparatively simple optical bench, is to use an alternative spectral source and interference filters to isolate the desired wavelength. This method overcomes the disadvantage of using inefficient phosphors. The use of filters is, however, only possible when the source lamp emits a line spectrum as interference filters do not have a sufficiently narrow band pass to provide monochromatic light from a continuum. The emission spectrum of the so-called medium or high-pressure mercury lamp has been used commercially for this purpose as the emission lines are well separated and interference filters enable a range of essentially monochromatic lines to be employed for detection purposes. These lamps do generate a considerable amount of heat and some means of heat dissipation must be provided to avoid an excessively rise in the temperature of the flow cell.

The logical way of providing a photometric detector with the option of varying the wavelength so as to optimise the response towards a particular compound is to employ an optical monochromator functioning in a manner similar to spectrophotometers. In this situation the spectral source provides a continuous light emission over a wide range of wavelengths and the desired wavelength is isolated by a diffraction grating and/or prism. Commercial general-purpose spectrophotometers are not very suitable for use as monitors for high-efficiency LC separations as the energy output at any discrete wavelength is low and the light is not focussed into a narrow, say 1-mm diameter, beam which is ideally required for good passage of energy through a low-volume flow cell. A compromise between energy throughput, flow cell volume and band width of the light has to be made. The band width of the light is one area where some sacrifice is possible enabling a higher energy throughput with some slight deterioration in the linearity at high absorbance values. Spectrophotometers specifically designed as LC monitors have band widths in the region of 5 nm rather than 0.1 nm, which is typical for an analytical spectrophotometer. A very practical feature, which is offered with at least one spectrophotometric detector (the Siemens/Zeiss, Model PM4 CHR), is the possibility to select band widths to suit the application, the choice being 5, 10, and 20 nm. Thus, if minor components must be detected, a high-energy throughput is obtained by utilising the widest band width, whereas precise quantitation is achieved with the narrowest spectral band width, as this will give the greatest linear dynamic range.

When working with differential photometric detectors one is sometimes faced with the choice between using a single flow cell plus an air reference and a dual flow cell arrangement, a reference system being employed where compensation for the characteristics of the mobile phase is possible. In practice a single flow cell plus an air reference is quite adequate when using solvents having good (greater than about 75%) transmission at the operating wavelength and when mobile phase composition is not changing due to a programming technique such as gradient elution. When working with solvents having higher background absorbance or where the mobile phase is changing, some compensation of the change in background absorbance can be achieved by using a dual flow cell differential system. In this arrangement the mobile phase entering the analyser is split at a T-piece immediately before the sample injector. One liquid flow path by-passes the injector, and then passes through a dummy or reference column to a reference flow cell in the detector. The other liquid stream continues from the T-piece through the sample injector, the separating column and the measuring flow cell. When the flows through the two columns are closely matched, the detector baseline stability can be improved over that obtained with the single-cell version. It should be appreciated that in practice exact compensation of baselines during say a gradient elution run where the solvents forming the mobile phase absorb to different extents at the wavelength at which the detector is operating, can require careful setting-up, particularly if the detector is to be operated at high sensitivity.

Although much work is performed with detectors which operate in the UV region of the spectrum, a good deal is also practiced in the visible region. Perhaps the most widely known is the now classical ninhydrin colour reaction by which amino acids and other structurally related compounds are detected by monitoring the intense blue colour developed in the reaction. For these and similar analyses sample detection relies on the measurement of the increase in absorption when a colorimetric reagent is mixed with the column effluent; accurate detection of the sample components depends on measuring the absorbance of the "coloured" species, corrected for any change in the background absorbance due to depletion of the reagent[3]. In a somewhat similar manner, other selective detection methods can be envisaged by using different types of post-column colour reactions in combination with this type of differential photometric detector.

There has been some interest in a photometric detector operating in the infrared region of the spectrum. Although there is at least one commercial model available (from Wilks Instruments), there appears to be little interest at this stage due largely to the severe restriction on the solvents that may be employed as mobile phases.

FLUORESCENCE DETECTION

Fluorimetry as an analytical method is well known for its very high selectivity and sensitivity to very small quantities of some samples, while being completely insensitive to many other materials. Interest in this principle of detection for LC has been generated since many important biological substances, *i.e.*, drugs, vitamins and steroids, fluoresce quite strongly under conditions which give rise to little interference from the complex co-extractives occurring in many biological fluids.

In this method, the eluting compound passing through the flow cell absorbs radiation from an intense spectral source (usually ultraviolet) and then fluoresces, emitting light of a longer wavelength. This emitted radiation is measured by some light-sensitive device, usually a photomultiplier. The success of any design of detector depends on maximising the fluorescent radiation reaching the photomultiplier while blocking the excitation

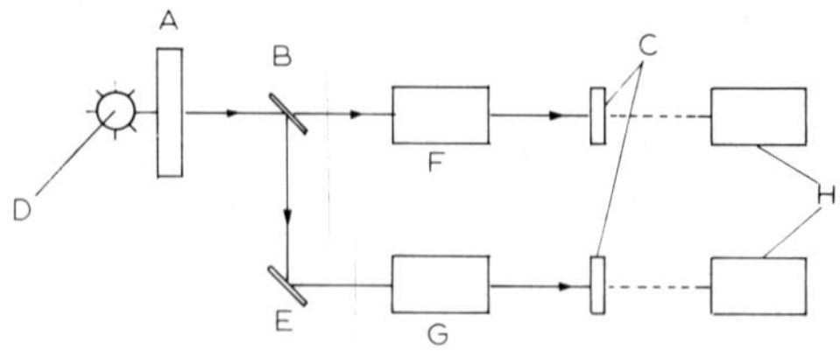

Fig. 5.2. Optical layout of an in-line fluorimeter. (A) Excitation filter; (B) beam splitter; (C) emission filters; (D) spectral source, *e.g.*, medium-pressure mercury lamp; (E) mirror; (F) analytical cell; (G) reference cell; (H) photo cells. (Reproduced by courtesy of Laboratory Data Control.)

(source) radiation. The emitted (fluorescent) light invariably has a significantly longer wavelength compared with the excitation light. Light from the excitation source is prevented from entering the photomultiplier by optical filters. One filter is fitted to the source to block any wavelengths which are longer than needed for optimum excitation and a second on the window of the photomultiplier to eliminate any of the excitation lines. Fluorescence detectors may be constructed in two optical arrangements, depending on whether the measuring photomultiplier is in line with the spectral source or positioned at right angles to the light beam. Both forms are available commercially. In a similar way to photometric detectors some fluorescence monitors may be fitted with a dual flow cell system enabling compensation for any fluorescence of the mobile phase. Fig. 5.2 illustrates diagrammatically the arrangement of a dual flow cell, in-line fluorimetric detector and Fig. 5.3, a single flow cell, right-angle fluorimeter which also has the capability of simultaneously monitoring the absorbance of the column effluent.

Conventional analytical methods involving fluorimetry often show much higher sensitivity than the corresponding absorbance methods; with LC–fluorescence detection the gain in sensitivity is comparable. In many instances it is possible to analyse picogram amounts of samples, *i.e.*, that quantity actually injected into the LC column as distinct from the mass of sample occupying the detector flow cell. The gain in selectivity using

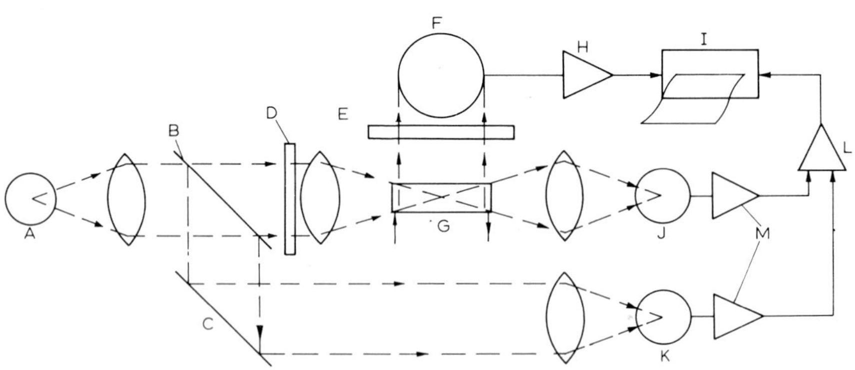

Fig. 5.3. Optical layout of a single flow cell photometer for simultaneous absorbance/fluorescence detection. (A) Lamp; (B) 10% mirror; (C) mirror; (D) excitation filter; (E) emission filter; (F) photomultiplier; (G) flow cell; (H) linear amplifier; (I) recorder; (J) sample phototube; (K) reference phototube; (L) linear amplifier; (M) log amplifier. (Reproduced by courtesy of DuPont.)

fluorescence monitoring is also very substantial, so that there are many instances where the overall sensitivity of a method is gained by analysing a larger sample where many of the components of little interest are non-fluorescent, making detection of fluorescent impurities a straight-forward matter. The versatility of this method of detection can be increased considerably by formation of derivatives of the sample using fluorigenic reagents. This may be accomplished before the chromatographic separation or afterwards by feeding the reagent in at a T-piece located between the column outlet and the detector flow cell. Possibly the most established methods based on these procedures are the formation of the fluorescent dansyl derivatives of amines and phenols using the reagent 5-dimethylamino-1-naphthalenesulphonyl chloride prior to chromatographic separation[4] and the use of fluorescamine as a fluorigenic reagent for amino acids after their separation by ion-exchange chromatography[5].

The intensity of fluorescent emission is dependent on the intensity of the excitation radiation. Since this radiation is of necessity absorbed by the compounds present in the flow cell, the effective intensity of the source decreases when strong absorption occurs, leading to an apparent non-linearity of the detector. Consequently, quantitation by fluorimetric methods is best performed with very dilute solute solutions and with UV transparent mobile phases.

When considering this method of detection, it is well to realise that some chemicals, particularly anions, possess marked fluorescence quenching characteristics (one of which is water – probably the most common liquid employed as a mobile phase) and also that decreased temperature or increased solution viscosity enhances fluorescent emission by reducing the chances of deactivating collisions. Unfortunately, the range of temperature over which a LC detector can be operated is quite small in relation to that required to effectively reduce the number of molecular collisions.

REFRACTIVE INDEX DETECTORS

Detectors based on differential refractometry, that is giving an output signal proportional to the difference in refractive index of liquids contained in two flow cells, are the most widely used detectors which are essentially non-specific. Although the absolute refractive index values of substances differ, the range of possible values is quite small relative to the very large differences that exist in the UV absorption or fluorescence characteristics of different compounds. Refractive index monitors have the distinct advantage that they are capable of detecting virtually all compounds provided the refractive index of the sample is not identical to the refractive index of the mobile phase. However, the systems only offer moderate sensitivity, *i.e.*, a limit of detection in the order of 10^{-6} g/ml of column effluent. The principal disadvantages of detectors of this type are that they are very sensitive to small changes in temperature and to pressure fluctuations, the former being a characteristic of the intensive physical property on which the detector is based rather than being the result of an instrumental design fault.

The temperature coefficient of refractive index is such that, when working at a sensitivity where differences in refractive index as small as 10^{-6} refractive index units are to be chromatographically significant, the temperature difference between the measuring and reference streams in the cell must be less than 0.001°C. This marked dependence on temperature is shared with other types of bulk (intensive) property detectors such as vapour pressure[6]. It is of major importance to eliminate any temperature difference between the two flow cells. All commercial differential refractometers incorporate some form of heat exchanger which enables the temperature of the two liquid streams to be closely matched. This usually comprises fine-bore capillaries which take the liquids to and from the detector flow cell and which are in intimate contact with a body of high thermal mass. This thermal mass may be a large metal block (heat sink) or a water-filled chamber. Heat exchangers of these types can adequately match the temperature of the two liquid streams and, due to the high thermal mass, provide some stabilisation against laboratory temperature fluctuations. Careful attention must be given to the design of heat exchangers, in particular to the internal volume of the capillaries, which, if large, can lead to excessive peak broadening. Maximum freedom from drift over a long period of time can only be obtained by working with the detector temperature controlled. This is achieved by working in a constant-temperature environment or providing the detector with an electrical or circulating liquid (normally water) thermostat. This latter arrangement is often employed where the heat-transfer liquid is continuously pumped through the heat exchanger in a closed loop. This compensates effectively against long-term drift as these thermostats are capable of maintaining the temperature within 0.01°C of a pre-set value. When operating the differential refractive index detectors at high sensitivity, however, it is often observed that the detector will respond to the on–off cycling of the heater in the thermostat giving rise to long-term noise. In most instances this can be eliminated by introducing some capacity or mixing volume into the line carrying the heat-exchange liquid to the detector. This mixing volume can be simply a large container, *i.e.*, a 5-l glass bottle with an inlet and outlet tube sealed into the stopper. Only one tube, the inlet, should reach to the bottom of the container.

Since these detectors are essentially non-specific, their sensitivity applies equally to variations in the mobile phase composition as it does to eluting samples. It is for this reason that it is common practice to operate the detectors in a truly differential mode, *i.e.*, two columns, two flow cells, etc., as described for photometric detectors and to inject samples into one column system only. It is possible to operate detectors with a static liquid in the reference cell but, in general, the stability of the detector in terms of drift is not as good as when there are two flowing liquid streams. There is also an imbalance of pressure in the two cells which can lead to an unacceptably high baseline offset or solvent leakage (described in more detail later). The sensitivity of the refractometric detectors to the slightest change in mobile phase composition rules out their use for monitoring separations involving gradient elution, since it is almost impossible to arrange for an exactly equivalent mobile phase composition to be in both flow cells at the same instant during a gradient elution programme.

Although several types of differential refractive index detectors have been described, the two most popular are known as the reflectance refractometer, based on the Law of Fresnel, and the deflection refractometer.

Reflectance (Fresnel) type of refractive index detector

In this version, shown in Fig. 5.4, the dual flow cell is formed by a very thin PTFE gasket held between a glass prism and a stainless-steel plate containing four ports for the inlets and outlets of the two liquid streams.

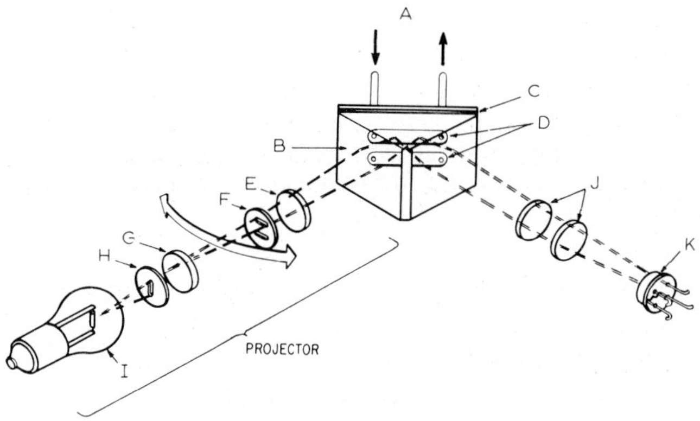

Fig. 5.4. Optical lay-out of a reflection type of differential refractive index detector. (A) Sample and reference stream flow; (B) prism; (C) base plate; (D) cells; (E) collimating lens; (F) aperture mask; (G) infrared blocking filter; (H) source mask; (I) source lamp; (J) detector lens; (K) dual detector. (Reproduced by courtesy of Laboratory Data Control.)

This design relies on measuring refractive index differences at the critical angle of the light reflected from the metal surfaces. The principle of detection is based on Fresnel's Law of reflection. This law may be stated as follows: the fraction of light reflected (or transmitted) at a glass–liquid interface varies with the angle of the incident light and the refractive indices of the two substances. Detector flow cells of this design are particularly attractive in that the cell volume is very low, in the order of 3 μl, and the cells are very efficiently swept. For this reason, such detectors are ideally suited for monitoring effluents from high-efficiency columns. These cells, comprising a thin layer of liquid between the prism and the back plate, are rather susceptible to a solid film forming on the surface of the cell, which must be removed if any imbalance should occur. This is achieved by removing the prism and cleaning the surfaces with a moist tissue. The task is more delicate than time consuming. Particular attention must be given to the alignment of the PTFE gasket forming the cells, for, if misplaced, leakages of mobile phase can occur. Similarly, the detector should be operated with two flowing streams rather than a stationary reference, otherwise the imbalance of pressure in the two cells can displace the fine centre part of the gasket. Two prisms are required to cover the entire range of refractive indices of possible mobile phases. One prism is satisfactory when working with liquids of refractive index 1.31–1.45 and the other for the refractive index range 1.40–1.55.

Deflection type of differential refractometer

With this model of differential refractometer, shown diagrammatically in Fig. 5.5, the

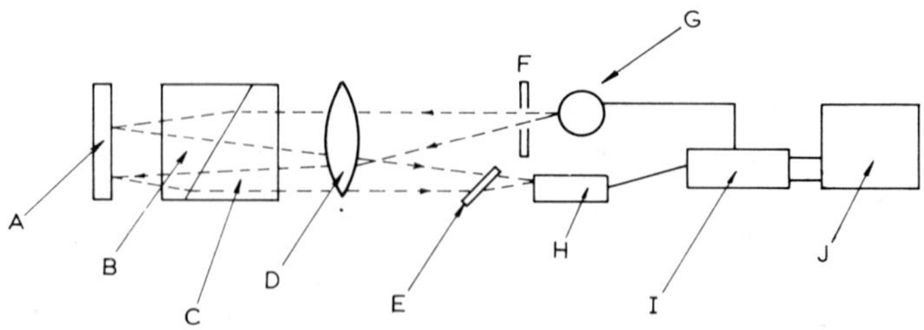

Fig. 5.5. Optical lay-out of a deflection type of differential refractive index detector. (A) Mirror; (B) sample; (C) reference; (D) lens; (E) optical zero; (F) mask; (G) light source; (H) detector; (I) amplifier and power supply; (J) recorder. (Reproduced by courtesy of Waters.)

cell consists of two wedge-shaped sections through which the sample and reference liquid streams flow.

The light beam is transmitted through the dual cell, reflected by a mirror so that it is passed back through the flow cell a second time, and focussed on a light-sensitive detector. This design is easier to use than the Fresnel type of refractometer as it is not necessary to change the optics for liquids of different refractive index and is less affected by contamination. It has also been suggested that this type of refractometer gives a superior linear range of response to an increasing mass of sample. The flow cells are, however, of significantly larger volume and not as efficiently swept as those of the Fresnel type.

PHASE TRANSFORMATION DETECTORS

The influence of the ideas of gas chromatographers on the development of LC is large and quite apparent. However, the concept of the phase transformation or solvent detectors must rate as one of the most significant. The previous sections have mentioned a number of limitations of LC detectors created because they respond to variations in composition of mobile phase, temperature, air bubbles, etc. The solvent transport type of detector sets out to eliminate these problems by providing a system whereby the column effluent, *i.e.*, the mobile phase plus any sample components, is fed on to a moving belt, wire or chain where the relatively volatile mobile phase evaporates, leaving a residue of the less volatile sample component. This is in turn removed from the transporting system by pyrolysis or oxidation at high temperature and the gaseous products are fed directly or indirectly to a GC type of detector. Thus in this process, provided the mobile phase is totally volatile, *i.e.*, leaves no residue, and the sample is relatively non-volatile, the detection system cannot suffer from the problems associated with the photometric or refractometric detectors. Gradient elution operation has no effect on the stability of the recorded baseline produced by the phase transformation detector provided the solvents employed volatilise readily and leave no residue. The apparent simplicity of this process led many companies to investigate LC detection systems based on this principle. Currently, however, only one manufacturer, Pye Unicam, continues to produce a detector based on this principle.

PHASE TRANSFORMATION TO FLAME IONISATION DETECTOR

The concept of the phase transformation detection system is readily deduced from the diagram produced in Fig. 5.6.

The column effluent flows through a coating block where a proportion of the liquid is taken up on a moving wire, the remaining column effluent passing to drain or to a fraction collector. The fraction of effluent picked up on the moving wire passes a small oven, where the solvent evaporates. The non-volatile residue is then carried by the moving wire to a high-temperature oxidation furnace which converts any carbon in the sample to carbon dioxide. This gas is swept in an oxygen stream to a molecular entrainer which effectively transfers the carbon dioxide to a hydrogen stream. This then passes through a catalyst, converting the carbon dioxide to methane before finally passing on to a flame ionisation detector.

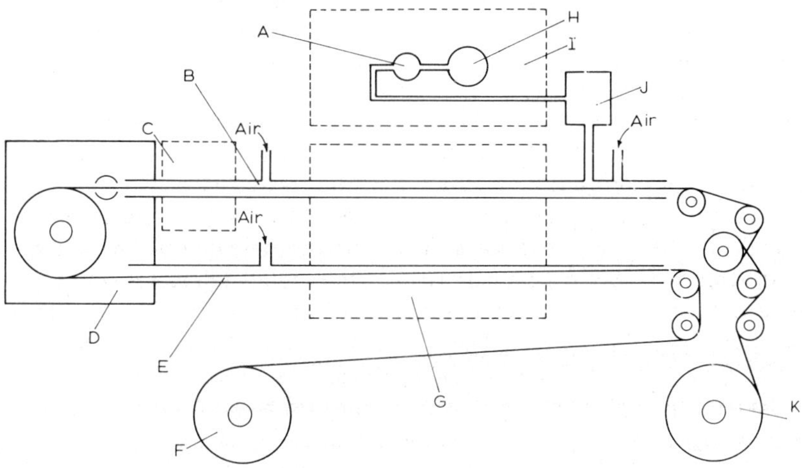

Fig. 5.6. Schematic of a phase transformation detector. (A) Catalyst chamber; (B) evaporator/oxidiser glassware; (C) evaporator oven; (D) coating block; (E) cleaner glassware; (F) feed spool; (G) cleaner/oxidiser oven; (H) flame ionisation detector; (I) flame ionisation detector/reactor oven; (J) molecular entrainer; (K) collecting spool. (Reproduced by courtesy of Pye Unicam.)

In this arrangement the detector response is directly proportional to the carbon content of the separated components and is linear over approximately five orders of sample size. The principal drawbacks to this system are the rather complicated multi-stage process, which requires a rather bulky unit, susceptibility to contamination, and a fairly modest sensitivity – the limit of detection claimed is in the order of 2×10^{-6} g/ml of column effluent. If an increased proportion or even the total effluent could be transported to the GC detector, the gain in sensitivity of this widely applicable device would be extremely useful. However, the essentially instantaneous evaporation of the total column effluent in close proximity to high-temperature oxidation furnaces could well pose some safety problems for the instrument manufacturer, particularly when using inflamable solvents such as hydrocarbons or ethers as mobile phases.

For quantitive work considerable care must be taken to control the mobile phase flow-rate and the wire speed as the sensitivity of the detector is enhanced by an increase in wire speed and a decrease in mobile phase flow-rate.

OTHER DETECTION DEVICES

The LC detectors described earlier in this chapter represent the most popular detectors in current use. Many other systems have been suggested as means of monitoring the effluents from LC columns, of which only a few have resulted in commercial products. Many of these rely on measuring a bulk property of the column effluent, for example electrical conductivity or heat of adsorption, and resemble refractive index detectors in their sensitivity to temperature changes and variations in the mobile phase composition. Each of these lesser used detection methods is now briefly described.

Electron capture detection

This detection principle is without doubt one of the most sensitive of GC detectors. For many years it has proved invaluable for the detection and quantitation of minute quantities of substances of biological importance, *e.g.*, chlorinated pesticides and derivatised steroids. The high sensitivity of this detector is also coupled with a very high degree of selectivity. Some studies have been made using an electron capture detector mounted on a solvent transport (moving wire) system with the aim to produce a detector with a similar sensitivity and selectivity of response, yet suitable for monitoring effluents from LC columns. More recently, Willmott and Dolphin[7] have described a much simpler system where the entire column effluent, including the mobile phase, is vaporised into a gas stream and passed through an electron capture detector. A very high sensitivity of detection to favourable compounds is reported for this device, for example a detection limit for aldrin of less than 10^{-10} g. Although this device is currently undergoing development, it could prove an interesting complementary detector to UV photometry and fluorescence for trace analysis. The principle limitation with electron capture detection is that the magnitude of the response is a function of the electronic structure of the molecule. Since LC is more concerned with the separation of thermally labile, or non-volatile, species, the likelihood of vaporising such components without fragmentation would seem small.

Electrical conductivity detectors

Conductivity detectors are available commercially (*e.g.*, from SpectraPhysics) and may be used to monitor changes in very poorly conducting media or measure the difference in the conductivity of two ionic solutions, *cf.* differential refractometry. Applications are restricted to aqueous or semi-aqueous (reversed-phase) systems where ionic species are being analysed. Conductivity detectors are susceptible to temperature fluctuations and impracticable for use with gradient elution.

Heat of adsorption

Chromatographic separation processes being governed by thermodynamic quantities are associated with a heat of reaction, for example, the exothermic reaction occurring when polar substances such as alcohols are adsorbed on the surface of silica gel. In an adsorbent-filled chromatographic column sample components are continually adsorbing and desorbing on and off the surface of the packing material during their passage through the column, as also are the molecules of the mobile phase. These reversible reactions are usually associated with an exothermic adsorption followed by an endothermic desorption. If a sensitive thermistor is buried in the column packing, a small yet significant temperature change is observed as a compound passes that point in the column bed. The temperature change is initially a sharp rise as the compound is adsorbed followed by a fall in temperature relative to the rest of the column as the sample is desorbed from the packing. Thus the detector produces a type of skewed sine wave response to a component eluting rather than the more familiar Gaussian type of peak. The need for strict temperature control, difficulty to quantitate and incompatibility with gradient elution operations have severely limited interest in this detection principle. Some more advanced designs have been reported[8] which are somewhat less sensitive to variations in ambient temperature yet have not led to any widespread revival of interest in its use.

Polarographic detectors

A number of interesting analyses have been reported using custom-built polarographic detectors[9,10]. These include both organic and inorganic applications. The technique of chromatographic separation of samples prior to polarographic detection is limited to aqueous or semi-aqueous systems as a high concentration of supporting electrolyte is necessary for satisfactory detector operation.

Two types of these detectors have been described, based either on the use of a dropping mercury electrode[9] or on a graphite-impregnated silicone rubber electrode[10].

Polarographic detectors exhibit rather unusual selectivity characteristics, the response depending on the nature of the base electrolyte and applied potential as well as the sample itself. These detectors are found to respond to a variety of compounds, not all of which can be rationalised on the basis that the compound contains an electrolytically reducible functional group. The limitations imposed by the necessity of having concentrated electrolytes in the mobile phase and lack of reliable commercially available units have both contributed to the general low level of interest in this principle of detection.

Radioactivity detectors

There are many applications in the studies of the metabolism of drugs, pesticides, etc., where radioactive samples are employed to enable the compounds of interest to be detected at very low concentrations. The use of radioactive detectors as sensitive monitors for fast LC separations poses an unavoidable compromise in that for the highest sensitivity a radioactive substance must reside in the "detector" for a long period of time, whereas for fast analyses the compound should reside in the detector for the shortest possible time.

Thus, if a short residence time is obligatory then, unless the sample exhibits very high activity, the detection is not particularly sensitive albeit highly selective. If high sensitivity is required, the best approach is to collect fractions of column effluent and measure the activity of the fractions in a scintillation counter. Various designs of flow through monitors for LC have been described and results of a very recent study on the feasibility and construction of such devices have been reported by Sieswerda[11].

Detectors based on converting the column effluent into a charged aerosol

A novel, nanogram-sensitive, detector for use with chromatographic systems involving water-based mobile phases has been described by Mowery and Juvet[12]. In this system, the column effluent is transformed into a charged aerosol by means of a stream of pressurised gas directed at a target electrode (the so-called Spray Impact Detector). An electrical charge is generated, dependent on the operating conditions and the nature of the column effluent. Changes in the magnitude of this charge are measured by an electrode system linked to a high impedance electrometer.

Only limited information regarding the performance of this detector has been reported, however, it is claimed to offer detection limits similar to those of the UV photometric detectors (low nanogram range) and to respond linearly with sample concentration over 3–5 orders of magnitude. Reported sample applicability includes the detection of fatty acids, detergents, amino acids and organic salts.

FINAL COMMENTS ON INSTRUMENT DESIGN

At various points in this text, and particularly in Chapter 3, the importance of eliminating dead space and badly swept regions within the chromatographic system has been emphasised. When trying to decide between two models of custom-built equipment or indeed comparing the likely performance of commercial instruments it is often of interest to assess these characteristics quantitatively as they represent the limit in performance that may be achieved with the apparatus.

The dead volume of any chromatographic system may be measured by connecting the injection device directly to the detector using the normal column connectors and the absolute minimum volume of other tubing, *i.e.*, no column fitted. With the pumping system delivering a typical flow-rate of mobile phase, say 1 ml/min, and the recorder operating with a fast chart speed, one may inject a small volume of a solvent known to give a response in the detection system while simultaneously marking the recorder chart.

The distance along the recorder chart measured from the point of injection to the first movement of the recorder pen from the baseline due to the injected solvent, converted into volumetric terms, is the dead volume of the chromatographic system. The profile of the detector response is usually a trailing peak. The rate at which the pen returns to the baseline relative to the position of the peak maximum gives an indication of how efficiently the system is swept. It should be emphasised that these simple tests must be performed with the electronic components of the equipment having fast (less than 1 sec) response times. In Chapter 3 it was mentioned that a perfect injection into an ideal chromatograph

TABLE 5.1

LIMIT OF COLUMN PERFORMANCE DUE TO DETECTOR DEAD VOLUME

Dead volume of the detector (μl)	Approximate minimum peak volume* (μl)
3	40
8	107
24	320

*Peaks eluting with a volume larger than this value will be faithfully recorded, *i.e.*, the resolution is unaffected by the detector dead volume.

where no dispersion occurred would result in a rectangular "peak" being produced on the strip chart recorder. In reality a trailing peak is obtained, and this, when the time scale is adjusted to that of a typical analysis, is a representation of the narrowest "peak" which may be obtained with the apparatus being tested irrespective of the efficiency of the chromatographic column employed for subsequent analyses. A more detailed and mathematical discussion of these factors can be found in the work of Sieswerda[11].

The best system from the viewpoint of providing highest resolution and sensitivity will be that which enables detection of an eluting component in the smallest volume, without the detector itself contributing to the broadening of the peak.

The work of Oster and Ecker[13] has indicated that for a detection system to have a negligible effect on the resultant chromatograms the volume of the flow cell should be less than 0.3 of the standard deviation of the eluting peak. Table 5.1 transposes this expression into practical terms by giving the minimum volume of an eluting peak which can be detected without significant band broadening occurring in the detector flow cell. The values presented are calculated for flow cells having volumes close to those in current use and for peaks of Gaussian form, where the basewidth can be taken as approximately equal to four times the standard deviation of the peak.

REFERENCES

1 J.J. Kirkland, *J. Chromatogr.*, 83 (1973) 149.
2 D.R. Baker, R.C. Williams and J.C. Steichen, *J. Chromatogr. Sci.*, 12 (1974) 499.
3 P.B. Hamilton, *Rev. Sci. Instr.*, 38 (1967) 1301.
4 R.M. Cassidy, D.S. Legay and R.W. Frei, *J. Chromatogr. Sci.*, 12 (1974) 85.
5 S. Udenfried, S. Stein, P. Böhlen, W. Dairman, W. Leimgruber and M. Weigele, *Science*, 178 (1972) 871.
6 R.E. Poulson and H.B. Jensen, *Anal. Chem.*, 40 (1968) 1206.
7 F.W. Willmott and R.J. Dolphin, *J. Chromatogr. Sci.*, 12 (1974) 695.
8 T.B. Davenport, *J. Chromatogr.*, 42 (1969) 219.
9 J.G. Koen, J.F.K. Huber, H. Poppe and G. den Boef, *J. Chromatogr. Sci.*, 8 (1970) 192.
10 P.L. Joynes and R.J. Maggs, *J. Chromatogr. Sci.*, 8 (1970) 427.
11 G.B. Sieswerda, *Thesis*, University of Amsterdam, 1974, p.21.
12 R.A. Mowery and R.S. Juvet, Jr., *J. Chromatogr. Sci.*, 12 (1974) 687.
13 H. Oster and E. Ecker, *Chromatographia*, 3 (1970) 220.

FACTORS INFLUENCING CHROMATOGRAPHIC SELECTIVITY

Chapter 6

Nature of the mobile phase

INTRODUCTION

In LC, unlike in GC, the retention characteristics of sample components within a given column are extremely dependent on the chemical composition of the mobile phase. The situation in LC is very different from that in a GC system, where, within quite wide limits, almost any stable compound which is capable of vaporisation will eventually elute from the column – the rate of elution being primarily a function of the column temperature. When working with LC, the novice is often concerned with peaks that elute only after considerable retention and fails to appreciate that a small change in the composition of the liquid mobile phase can cause a drastic change in the sample retention. This latter feature represents the most powerful parameter available to the liquid chromatographer who wishes to develop and optimise the separation of chemical mixtures.

The exact chemical characteristics of the mobile phase required when performing analyses on the various separation techniques, *i.e.*, adsorption, ion exchange, partition or steric exclusion, depend markedly on the sample and the type of chromatographic packing being employed. In this chapter features concerning the characteristics of the mobile phase common to the separation techniques are described, leaving discussion of the finer details of the mechanism and usage to the chapters devoted exclusively to the individual separation methods.

A systematic approach to the selection of the most appropriate chromatographic conditions for any sample is proposed, indicating some of the pitfalls likely to be encountered. There have been many attempts to rationalise the choice of chromatographic conditions based on the characteristics of the sample, *e.g.*, solubility, mobile phase, and column type. The treatment outlined here attempts to base a scheme on practical experience, if only from an empirical viewpoint, which is probably of more value to the inexperienced chromatographer.

The final sections of this chapter describe the methods employed to increase the capacity of a chromatographic system to achieve separations of complex mixtures while simultaneously attempting to reduce the time required to achieve the separation. Such techniques include column switching, pressure programming, and gradient elution.

Before proceeding to a discussion of the characteristics of the mobile phase it is considered helpful to explain in very general terms the nature of the modes of separation which are employed in LC. For a detailed discussion of the individual separation methods see Chapters 7–10.

A common approach may be applied to selecting the composition of the mobile phase for all systems which depend, for separation, on the selective retardation of components of the sample by the column packing material or a coating thereon.

METHODS OF SEPARATION IN THE LIQUID PHASE

Liquid–solid (adsorption) chromatography

Separations achieved by liquid–solid (adsorption) chromatography are based on the competition for sites on an active adsorbent surface, such as silica gel or alumina, between molecules of the sample and molecules of the mobile phase (or a component thereof). The mobile phase used in a typical adsorption system would comprise hexane or dichloromethane as a principal solvent, to which is added a second, modifying solvent. This may be a polar solvent, such as water, an alcohol or dimethyl sulphoxide and is added in relatively minor proportions, *i.e.*, less than 5%.

When such a mobile phase is passed through a column, part of the modifying solvent is adsorbed on to the surface of the chromatographic support, thus altering its adsorptive activity. Variations in the level of the modifying solvent in the mobile phase give rise to considerable changes in the retentive power of the column, a higher percentage of modifier leading to earlier elution. Traces of water, even that present in water-immiscible solvents such as hexane or chloroform, will modify the activity of an adsorbent and for maximum reproducibility the level of water present must be controlled. It should be appreciated that the modifying solvent is often only slightly soluble in the mobile phase and consequently a deliberate change in the activity of the adsorbent column packing material will be achieved only after a prolonged passage of the new mobile phase. After a sample has been introduced into the system, there is a competing reaction for the active sites on the adsorbent surface. If the affinity of the column packing for the sample molecules is greater than its affinity for the mobile phase, then the sample will be retained and the previously adsorbed solvent molecules displaced. Conversely, a stronger affinity for the mobile phase will lead to rapid elution of the sample. In practice it is necessary to find an intermediate condition, by changing the chemical composition of the mobile phase to give a certain degree of retention rather than either complete or zero retention of the sample components.

Liquid–liquid (partition) chromatography

In this method a comparatively inert chromatographic support is used, the surface of which is coated with a "liquid film" or stationary phase in which the sample components are soluble. The liquid film forming the stationary phase may be a true liquid, a polymeric material or a chemically bonded layer on the surface of the support. In the simplest case, the first step in selecting the optimum mobile phase composition is to choose one solvent in which the sample has limited solubility. When this solvent is used as the mobile phase, total retention of the sample would be expected. The eluting power of this primary solvent is then modified by the addition of a second solvent, which is a good solvent for the sample and which would, if present in excess, cause rapid elution of the sample components from the column. The proportion of the two solvents necessary for optimum resolution is then decided by experiment. Typical solvent pairs which are often used are: hexane with chloroform and dichloromethane with methanol. The most common stationary phases possess either nitrile or hydroxyl functionality.

Care should be exercised when working with column packings having rather labile stationary phases, *e.g.*, those with a simple liquid coating. With these materials it is important to ensure that a change in the mobile phase composition does not lead to dissolution of the liquid coating; this problem can be avoided by carefully saturating the mobile phase with the stationary liquid before passing it through the column. The use of a column packing where the stationary phase is bonded chemically to the support material is ideally suited to this approach of developing methods as the nature of the mobile phase may be changed over a wide range without disrupting the stationary phase.

Reversed-phase chromatography

Although strictly just a special case of partition chromatography, reversed phase chromatography is often regarded as a separate category. The expression has been adopted to describe a partition system where the mobile phase is more polar than the stationary phase. The most common example of a reversed-phase system is one in which the stationary phase is a C_{18} hydrocarbon usually introduced on to a support by the action of an octadecylchlorosilane, *i.e.* a bonded phase. Mobile phases used in this case are based on water to which a water-miscible organic solvent is added to modify the elution characteristics of samples. Compounds elute more rapidly when the proportion of organic solvent in the mobile phase is increased. Reversed-phase solvent systems usually employ water mixed with methanol; however, in some applications the use of acetonitrile–water mixtures offers an additional degree of selectivity.

Ion-exchange chromatography

The basic concept of ion-exchange is somewhat analogous to adsorption chromatography, *i.e.*, the sample interacts with an active surface, only in the present case the surface carries a charge. An anion exchanger possesses positively charged sites, most commonly derived from quaternary ammonium groups. Cation exchangers bear negatively charged sites and are often produced by incorporating sulphonate groups.

In a "true" ion-exchange system, the degree of retention of a sample is decided by the pH of the mobile phase, the concentration of the buffer solution, and the presence of any counter ions which could compete with the sample for the active sites on the ion-exchange surface. Many of the reported separations using modern ion-exchange packings cannot be explained by the straightforward ideas of ionic equilibria. This situation arises as most packing materials interact with samples via some secondary mechanism of adsorption, partition or hydrogen bonding effects. A hybrid mechanism is then found to govern the order of elution of sample components, making chromatographic behaviour hard to predict.

Steric exclusion chromatography

This method differs from all those previously described in that steric exclusion does not involve the retention of a sample on a column packing. The mechanism of separation relies on the different rates of diffusion or permeation of molecules of different size through a porous matrix. Very large molecules, being unable to enter narrow pores, elute first as they can travel through the column only by way of the spaces between the gel particles in the column. Smaller molecules can enter (permeate) the pores of the gel and elute later. A separation is achieved where the largest species elute first followed by progressively smaller species. It is important to realize that the separation is according to molecular size and not molecular weight. In some cases, particularly in the field of high polymers, the shape of the molecules has an influence on the elution characteristics, as does any solvation of the molecules. In this method it is important to eliminate any possible interaction between the sample components and the surface of the gel. This condition is usually met by selecting a mobile phase with similar characteristics to the gel and/or which is an excellent solvent for the sample being studied. Unlike retentive chromatographic systems, *e.g.*, partition and adsorption, in steric exclusion one only needs to optimise the mobile phase so that it is compatible with the detection system and eliminates any possible adsorption effects. This procedure can often be predicted with comparative certainty without recourse to experiment.

CLASSIFICATION OF MOBILE PHASES

The term polarity has for many years been the yardstick of most chemists, particularly chromatographers, for the qualitative classification of organic solvents and samples. Solvents such as low-molecular-weight alcohols, water, acids and bases are considered to be highly polar, whereas normal paraffins, *i.e.*, *n*-pentane and *n*-hexane, are regarded as non-polar. This description originates from classical methods of determination of dipole moments or dielectric constants of different substances. Any text book of physical chemistry will contain descriptions of the fundamental principles and methods of measurement of dipole moments.

Data derived from many experimental measurements of dipole moments enabled lists of solvents to be produced in some relative order of increasing or decreasing polarity. From the early practice of classical column chromatography using adsorbent packings such as silica gel and alumina, it was realised that the eluting power, *i.e.*, the ability to displace a sample component from a column, of solvents used as mobile phases approximately parallelled the polarity of the solvent. A highly polar solvent, such as an alcohol, is very effective at displacing components from the column. In a similar manner to the measurement of dipole moments, based on experience, solvents were tabulated in order of their ability to elute compounds from the adsorbent-filled column. These tables of solvents are known as eluotropic series and as mentioned earlier their order resembles the order in lists of dipole moment measurements.

There have been a number of different eluotropic series proposed, all of which are essentially similar, but vary in the solvents studied and sometimes in the relative positions in the list of two solvents which possess rather similar characteristics. The apparent discrepancy should not be considered a limitation due to experimental error but more a variable originating from the nature of the samples chosen as "test compounds" for the various comparative elution tests. Similarly, the choice of adsorbent packing employed, *i.e.*, whether silica gel or alumina, and if the solvent is electron withdrawing (*e.g.*, methanol) or electron donating (*e.g.*, acetonitrile) will impose certain different selectivity effects.

One typical eluotropic series of solvents derived from data reported by Hais and Macek[1] is shown in Table 6.1. Information regarding the usefulness of these solvents with refractive index, photometric (UV cut off) and solvent transport (boiling points) detectors and their viscosity at 20°C is also included in the table. The solvents listed represent those most commonly used as constituents of mobile phases employed in modern LC.

In recent years efforts have been made to establish polarity or solvent strength on a more quantitative basis by taking into account a number of characteristics of the solvents including solubility data and proton acceptor/donor characteristics. The work and publications of Snyder[2] are probably the most authoritative on this subject, full details of which are considered beyond the scope of this text.

TABLE 6.1

PROPERTIES OF SOLVENTS COMMONLY USED AS MOBILE PHASES IN MODERN LC

The order of the solvents is based on data reported by Hais and Macek[1].

Solvent	R.I. ($n^{D}_{20°}$)	UV cut off (nm)***	B.p. (°C)	Viscosity**
Heptane – Least polar	1.39	200	98	0.42
Hexane	1.38	200	69	0.32
Cyclohexane	1.43	202	81	0.93
Carbon disulphide	1.63	380	47	0.37
Carbon tetrachloride	1.46	265	76	0.97
Toluene	1.49	285	111	0.59
Propyl chloride	1.39	225	47	0.35
Benzene	1.50	280	80	0.65
1,2-Dichloroethane	1.44	230	83	0.79
Chloroform	1.45	245	61	0.57
Dichloromethane	1.42	233	40	0.44
Dibutyl ether*	1.40		143	
Nitromethane	1.38	380	101	0.67
n-Butyl acetate	1.40	255	125	
Diethyl ether	1.36	202	35	0.23
Ethyl acetate	1.37	260	77	0.45
n-Butanol	1.40		117	
Methyl ethyl ketone	1.38	330	80	
Tetrahydrofuran*	1.41	230	66	0.51
Dioxane*	1.42	215	101	1.54
Acetone	1.36	330	56	0.32
Isopropanol	1.38	207	82	2.30
Ethanol	1.36	205	78	1.20
Acetic acid	1.37	230	118	1.26
Methanol	1.33	208	65	0.60
Acetonitrile	1.34	212	82	0.37
Formamide	1.45	210	210	
Water – Most polar	1.33	200	100	1.01

*These solvents often contain stabilisers which are strong UV absorbers.

**Viscosity measured in centipoise at 20°C.

***The approximate wavelength below which the transmission is less than 10% in a cell with a path length of 10 mm.

The selection of the most appropriate mobile phase for the separation of a completely unknown sample by an adsorption or partition system is based on a great deal of trial and error. However, the mobile phase composition can often be anticipated quite closely if additional data such as solubility characteristics or some information regarding the chemical nature of the sample are available. When dealing with any given chromatographic packing material, it is useful to seek to establish the composition of two mobile phases, one which will result in complete retention of the sample on the column packing and the second which, if used alone, would elute the entire sample with no retention. For example, if an adsorbent-filled column is employed, *i.e.*, silica gel, many substances would be completely retained if a non-polar solvent, such as hexane, was used as the mobile phase yet the same samples would elute without retention if the mobile phase was changed to a very polar liquid such as ethanol. Similarly in ion-exchange work, one often finds that the pH of one buffer solution will give complete retention while a mobile phase having a different pH gives no retention. In both cases once the two extreme mobile phases have been established the study may be continued with mobile phases formed by mixing the two solvents in different proportions and observing the effect on sample retention.

Several practical points can be suggested which may assist the inexperienced chromatographer. Firstly, it is usually quicker to carry out such a study of mixed mobile phases using the most strongly eluting solvent initially and progressively decreasing its strength by addition of the second solvent after each successive test run has been completed. In this way one can often assess the onset of any retention by observing the detector response near to the solvent front, *i.e.*, in the very minimum of time. If the weakest eluting solvent is used first, one must wait an extended period of time to determine whether or not the compound is completely retained; also, after subsequent modification of the mobile phase, components from earlier injections will begin to elute, possibly leading to confusion. Secondly, when carrying out these exploratory tests it is advisable to inject an equivalent volume of the solvent used to dissolve the sample; in this way the extent of the response due to the solvent of the sample can be differentiated from any peaks originating from the sample. A final practical point when carrying out any exploratory work is to ensure that the chromatographic system has reached an equilibrium state with the new mobile phase. This can sometimes take a considerable time, particularly if employing totally porous column packing materials. Repeated injection of the sample allowing ten column volumes of mobile phase to pass through the column between successive injections is perhaps the most straightforward way of assessing if equilibrium has been attained; any non-equilibrium will show as a change in the retention characteristics of the system from one injection to the next.

Having established a mixture of, maybe, two solvents which produce reasonable retention of the components, *i.e.*, most peaks elute with capacity factors in the range $k' = 1$ to $k' = 10$, it is sometimes observed that two or more components are incompletely resolved. This is an example of the relationship between the number of effective plates available in the column (*i.e.*, its efficiency) and the selectivity of the phase system – as described in Chapter 2. To improve the analysis one must either increase the efficiency of the chromatographic column by using a longer column or one packed with smaller particles, or one may change the selectivity of the system. This latter characteristic can sometimes be accomplished without changing the column but by using an alternative mobile phase, for example by employing a solvent of intermediate polarity as distinct from a mixture of two solvents of widely different polarities or substituting an electron-donating for a proton-donating solvent of similar polarity, *e.g.*, acetonitrile in place of methanol.

Other characteristics of solvents will also govern their selection as potential mobile phases, particularly with respect to the type of detection system employed, *i.e.*, most commonly photometric, refractometric or solvent transport (phase transformation) detectors. Each of these detectors have their own criteria for an acceptable solvent for use as a mobile phase. Clearly, for photometric detectors the mobile phase should not absorb strongly at the wavelength at which the detector operates. Most detectors of this type are, however, capable of at least one absorbance unit off-set of the signal, thus solvents with as little as 10% transmission are just acceptable. Detectors fitted with both analytical and reference flow cells should, in principle, be able to compensate for even greater absorption of the mobile phase; unfortunately, the considerable reduction of light energy frequently leads to non-linear behaviour and poor baseline stability and as such cannot be recommended for quantitative studies.

When using refractometric detectors it is advisable to keep the overall refractive index of the mobile phase as low as possible, thus giving the maximum difference in refractive index between the sample and the mobile phase, hence optimum sensitivity. In instances when different selectivity is being sought it may be useful to examine the characteristics of highly refractive mobile phases such as the aromatic or halogenated solvents. Under these conditions many sample components will often be observed as negative peaks. The use of highly refractive mobile phases is not strongly recommended as the most likely liquids are more toxic and expensive than most other solvents.

Phase transformation detectors are essentially insensitive to the nature of the mobile phase providing that the solvents used are relatively volatile, *i.e.*, boiling point less than about 100°C, and are free from non-volatile impurities. Redistilled solvents are virtually essential when working with these detectors.

When optimum performance and minimum inlet pressure are being sought, the viscosity of the solvents forming the mobile phase should be considered. A low-viscosity solvent will tend to give a higher column efficiency as the kinetic processes within the column are improved. Careful selection of solvents of approximately the desired polarity according to their viscosity characteristics holds some advantage, for example, a choice between heptane, hexane, pentane, and cyclohexane. However, it should be appreciated that the viscosity of a liquid decreases markedly with increasing temperature so that operation above ambient temperature can lead to an enhancement of the column performance provided the column and sample under examination will tolerate an increase in operating temperature.

For many liquids the viscosity decreases with temperature at a rate which itself decreases with an increase in temperature. Graphs constructed by plotting viscosity against temperature for different liquids are often similar in shape, and may be superimposed by a shift in the temperature axis[3]. Fig.6.1 gives the viscosity data for water (in centipoises) against temperature, showing the general form of the relationship.

Working at an elevated temperature can improve the mobile and stationary phase mass transfer, the solubility of the sample components, and lead to a reduction in the inlet pressure for a given combination of linear velocity, mobile phase, and column packing.

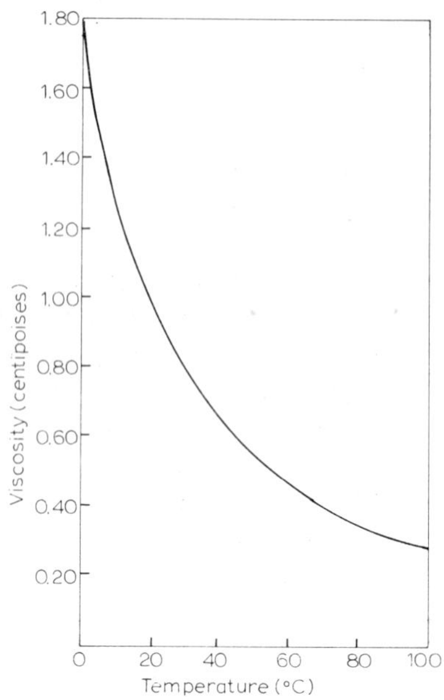

Fig. 6.1. Temperature dependence of the viscosity of water.

This advantage of elevated temperature operation is perhaps of greatest value when working in the fields of reversed phase and ion exchange, where the mobile phases contain a high proportion of water, a comparatively viscous solvent. Steric exclusion studies of high polymers are also frequently performed at elevated temperatures – in the case of polyolefin samples usually at temperatures in excess of 100°C – to enhance the solubility of the polymer and reduce the viscosity of the resultant solutions.

Studies on the influence of operating temperature on the efficiency of the chromatographic column by Schmit and co-workers[4] have demonstrated that in reversed-phase systems the plate height decreases with increasing temperature in a manner closely resembling the change of viscosity with temperature, as was illustrated in Fig. 6.1. Over the temperature range from ambient to 75°C the plate height decreases by a factor of approximately two, *i.e.*, the column efficiency is doubled. This gain in the separating power of a column can be of quite significant value when striving to resolve the components of a particularly difficult sample.

DEVELOPMENT OF CHROMATOGRAPHIC METHODS

Deciding the best method of separation

It is often the wish of those with limited experience of LC to be able to decide in a rational manner the most appropriate chromatographic column packing and mobile phase combination for any sample mixture which they may be required to separate. The likelihood of ever being able to devise a scheme that will enable this to be accomplished with a 100% success rate is very small, as in many instances any one chemical substance may be amenable to several different chromatographic methods.

The problem is best illustrated with an example. Let us consider the case where the most important constituent of a sample, if pure, is found to chromatograph either on a reversed-phase chromatographic system using, say, an aqueous alcohol mobile phase or on an adsorptive column with a mobile phase of chloroform. The decision of which of these two procedures to use then lies with an understanding of the nature of the rest of the sample and likely interferences. If appreciable quantities of lipophilic material are present, *e.g.*, a greasy base to an ointment where the compound of interest is some pharmaceutical product in the base, the sample would probably be best analysed by the adsorption method, as the solvent, chloroform, would readily dissolve the greasy base material and the lipophilic material would not precipitate in the chromatographic column. If the reversed-phase method was employed the lipophilic substances would have little or no solubility in the mobile phase and would be strongly retained in the chromatographic column, giving rise to very slowly eluting peaks which would interfere with subsequent analyses. Alternatively, it may happen that there are many components originating from the sample which elute with very similar retention to the component of interest so that quantitation of the peak is impracticable. In this instance it may be preferable to solvent extract the component of interest by simple liquid–liquid partition in a separating funnel using aqueous alcohol and chloroform as the two liquids. The proportions of water to alcohol required to achieve a satisfactory distribution coefficient must be determined by experiment. The net result, however, could be that the component of interest is contained in a water-miscible phase to an extent dependent on the distribution coefficient, which may itself be determined if the nature of the analytical problem demands it. In many instances there will also be some co-extractives, but the procedure does ensure that the solution containing the component of interest is completely miscible with the mobile phase used in the reversed-phase procedure and that strongly lipophilic species which could otherwise cause the greatest concern, in terms of possible column contamination, are now absent or at least their concentration is substantially reduced. The example given here is typical of the problems encountered in modern LC and illustrates that although initially it may appear confusing to have several possible separation procedures to choose from, the situation reflects the power of the technique in its ability to solve problems common to everyday chemical analysis.

In the following paragraphs factors leading to a systematic approach to the selection of column type and mobile phases are discussed. This approach should be considered in the light of the comments made earlier that in many instances a given compound may be satisfactorily chromatographed in the pure state on more than one column. Also that polarity, whether referring to a solvent, stationary phase or a sample, is in reality continuous, *i.e.*, the division between "moderately polar" and "polar" cannot be rigidly defined.

The first requirement when commencing a study of a completely unknown sample mixture is to establish the approximate molecular weight range of the components in the sample mixture. In most work the approximate molecular weight of a sample will be known from independent data. If there is a genuine possibility that the sample may contain species of a wide range of molecular weight, it should be determined by examining the sample by steric exclusion chromatography.

It was noted earlier that procedures based on this method permit the separation of sample components in such a manner that the largest molecular species elute first and the smallest last. The field of steric exclusion is usually subdivided into two categories, depending on the solvents used in the method. Those separations which are performed in aqueous media are often referred to as gel filtration methods. When organic solvents are employed the technique is usually described as gel permeation chromatography (GPC). Both methods are capable of handling samples up to a molecular weight of several millions.

With regard to deciding the most appropriate separation method, the point of interest is whether the molecular weight of the sample is above or below about 2000. If it is greater than 2000, it is probable that the steric exclusion techniques, *i.e.*, gel permeation or filtration, will hold the greatest promise. But although these methods are capable of separating species of smaller molecular weight, in this range the other LC techniques, such as ion exchange, partition, and adsorption, are generally more rapid and more selective.

Having decided that the sample contains only species with a molecular weight less than 2000, further classification of the sample is necessary to decide the approximate range of polarity. Again, background information concerning the sample origin can help considerably. If this is not available some understanding may be derived from simple qualitative, or at best semi-quantitative, solubility tests. A highly polar compound, which is readily soluble in water, acids or bases, indicating the presence of some ionic or potentially ionic functional groups, would almost certainly show little or no tendency to dissolve in low-polarity solvents like hexane or toluene. A compound of this type would certainly appear to be a good candidate for an ion-exchange method of separation. With a few additional tests, an indication of whether the sample is acid, neutral or amphoteric will be obtained.

At the opposite end of the polarity scale many samples are completely insoluble in water but have some affinity for solvents of lower polarity, *e.g.*, hexane, toluene, and chloroform. Additionally, many of these compounds dissolve quite readily in solvents of intermediate polarity, such as alcohols, ethers, and ketones. Samples at this end of the polarity range can normally be satisfactorily chromatographed by reversed-phase chromatography.

A working rule that is used by many chromatographers is that for the selection of a stationary phase in liquid (or even gas) chromatography one should consider the simple relationship that "like dissolves like". In other words, a reversed-phase procedure is most suitable for a lipophilic sample, a normal partition or adsorption procedure for a sample which is lipophobic but is not so polar that it is an ionic substance. Unfortunately, life is not always quite that simple. The concept, however qualitative, does suggest that if one is seeking to analyse a compound of intermediate polarity, some moderately polar stationary phase should be employed, *i.e.*, maybe a phase with an ester or ether functionality.

The discussion so far has implied that separations of small molecular species which are not ionic will be achieved by some partition process. This situation is by no means completely true. Many successful chromatographic analyses are performed by adsorption, often complementing a separation achieved by partition chromatography. In the first approximation one could consider that all separations achieved by liquid–solid (adsorption) chromatography using, say, silica gel as the chromatographic column packing are all performed with the same polar adsorptive stationary phase, *i.e.*, one with inorganic hydroxyl functional groups. A more practical way of considering the role of adsorbent column packings is to appreciate that in reality one uses the adsorbent at different levels of activity, depending on the separation problem in hand. In TLC, the well established Brockman scale of activities is used to classify alumina into grades of retentive power. A similar range of activities is possible with all adsorbents but in practice it is possible to vary the activity in a continuous, as distinct from a stepwise, manner. Thus in adsorption chromatography of samples of low polarity, *e.g.*, hydrocarbons, a very high degree of activity is required to effect retention of the sample on a column loaded with silica gel, whereas polar samples are only eluted from a column which possesses a significantly lower activity. The principal difference between adsorption and partition chromatography is that in the former technique the retentive power of the "stationary" phase is decided by the composition of the mobile phase since polar constituents from the mobile phase are

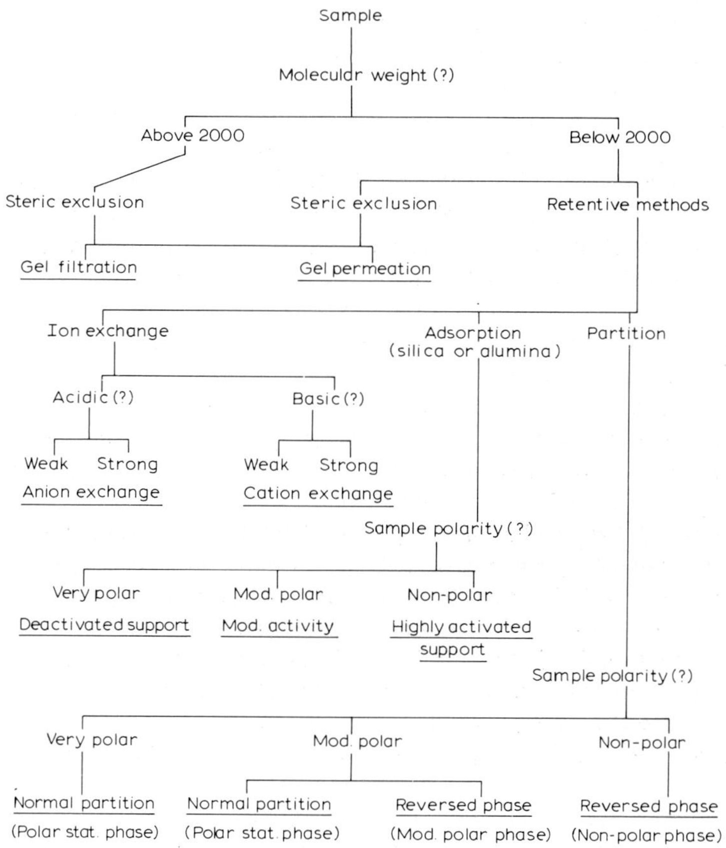

Fig. 6.2. Selection of column type.

initially adsorbed on the surface of the support, reducing its adsorptive power. In partition chromatography the support should have little or no retentive power in its own right, as the stationary phase coating on its surface is responsible for the selectivity and retention characteristics of the column packing. In all cases the nature of the mobile phase will govern the degree of interaction that the sample will experience with the column packing or its stationary phase.

The classification of the different separation methods and how these relate to sample type are outlined in Fig.6.2. This scheme indicates the main classes of column packing materials that are most commonly used in modern LC. There are other LC separation methods known by names such as ion-pair and affinity chromatography. These techniques tend to be less commonly employed, as they have, at the present time, only been examined for a few specialised applications. The potential range of application of these methods is, however, quite extensive, and a brief summary of the principles and background of these techniques is included in the most relevant chapter which describes the more common separation methods.

Fig.6.2 provides some general indication of the types of stationary phase and adsorptive packing that may be employed in LC. A more detailed analysis is given in the chapters dealing with the specific method of separation together with documentation concerning the range of commercially available column packing materials.

Deciding the best mobile phase

The selection of the mobile phase that is to be used for a particular separation follows the guidelines set out in earlier sections of this chapter. In the present section the method of selecting an appropriate mobile phase is considered. Initially the emphasis is placed on partition and adsorption chromatography, *i.e.*, where the stationary phase is more polar than the mobile phase. In reversed-phase chromatography the logic is similar, but the effect is the opposite, in that the use of a less polar solvent as the mobile phase will lead to sample components eluting earlier, *e.g.*, water will tend to give the strongest retention of non-polar components on the column packing whereas alcohol will result in only weak retention of components.

One of the drawbacks of the method of selection of mobile phase composition that has been described is that, although logical, a certain amount of trial-and-error experimentation is necessary. Some workers prefer to derive the same information by injecting the sample into the column packing using a mobile phase composition that will ensure as far as practicable complete retention of all components and then programming the solvent composition over a wide range of solvent polarity, *e.g.*, an adsorptive type of column packing and operating a gradient from hexane to ethanol.

The procedure will result in the sample components eluting at some stage during the gradient programme, the degree of hold-up on the column indicating the approximate order of mobile phase polarity that might be necessary for a separation when the carrier composition is held constant. Care should be exercised when using this approach (i) because the column system is not in equilibrium, (ii) because of the unavoidable error that the mobile phase entering the system as the sample elutes is not of the same solvent strength as that which caused the sample to elute (in fact, a weaker solvent must have

TABLE 6.2

SOLVENTS USED FOR INCREMENTAL GRADIENT ELUTION

(After R.P.W. Scott and P. Kucera, reproduced from *J. Chromatogr. Sci.*, 11 (1973) 83, with permission)

1	*n*-Heptane*	9	Methyl ethyl ketone
2	Carbon tetrachloride	10	Acetone*
3	Heptyl chloride	11	*n*-Propyl alcohol
4	Trichloroethane*	12	Isopropyl alcohol
5	*n*-Butyl acetate	13	Ethyl alcohol*
6	*n*-Propyl acetate	14	Methyl alcohol
7	Ethyl acetate*	15	Water
8	Methyl acetate		

*Solvents used for reconditioning the column between sample injections.

sufficed), and (iii) because when repetitive work is considered, the time necessary for the column to return to true equilibrium with the initial mobile phase can be unacceptably long. In the case of columns with adsorptive packings, volumes of solvent in excess of one hundred times the column volume may have to be flushed through the system before the initial starting conditions will have been restored.

This procedure for deciding the range of polarity of the components of a sample has been extended considerably by Scott and Kucera[5] in that they employ a solvent gradient system, termed incremental gradient elution, in which a range of fifteen different solvents is delivered in order of increasing polarity to a chromatographic system using a silica gel, adsorptive column. The procedure is to inject the sample into the head of the column and then to pass a definite quantity of each of the solvents named in Table 6.2, in turn, through the column.

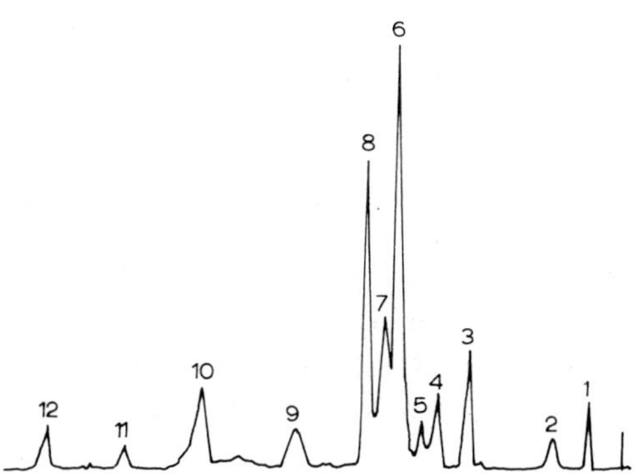

Fig. 6.3. Chromatogram showing the separation of compounds of widely different polarity using incremental gradient elution. Operating conditions: column, 0.5 m × 5 mm I.D.; packing, Bio-Sil A; mobile phase, 12 ml of each solvent given in Table 6.2; sample size, 1.8 mg; detector, phase transformation (moving wire to FID); total separation time, 150 min. 1 = Squalane; 2 = anthracene; 3 = methyl stearate; 4 = octadecanol; 5 = vitamin A acetate; 6 = corn oil glycerides; 7 = dihydrocholesterol; 8 = 11-keto-progesterone; 9 = benzoic acid; 10 = chlordiazepoxide; 11 = phenylalanine; 12 = glucose. (Reproduced with permission from R.P.W. Scott and P. Kucera, *J. Chromatogr. Sci.*, 11 (1973) 83.)

The principal idea behind this approach is to employ a series of solvents covering the entire polarity range while minimising the excess free energy of adsorption between the successive solvents. The use of such a system is to enable any unknown sample mixture to be studied by the one procedure and thus obtain basic information on the most appropriate solvent polarity for the mobile phase in the subsequent optimised separation. These authors have demonstrated the concept with a separation, in a single analysis, of a complex sample containing components ranging in polarity from the non-polar squalane to the highly polar glucose. The chromatogram obtained is reproduced in Fig. 6.3.

One of the main disadvantages of this technique, based on data published at the present time, is that the solvents used are compatible with only one system of detection, the solvent transport detector, which places a most definite limit to the sensitivity of detection that may be achieved. Present data also indicate a quite lengthy time scale for individual runs, *e.g.*, fifteen solvents at 10 min each, followed by column reconditioning with five solvents.

The selection of chromatographic conditions, based on experimental results and the more commonly used solvents can be summarised as follows.

Reversed-phase chromatography

Hydrocarbon types of stationary phase are used for this method, which finds application in the separation of non-polar compounds and compounds of low polarity. Typical examples are steroids, sterols and hydrocarbons, which are (1) insoluble in water, (2) partially soluble in methanol, or another water-miscible solvent, and (3) have molecular weights below about 2000.

Water is used as the principal or primary solvent. To obtain optimum retention of the sample components, that is capacity factors falling between 1 and 10, the water is modified using solvents such as:

Methanol – the most useful.

Isopropanol – if greater modification is required to reduce retention.

Acetonitrile – offers a somewhat different selectivity.

Methanol + 5–10% dichloromethane – used when the sample components are otherwise very strongly retained, avoiding any immiscibility of the liquids water, methanol, or dichloromethane.

Retention times are increased by increasing the water content of the mobile phase, conversely, an increase in the organic modifier concentration causes a decrease in sample retention. If insufficient retention is attained, even when using the least powerful mobile phase (100% water), *i.e.*, the capacity factors of all components are less than 2, one should consider alternative separation methods.

Normal partition chromatography

This method employs polar stationary phases, such as polyglycols, ethers or nitriles. Samples which are successfully separated by this method are most commonly polar species, *e.g.*, phenols, amines and heterocyclic compounds. These have (1) very low solubility in hexane (or other alkanes) and (2) good solubility in polar solvents.

Here hexane is used as the principal solvent and the retention of sample components is adjusted by the addition of organic modifiers to the mobile phase. Where liquid-coated packings are employed, care should be taken to ensure the stability of the stationary phase layer (see Chapter 8). Typical modifiers are:

Ethanol – very powerful modifier, often only needed in low concentrations.

Tetrahydrofuran – slightly less powerful, but offers some distinct selectivity differences to alcohol.

Chloroform – although only of moderate strength as a modifier, it is useful in that an appreciable volume is required, making the proportions of the mobile phase mixture more easily reproduced.

In a similar manner to reversed-phase chromatography, an increase in the concentration of these modifiers in the mobile phase will lead to more rapid elution of the sample components. If retention of the sample is insufficient, *i.e.*, k' is less than 2, in some instances a heavier loading of stationary phase may be applied to the support material. Alternatively, reversed-phase chromatography (q.v.) or adsorption chromatography should be used.

Adsorption chromatography

When column packings such as silica gel and alumina are used, the columns generally exhibit stronger retention towards polar samples than normal partition systems; consequently, more powerful solvents are required to cause elution of components from the column. Most non-ionic samples can be retained to some extent on adsorptive packings, those samples which are moderately polar, for instance, phenols, heterocyclics and esters being typical. Compounds of these types show (1) a fairly low solubility in hexane, (2) good solubility in most moderately polar solvents, and (3) low solubility in water.

Hexane is a useful starting solvent, the retention of compounds being decreased by the addition of an organic solvent which is more powerful in eluting strength, *i.e.*, one which occupies a lower position in the eluotropic series reproduced in Fig. 6.1. While modifying the composition of the mobile phase it is important to be aware of any influence the new mobile phase may have on the activity of the adsorbent packing material. The control of the level of activity of an adsorbent surface is detailed in Chapter 7. The most common organic modifiers used in mobile phases are:

Dichloromethane and chloroform – moderate modifying power, often used in high concentrations and useful as the principal solvent when studying quite polar samples; these two solvents frequently show distinct differences in selectivity.

Diethyl ether – more powerful modifier; its high volatility can cause changes in the composition of the mobile phase if the reservoir has not been closed.

Ethyl acetate – similar eluting power to diethyl ether, but less volatile; not useful at wavelengths below 260 nm.

Isopropanol and methanol – powerful modifiers, often used in trace amounts (less than 1%), hence precise control of actual concentration is essential.

By an appropriate selection of solvents it is possible to obtain a similar eluting strength of mobile phase using an almost pure solvent of moderate polarity or a mixture of two solvents of different polarity. As a general rule greater differences in selectivity, therefore greater resolution of components, are normally obtained when a mixture of solvents having widely different polarity are used, for instance hexane and alcohol, rather than using a single solvent such as diethyl ether.

In all of these forms of chromatography, any tendency for the sample components to dissociate, *i.e.*, a weak acid or base, frequently leads to excessive peak broadening or a tailing peak. The addition of small quantities (one or two drops per litre) of acetic or phosphoric acid – in the case of a weak acid – or ammonia solution – in the case of a weak base – to the mobile phase will suppress the dissociation, giving a much improved peak shape.

ELUTION BEHAVIOUR OF COMPLEX MIXTURES OF DISSIMILAR COMPOUNDS

As soon as one starts to carry out studies to establish the ideal composition of mobile phase for the separation of components present in anything but a simple mixture it is often found that not all of the components can be eluted as separate peaks by using a single mobile phase. It often happens that a mobile phase which is capable of eluting all the components does not allow sufficient selectivity for resolution of the individual components. This is a general result of the situation considered earlier, *i.e.*, resolution is a function of column efficiency, capacity and the selectivity of the phase system. However, if the mobile phase composition is changed to make the chromatographic system more selective (and inevitably more retentive), early eluting peaks are resolved to a greater extent but at the expense that components that were appreciably retained with the previous solvent system are now completely retained on the column. This situation has been referred to by Snyder[6] as the general elution problem and is common to all forms of retentive chromatography, *i.e.*, all LC methods excepting steric exclusion chromatography.

There are a number of methods by which this elution problem may be overcome. These rely on operating the LC system in such a manner that one can alter the selectivity, the capacity, the resolving power or simply speed up the velocity of the mobile phase in a repeatable and systematic manner during the course of a chromatographic separation. These methods all require equipment which is somewhat more complex than that needed for a simple separation system.

Separation methods involving a change in column selectivity

In earlier chapters it was indicated that the selectivity of a chromatographic system is a function of the chemical composition of the mobile phase, the stationary phase, the temperature and the nature of the surface layer on the chromatographic support, *e.g.*, whether silica, alumina, cation exchange or anion exchange, etc. These factors influence the solubility of the components in a phase and also the extent of any interactive forces. Of these factors which influence the selectivity, only those concerned with changing the chemical composition of the mobile phase and the temperature are capable of being changed in a repeatable manner during the course of a separation. A change in the chemical composition of the column packing or stationary phase is impracticable, if not impossible. If the level of stationary phase is changed, it will effect the capacity of the column system, not the selectivity. The method of carrying out a controlled change in chemical composition of the mobile phase is known as gradient elution or solvent programming. In an analogous manner varying the column temperature is referred to as temperature programming.

Gradient elution

The very definite effect that even slight changes in mobile phase composition can have on the retention characteristics of sample components has been indicated and inferred in many places in this text. The dependence of retention on the nature of the mobile phase has largely been responsible for much of the success in achieving very highly selective phase systems in LC, which reduces the need of always having to work with columns of exceedingly high efficiency. Various apparatuses that may be used to provide a programmed change in mobile phase composition have already been described in Chapter 4.

In the simplest case of operating a gradient elution system, one uses two solvents which are miscible and differ in their eluting power with respect to the sample being studied. The solvent used as the mobile phase at the beginning of the separation is that which gives strongest retention of the sample components on the column packing being employed. The second solvent, selected as one which will, if used alone, cause the sample to elute without retention, is blended into the first solvent during the course of the separation. This action leads to the effect that components which would tend to elute early in the chromatogram experience a mobile phase which permits maximum interaction with the column packing. The low solvent strength of the mobile phase attempts to shift the equilibrium distribution of the sample in favour of the stationary phase, thus increasing the capacity factors and the chances of achieving a separation. If this composition of mobile phase was continued indefinitely, other components which show greater affinity for the stationary phase would not be eluted from the column in an experimentally acceptable time. To increase the speed of the elution, the second solvent is bled into the mobile phase in ever increasing proportions to cause the distribution coefficients of the sample components to change in favour of the mobile phase, resulting in elution of the components. Since the distribution coefficients of substances differ quantitatively, the point where the distribution of each component reaches a value which causes it to be eluted from the column will vary from one substance to another, hence giving rise to a separation.

The above description outlines the situation in the simplest case, in practice there are a number of features which can lead to experimental difficulties. The first of these is that some detectors used to monitor the system respond to the change in mobile phase composition. This response can be so serious a problem that it completely rules out the use of detectors which respond to bulk properties of the column effluent such as refractive index. Most types of selective detectors may be used in gradient elution studies. Optically transparent solvents must be employed when working with photometric and fluorescence detectors; however, in practice this does not pose a severe restriction on the types of elution performed as, in many applications, it is possible to find a transparent solvent with similar characteristics to those which do absorb in the UV region. If a non-selective, albeit less sensitive, detector is required in gradient elution work, the phase transformation detector is the only practical choice.

A second feature is associated more closely with chromatographic behaviour, *i.e.*, that solvent demixing, or dehomogenisation of the mobile phase, can occur if the solvent being added by the gradient system varies considerably in polarity from the initial solvent. This effect is caused by the secondary solvent being retained by the chromatographic support giving an initial depletion of the solvent in the mobile phase. Thus in the early stages of a gradient elution run molecules of the second solvent are retained by the column packing; this results in a decrease in the concentration of this solvent in the mobile phase until the capacity of the column packing for this particular solvent is satisfied. At this point the concentration of secondary solvent in the mobile phase will rise sharply causing a sudden change in polarity which has the effect of accelerating any fairly early eluting components through the column giving rise to a sharp peak on the chromatogram at the point where the first breakthrough of the secondary mobile phase occurs. This effect is illustrated in Fig. 6.4, where the spurious peak can be clearly differentiated by its shape from those of components eluting from the column in a normal manner. This problem is liable to occur in adsorption and ion-exchange systems and may be overcome by having a small proportion of the secondary solvent in the initial solvent at all times so that the affinity of the column for the secondary solvent is always satisfied. A more detailed discussion on the dehomogenisation of the mobile phase can be found in the work of Liteanu and Gocan[7].

Once aware of how to avoid these operational problems, gradient elution is by far the most powerful method by which one can vary the retention characteristics of sample components to effect a separation in a realistic time. By varying the rate at which the second, modifying, solvent is added to the mobile phase, the extent of the reduction in retention of a component may be controlled. Under optimised gradient elution conditions it is possible to obtain a chromatogram where each component peak is sharp and has little,

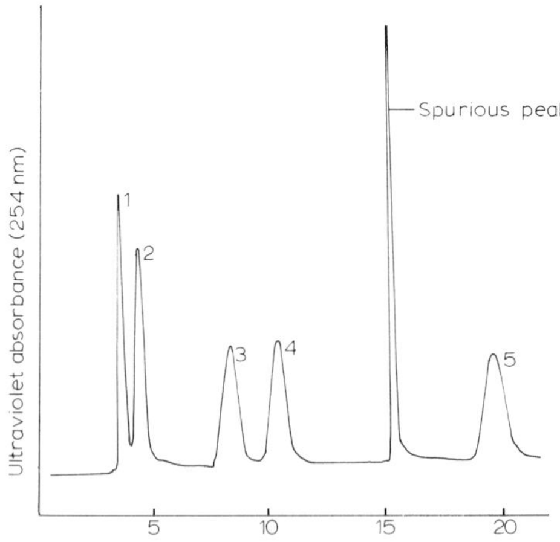

Fig. 6.4. Spurious peak during gradient elution due to dehomogenisation of the mobile phase. Operating conditions; column, 1 m × 2.1 mm I.D.; packing, Zipax SAX, strong anion exchanger; initial mobile phase, 0.1% ammonia in water; modifying mobile phase, 0.1% ammonia + 0.1 *M* sodium perchlorate in water; flow-rate, 1 ml/min. 1 = Sulphaguanidine; 2 = sulphanilamide; 3 = sulphanilylurea; 4 = sulphanilic acid; 5 = sulphacyanamide.

if any, tailing. This results in considerable improvement in the detectability of what would normally be slowly eluting minor peaks, thus increasing the apparent sensitivity of the method. The concentrations of the component bands also increase the chance of more peaks being resolved in a chromatographic column. Fig. 6.5 shows the separation of components in a particularly complex mixture, a commercial distillation residue, analysed

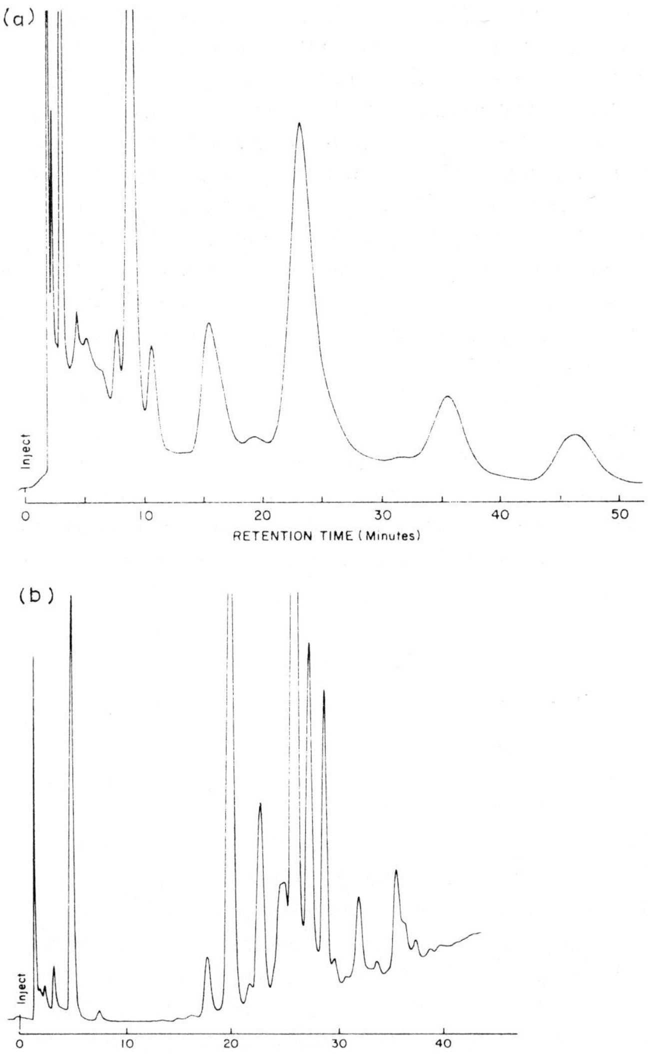

Fig. 6.5. Comparison of (a) isocratic and (b) gradient elution for a complex terephthalate mixture. Operating conditions: (a) Column, 1 m × 2.1 mm I.D.; packing, Permaphase ODS; column temperature, 50°C; mobile phase, 25% methanol in water; inlet pressure, 650 p.s.i.; detector, UV photometer, 254 nm; (b) Column, 1 m × 2.1 mm I.D.; packing, Permaphase ODS; column temperature, 40°C; mobile phase, linear gradient 10% methanol–90% water to 100% methanol, at 2%/min; inlet pressure, 1200 p.s.i.; detector, UV photometer, 254 nm. (Reproduced by courtesy of DuPont.)

under isocratic conditions, *i.e.*, with a mobile phase of constant composition (a) and under gradient elution conditions (b), illustrating the improved separation of the individual components, relatively constant peak widths and much improved detection of late eluting components when using gradient elution.

The use of solvent gradients other than a linear change with respect to time has the effect of enabling component peaks to be affected to a greater or lesser extent by the solvent gradient. Thus, relative to the effect of a linear gradient the peaks may be spread out more at the early or later part of the gradient run depending on the shape of the gradient profile. Some gradient elution devices permit composition *versus* time profiles to be a smooth continuous curve, *e.g.*, a logarithmic, exponential or linear function. With other devices the gradient profile may be tailored to suit a specific sample by selecting, for example, an initial exponential gradient which at some point changes to a linear or a logarithmic function. This latter type of system is in principle more versatile, but considerable preliminary work must be performed on a trial-and-error basis if a separation is to be completely optimised in this manner.

On a semi-theoretical basis it is generally considered that an exponential increase in the volumetric concentration of the modifying solvent in the mobile phase is the most suitable for adsorption chromatography. A linear increase in the modifier concentration with respect to time is similarly considered most useful in applications involving partition chromatography. In practice, however, the optimum gradient profile is invariably decided by experiment.

Gradient elution is sometimes performed by changing the composition of the mobile phase in a step-wise manner rather than by a continuous smooth change; the apparatus required in this case is less complex. The technique, however, often leads to spurious peaks being recorded at the breakthrough point of the new mobile phase due to the solvent demixing effect unless the difference in polarity of the solvents is small. The two different methods may be rationalised by considering a continuously changing type of gradient as a series of infinitely small step changes and thus as a series of step gradients.

In all forms of gradient elution one is faced with having to return the chromatographic packing to its initial form, *i.e.*, reconditioning of the column, before another separation can be attempted. In most two-solvent gradient systems this consists simply in switching back to the initial solvent and flushing the column for a suitable period of time (discussed below). If the number of solvents employed is greater than two, it may occur that the initial and final solvents of the mobile phase are not miscible so that a series of solvents has to be used to overcome this problem. Any immiscibility would lead to one solvent remaining on the packing material almost indefinitely, modifying the chromatographic characteristics of the column somewhat, as the retained liquid would act as a stationary phase. Some chromatographers are of opinion that reconditioning of the column after a gradient run should be achieved by retracing the gradient profile. This approach would certainly overcome any problem of immiscibility but can be wasteful with respect to time and solvents. Although there may be some merit in this approach, most commercial gradient systems deliver only mixtures of two solvents which are completely miscible. In these circumstances there seems to be a lack of well established experimental evidence to substantiate the need to routinely retrace the gradient profile.

The time taken to re-equilibrate the column packing with the initial mobile phase varies widely with the types of packing employed and the extent of the solvent change. Totally porous adsorbent materials, which have a very high surface area, such as silica gel and alumina, may require several hundred column volumes of the mobile phase flushed through the column to have their initial condition restored. Porous layer supports take a very much shorter time to equilibrate, the thin surface layer of stationary phase or active layer being ideally positioned for maximum contact with the new solvents. The types of column packing which re-equilibrate most rapidly are probably those porous layer materials designed for liquid partition chromatography using a stationary phase which is chemically bonded to the surface of the support. These materials can be re-used in gradient elution work within a few minutes of returning from a previous gradient with little or no adverse effects on the reproducibility of retention data. For this reason porous layer chromatographic packings for either adsorption, partition and ion-exchange work find wide application in gradient elution studies, additionally this combination also enhances the total number of components that may be resolved in a single chromatographic operation.

Temperature programming

In GC temperature programming is one of the most important methods by which complex mixtures containing components of widely differing vapour pressure can be separated in a single analysis. In the liquid phase an increase in temperature will invariably increase the solubility of sample components in any liquid phase, whether it be the mobile or the stationary phase. However, in many liquid phase separations the solubility of the components (or lack of solubility) in the mobile phase is not the principal factor influencing the retention of a compound on the column packing. A rise in temperature influences the chromatography in different ways, depending on the separation method being used. For instance, in steric exclusion a rise in temperature may change the viscosity of the mobile phase but this, in itself, does not lead to an earlier elution of the sample assuming the flow-rate of the mobile phase remains constant. In adsorption chromatography a significant rise in the operating temperature of the column can have the effect of displacing polar species such as water or alcohol which may have been deactivating the adsorptive surface; in this instance one might expect an increase in retention as the column packing will behave as a more powerful adsorbent. This situation is complicated in that the mobile phase will, at least initially, contain an increased concentration of polar modifier which may tend to compete with the increased adsorptive power of the column and try to displace the components earlier. Maggs[8] has studied the effect of temperature on adsorptive column systems and concluded that in some instances it could be useful to consider temperature changes as a means to vary the activity of the adsorbent and hence the selectivity. Unfortunately, the dynamics of changing the activity of an adsorbent in this way are very slow and make the procedure unsatisfactory as a programming method.

It has been mentioned that a rise in temperature would most likely cause an increase in solubility of the sample in both the mobile and the stationary phase, thus in liquid partition work it is conceivable that the retention of a compound may increase or decrease as the temperature of the column is raised, depending on how the distribution coefficient varies with temperature. One must assume when discussing the effect of temperature on partition systems that the column packing and stationary phase loading are stable to a change in temperature. This virtually implies that the packing material is of the type which has the stationary phase bonded chemically to the surface of the chromatographic support. In practice, an increase in the operating temperature of the column will normally give rise to a decrease in the retention time of the sample, thus some form of temperature programming can be considered feasible. The behaviour of ion-exchange materials and ionic substances parallels this behaviour, as a rise in temperature will increase the degree of dissociation of a partially ionised sample, suggesting stronger retention at elevated temperature, yet the same increase in temperature may increase the solubility of the sample such that an earlier elution occurs.

In all cases of liquid phase separations, whatever the mechanism, an increase in temperature will decrease the viscosity and improve the mass transfer characteristics in both phases. This effect has more benefit to the overall analysis than attempting to exploit temperature programming, which can, in some cases, be somewhat unpredictable in its effect and slow in its response to change and this may affect re-establishing the initial condition after a temperature-programmed run. The method could hold some advantage in a limited number of applications but is nowhere near as powerful and versatile as in the gas phase or as gradient elution in the liquid phase.

Methods of changing column capacity

Column selectivity and capacity factor are very closely related in that the selectivity of a column towards two different components is determined by the ratio of the capacity factors of the two components. For this reason it may appear rather inconsistent to segregate methods which influence column selectivity from column capacity. In the preceding paragraphs the methods described have the ability to affect the capacity factor of each component to a different extent, *e.g.*, a change in mobile phase composition may affect the retention characteristics of one component a great deal yet hardly influence the retention of another, thus the ratio of the two capacity factors, *i.e.*, the selectivity will change. On the other hand, if one studies the retention of the same two components on two columns which differ only in the level of stationary phase that has been applied to the support material, the capacity factors will again be different, depending on the column used, but the ratio of the capacity factors, *i.e.*, the selectivity, will remain the same. These different characteristics are illustrated in Fig.6.6, which depicts, in the form of simple chromatograms, the effect of a change in selectivity or a change in the column capacity on a pair of chromatographic peaks.

Methods which enable the capacity of the chromatographic system to be changed will rely on methods which permit the relative quantity of mobile phase to stationary phase or surface area to be altered. This of necessity must involve a physical rather than chemical change to the system, *e.g.*, changing columns from those having limited capacity (low surface area or stationary phase) to ones having a higher capacity. This may involve simply substituting another column or column switching during the course of an analysis.

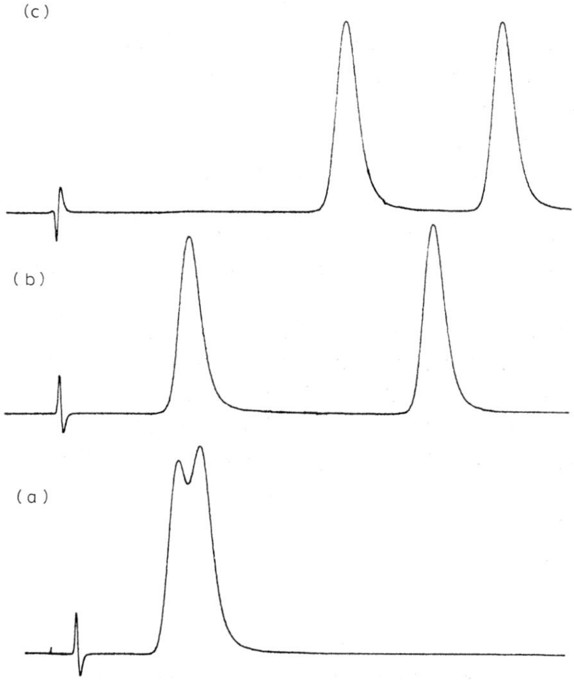

Fig. 6.6. Influence of selectivity and column capacity on a chromatographic separation. (a) Original incomplete separation; (b) improved separation due to increased selectivity; (c) improved separation due to increased column capacity.

Capacity characteristics of columns

The capacity of chromatographic columns is a function of the available surface (for adsorption), the level of stationary phase on the packing (for partition), the number of equivalents of exchangeable ionic sites per gram (ion exchange), and the pore volume per gram (for exclusion). All of these functions are, to the first approximation, related to the surface area of the chromatographic support, since a high proportion of stationary phase or ionic sites is only possible if there is a sufficient surface available on which to place the active coating. Since differences between specific chromatographic packings will be dealt with in later chapters, it is sufficient at this stage to describe the effect in general terms. As the diameter of the support material is reduced, the surface area per gram will increase, hence so will its capacity to retain a compound or support a stationary phase. Totally porous supports often have surface areas in the range of 100–400 sq.m/g whereas a superficially porous (porous layer) packing will have a surface area significantly lower than 50 sq.m/g. Thus, if a method originally used a porous layer type of packing, an increase in the capacity in order to effect and improve separation could be obtained by changing to a column packed with a totally porous support provided that the column efficiency is at least as good, for example by using a support of a mean particle diameter of less than 20 μm. The separation of several hydrocarbons shown in Fig.6.7 illustrates the

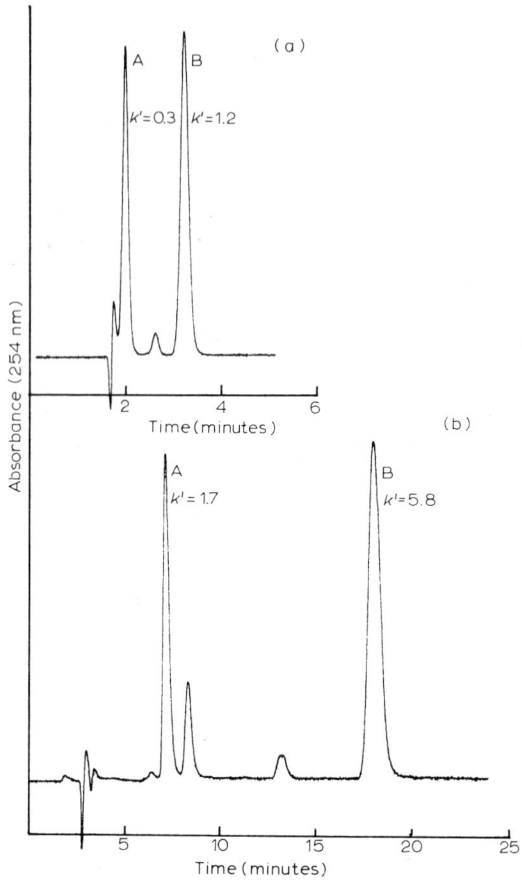

Fig. 6.7. Separation of hydrocarbons using solid-core and microparticulate porous column packings. Operating conditions: (a) Column, 1 m × 2.1 mm I.D.; packing, Permaphase ODS (solid core, 30 μm); flow-rate, 0.9 ml/min; inlet pressure, 60 bars (900 p.s.i.); mobile phase, methanol–water (75:25); temperature, 40°C. (b) Column, 0.25 m × 2.1 mm I.D.; packing, Zorbax ODS (porous, 4–6 μm); flow-rate, 0.25 ml/min; inlet pressure, 100 bars (1500 p.s.i.); mobile phase, as under (a); temperature, as under (a). A = Naphthalene; B = pyrene.

higher degree of retention and improved separation obtained by using a packing material having a larger surface area and a higher level of stationary phase.

The diameter of the internal pores will also govern the surface area of a support. For a given type of column packing the surface area is approximately inversely proportional to the pore diameter. The variation in surface area provides supports which will accept different loadings of stationary phase, allowing columns – of otherwise similar characteristics – to have different capacities to retain the sample while using the same mobile phase. If combinations of columns of this type are prepared, it is sometimes advantageous to employ column switching in order to optimise a separation.

Column switching

In this method two or more columns are linked together via a switching valve in such a manner that any component flowing through the first column can either be directed to the detector or into the second column in which further resolution can occur before the sample passes into the detector cell. The second column may be simply a longer version of the first column with a higher resolving power (in this case the resolving power of the system, not the capacity, is increased as the ratio of mobile to stationary phase remains constant) or a column having a higher loading of stationary phase, surface area or exchange capacity (in which case the capacity will be increased). The arrangement of instrument components necessary for this procedure is illustrated in Fig.6.8 and consists of a switching valve located between the two chromatographic columns, the outlet of each feeding into a T-piece immediately ahead of the detector.

In this system the column having the lower capacity is placed ahead of the column of higher capacity. Having injected a sample, the more rapidly moving components pass through "Column One" very quickly and enter "Column Two". Once this step has been achieved, the valve is actuated and any of the more slowly eluting components are passed directly to the detector and recorded while the less retained components of the sample are being further resolved in the second, high-capacity column. These components on elution flow through the detector and are recorded on the chromatogram after the more strongly retained components, which passed only through "Column One".

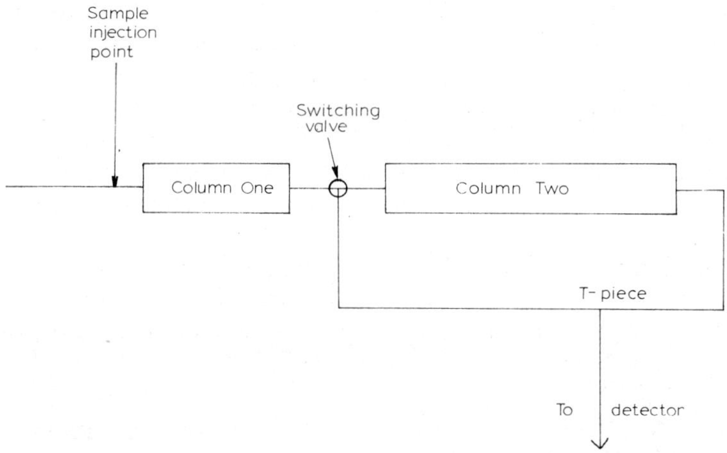

Fig.6.8. Schematic lay-out of apparatus for column switching.

For optimum control of this system it is necessary to have some previous experience of the behaviour of the sample components in both columns; however, it is possible to arrange the conditions such that strongly retained peaks do not enter the second column, thus reducing considerably the overall analysis time. In a similar manner a fraction of the sample eluting only partially resolved from the first column can be diverted into the second column to increase the resolution of the components. If the equipment used includes a

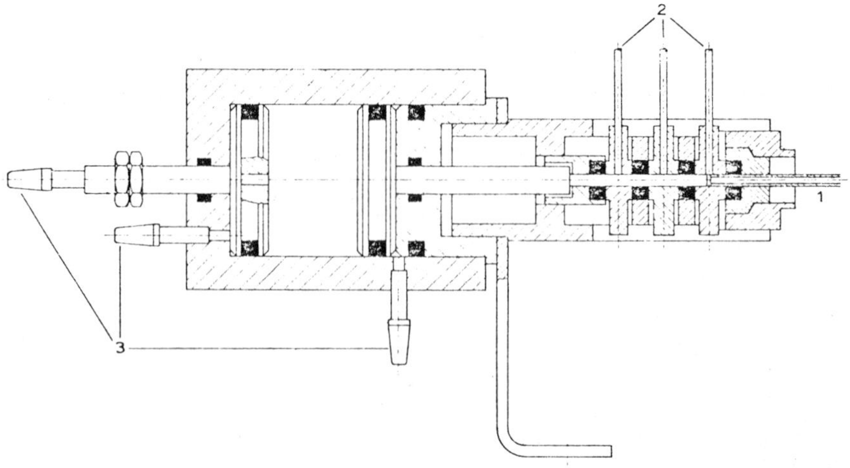

Fig.6.9. Pneumatic valve for column switching operations. 1 = Inlet; 2 = outlets; 3 = air supply. (Reproduced with permission from J.F.K. Huber, R. van der Linden, E. Ecker and M. Oreans, *J. Chromatogr.*, 83 (1973) 267.)

differential detector which has two flow cells capable of withstanding the operating pressure of the separation system, it can be advantageous to arrange one of the flow cells in the liquid flow path immediately ahead of the column switching valve so that the point at which to actuate the valve can be accurately determined. The success of the method relies very much on minimising the dead volume in the valve and interconnecting tubing in relation to the volume likely to be occupied by an eluting peak. A low dead volume valve that has been specifically designed for this application is shown in Fig.6.9 (ref.9).

This valve is a pneumatically operated device with one liquid inlet and can have either two or three outlets, depending on the model of valve. Air-activated valves are particularly useful for this type of operation as they enable the system to be automated without difficulty.

The method of column switching in LC has two advantages over gradient elution and the less useful temperature programming. The advantages are that the method involves considerably less expensive components and is operated with a mobile phase of constant

TABLE 6.3

SOME POSSIBLE PAIRS OF COLUMN PACKINGS FOR USE IN SWITCHING TECHNIQUES IN ADSORPTION CHROMATOGRAPHY

Column I	Column II	Difference in columns
Corasil I	Corasil II	The surface area of Column II is approximately double that of Column I (both are solid-core supports)
Pellosil HS	Pellosil HC	As for Corasil
LiChrospher SI-1000	LiChrospher SI-100	Porous microparticulate packings; the surface of SI-100 is more than ten times greater than that of SI-1000

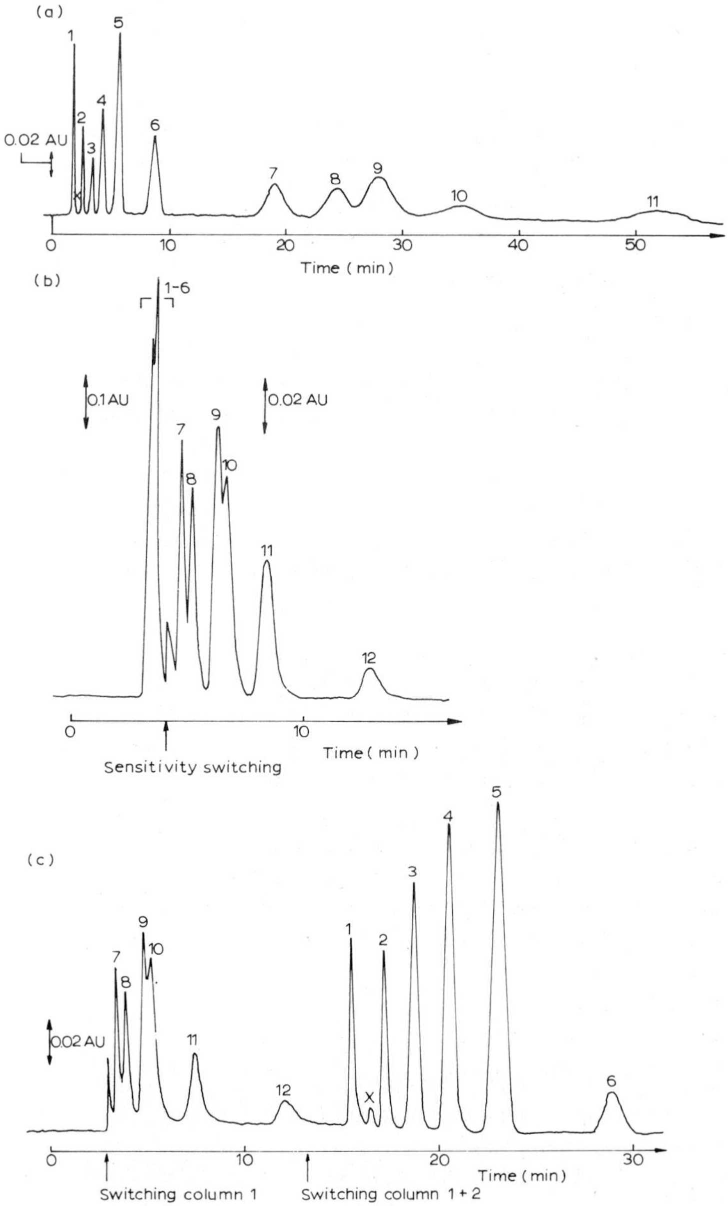

Fig. 6.10. Application of column switching to the separation of steroids. Sample (distribution coefficient): 1 = decylbenzene (0), X = impurity, 2 = progesterone (9), 3 = androstenedione (26), 4 = methyltestosterone (36), 5 = testosterone (65), 6 = andrenosterone (122), 7 = 16α-hydroxy-pregn-4-ene-3,20-dione (300), 8 = 19-hydroxy-androst-4-ene-3,17-dione (380), 9 = corticosterone (560), 10 = 11-dehydrocorticosterone (700), 11 = cortisone (1300), 12 = cortisol (2900); injection volume, 30 μl. Columns: liquid–liquid system water–ethanol–2,2,4-trimethylpentane; % (w/w), water-rich phase (stationary) = 25.5:71.5:3.0, water-poor phase (mobile) = 0.1:3.0:96.9; column 1, 250 × 2.7 mm, diatomite support, 2 m^2/g, 5–10 μm; column 2, 250 × 2.7 mm, silica support, 15 m^2/g, 5–10 μm. Detector: UV, 236 nm. (a) Columns 1 + 2; (b) column 1; (c) first part column 1, second part columns 1 + 2. (Reproduced with permission from J.F.K. Huber, R. van der Linden, E. Ecker and M. Oreans, *J. Chromatogr.*, 83 (1973) 267.)

composition, thus it is compatible with all types of LC detector. Notwithstanding these last points, column switching is not widely used in practice. The principal reason for its lack of popularity is probably the general difficulty of finding columns offering distinctly different selectivity or capacity characteristics while operating with exactly the same composition of mobile phase. Clearly, columns of different length which are packed with the same material are directly suited to this method. Candidate packing materials for use in the two columns where more pronounced differences are required can include those given in Table 6.3.

In liquid partition chromatography, the level of stationary phase held on the two columns will control their retention characteristics; however, the level of stationary phase is dependent on the nature of the chromatographic support contained in the columns. The use of supports which differ in surface area for columns I and II represents the most straightforward approach to this method when the stationary phase is applied to the columns by physical coating procedures. The ease of operation of the column switching technique may be increased considerably by employing two columns each containing a different stationary phase bonded chemically to its chromatographic support. With this approach it is most important to check the retention characteristics of all the sample components on each column independently so as to ensure that both columns contribute to the overall separation.

The chromatograms reproduced in Fig.6.10 illustrate applications of column switching reported by Huber *et al.*[9]. The separation relates to a group of steroids having widely different elution characteristics.

An alternative method of increasing the resolution of closely related components is to employ recycle chromatography, a method in which the sample is passed repeatedly through the chromatographic column or a set of columns until sufficient resolution is obtained.

Recycle chromatography

This method can be considered as a special case of column switching where the sample, after having passed through the chromatographic system without complete separation, is re-directed from the detector outlet back to the column inlet and passes a second time through the chromatograph. This process can be repeated a number of times until either sufficient resolution is obtained or the sample has spread to such an extent that it occupies the complete volume of the column(s).

Recycling a sample through a column system should, in principle, present a general method for improving resolution between components in any LC system. In practice, however, it is found to be applicable only to compounds which would elute with very little resolution after a single pass through the column, otherwise a situation is created where the leading edge of the first peak catches up with the trailing edge of the last peak from the previous pass through the column. Minimising intra- and extra-column band broadening of the sample components is critical to the success of the method. Intra-column effects cannot easily be modified once a column has been selected, but the extent of extra-column dead volume depends largely on the design of the chromatographic apparatus. A guide to the extra-column volume that is permissible in a recycle apparatus can be taken from the data pertaining to the influence of detector flow cell volumes given in Chapter 5, Table 5.1.

There are two somewhat different recycle procedures which have been employed, depending on whether the volume of the pumping system is negligible in relation to the volume of the chromatographic columns used.

Recycle chromatography with low-volume pumping systems

This method is especially useful for instruments fitted with reciprocating pumps which have an inherently low volume. The diagram illustrated in Fig.6.11 gives the essential features of this method.

The sample after injection passes through the column and detector in the normal way. After this point the effluent passes through a valve which enables the sample to be directed to the inlet of the pump, through the pump, injector and back to the head of the column. Once sufficient resolution has been obtained the valve is positioned to allow the components to flow out of the apparatus to a drain or fraction collector.

The success of this method relies on having an absolute minimum dead volume in the entire chromatographic system. Unfortunately, pumps that have the lowest internal volume are ones which, unless of special design, will give a pulsating flow and thus a pulse damper should be installed in the liquid flow path. The dead volume associated with a pulse damper can lead to very serious mixing of the sample components which are being separated. A similar problem sometimes occurs with pumps that possess an internal volume which is much larger than the displacement volume of the piston of the pump. When excessive broadening of peaks does occur, it may be useful to take a centre cut from the peak of greatest interest and recycle that part further; this technique holds some advantage when trying to isolate a component preparatively. An increase in the volume of the column relative to the extra-column volume, by increasing the size of the columns used, will help to reduce the influence of the band broadening.

An advantage of this approach is that it is possible to operate the detector flow cells at or near atmospheric pressure; also, since the system is effectively a closed loop, very little fresh mobile phase is required whilst actually recycling the sample.

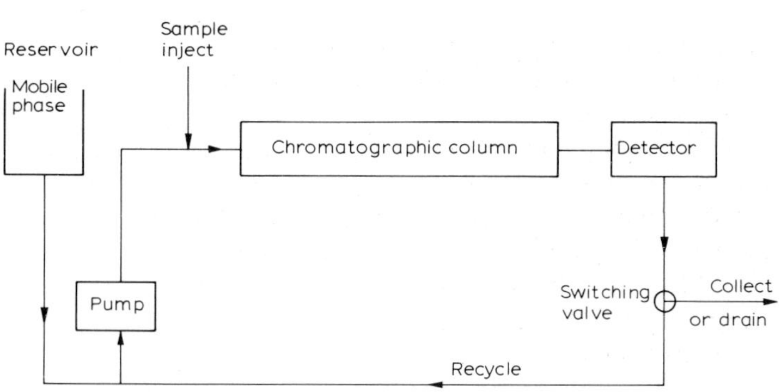

Fig.6.11. Schematic lay-out of apparatus for recycling sample through the column with the aid of a low-volume pump.

Recycle chromatography using the alternate pumping principle

The "alternate pumping principle" recycle action is achieved with two chromatographic columns and a six-port valve in such a way that the sample components being recycled do not have to flow through the pump, any pulse damper (if fitted) or the injector at each pass through the columns. This allows any type of pumping system to be employed irrespective of its volume. Moreover, the overall loop of the "recycle" part of the chromatograph can be designed in such a way that it gives very low dead space, leading to an improvement in the overall performance of the recycling action. A schematic drawing of the alternate pumping type of recycle chromatography is given in Fig. 6.12.

At first sight this method would appear to be wasteful of solvent as mobile phase is continually flowing through the pump and when the sample is recycled the mobile phase eventually leaves the apparatus without any sample components dissolved in it. This apparent loss of solvent can be easily eliminated by running the solvent back into the reservoir that supplies the pump at all times except when sample components are "tapped-off" from the recycle system. During each pass of the sample through a column the six-port valve must be actuated to switch the effluent leaving the second column either to drain or back into the first column. This action must be carried out within a fairly narrow limit of time after the sample has just passed through the valve. Deciding on this moment

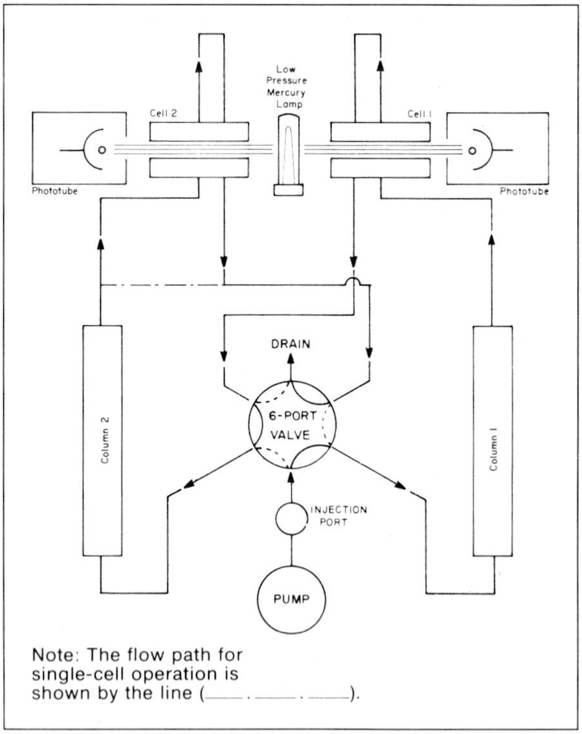

Fig. 6.12. Schematic lay-out of apparatus for recycle chromatography using the alternate pumping principle. (Reproduced with permission from R.A. Henry, S.H. Byrne and D.R. Hudson, *J. Chromatogr. Sci.*, 12 (1974) 197.)

can present some difficulty as, although the time to pass through two columns should be twice that taken in one, the columns might not be identical and also the peaks are continually broadening. The procedure is much simplified by using a detector which can be fitted with two flow cells, one at the outlet of each column. With detectors of the photometric or refractometric type the reference flow cell can be used to monitor one of the column outlets; thus when observing the chromatogram a peak eluting from one column will give a peak in one direction on the chart whereas one leaving the second column will give a peak in the opposite direction so that the location of the sample at any one instant can clearly be identified. Recycle chromatography in this form does require that the flow cells in the detector are capable of withstanding the high pressure within the system. This can create a problem with some designs of detectors, particularly with the differential refractometers.

It should always be borne in mind that the success of all forms of recycle and column switching techniques depend very much on minimising the extra-column band broadening. Chromatographic columns having internal diameters in the region of 2–3 mm have been used extensively for general analytical work for the past decade; columns of these dimensions cannot be used in recycle or switching methods without a very significant loss in resolution occurring as the sample passes through the pump or valves. Best results are obtained by working with columns of larger internal diameter, for instance, greater than 5-mm bore.

During the development of modern LC, recycle techniques have found greatest application in the field of steric exclusion chromatography, where columns have generally been of larger size and limited in resolving power. It is probable that interest in the method will decline as more chromatographers are making use of columns offering superior efficiency that have only become available comparatively recently.

Flow programming/pressure programming

These techniques are related closely with pressure programming in GC, whereby during the course of the separation the mobile phase velocity is increased by either applying a progressively higher inlet pressure or, in the case of mechanically driven pumps, by accelerating the drive mechanism in a systematic manner. This type of programming is accomplished with very little innovation with respect to instrument design; practically all pumping systems can be adapted for this mode of operation. Equally, as no change in the mobile phase composition is involved, the method can be employed with detectors such as the refractive index, which are incompatible with gradient elution methods. The major factor which limits interest in this technique is the range over which it can operate, for often the inlet pressure requirements are sufficiently high relative to the ultimate pressure capabilities of the equipment or column packing that the pressure cannot be increased by more than a factor of three or four, so that at best the technique will give a fourfold increase in speed of elution. Columns where the inlet pressure requirements are usually lower tend to be those whose efficiency is dependent on the mobile phase velocity so that a large increase in the flow-rate will lead to a decrease in column efficiency, thus destroying the object of the exercise. Wiedemann *et al.*[10] have described an apparatus for generating reproducible flow programmes in LC.

In conclusion, the only programming techniques which offer considerable potential in LC are gradient elution and column switching. The former is much more versatile and enables the optimum composition of the mobile phase to be found quickly by experiment, although the equipment required for a very reproducible system can be costly. Column switching, although involving much less capital outlay on equipment and although it may be used with all types of detectors, does require a good deal of preliminary knowledge of the chromatographic behaviour of the components present in the sample. This latter technique should be considered much less attractive relative to gradient elution as a tool for developing chromatographic separations. On the other hand, column switching does hold some potential in highly repetitive analyses where the qualitative composition of the sample is known, for instance in the monitoring of chemical plant processes.

REFERENCES

1 I.M. Hais and K. Macek, *Paper Chromatography*, Academic Press, New York, 1963, p.115.
2 L.R. Snyder, *J. Chromatogr.*, 92 (1974) 223.
3 W.K. Lewis and L. Squires, *J. Oil Gas*, 33, Nov.15th (1934) 92.
4 J.A. Schmit, R.A. Henry, R.C. Williams and J.F. Dieckman, *J. Chromatogr. Sci.*, 9 (1971) 645.
5 R.P.W. Scott and P. Kucera, *J. Chromatogr. Sci.*, 11 (1973) 83.
6 L.R. Snyder, *J. Chromatogr. Sci.*, 8 (1970) 692.
7 C. Liteanu and S. Gocan, *Gradient Liquid Chromatography*, Ellis Horwood, Chichester, 1974, p.19.
8 R.J. Maggs, *J. Chromatogr. Sci.*, 7 (1969) 145.
9 J.F.K. Huber, R. van der Linden, E. Ecker and M. Oreans, *J. Chromatogr.*, 83 (1973) 267.
10 H. Wiedemann, H. Engelhardt and I. Halász, *J. Chromatogr.*, 91 (1974) 141.

Chapter 7

Liquid–solid (adsorption) chromatography

INTRODUCTION

Of all the methods of separation possible in the liquid phase, adsorption chromatography is probably the widest used and has been practiced for the longest time. The original work of Tswett, considered to be the earliest application of LC, involved separation of substances in a column filled with powdered chalk which acts as a weak adsorbent. Most applications of classical column chromatography are based on the use of packing materials such as silica gel, alumina, charcoal and Florisil, all of which possess very definite, yet often quite different, adsorptive properties.

A great broadening of the application of this separation method came about with the advent of TLC. In this technique a thin layer of adsorbent, most often silica gel or alumina, is used as the medium on which a sample is applied as a spot and developed by the action of a liquid phase rising up through the adsorptive layer. In this text it is not important to describe TLC in any detail, but only to record the fact that through its use came the realisation that quite complex separations could be achieved on adsorbent materials, provided that the composition of the eluting liquid (mobile phase) was selected carefully. Modern liquid–solid chromatography (LSC) offers the same style of separation mechanism, only with greater resolving power, speed and ease of quantitation.

Although adsorption chromatography has been widely used over a considerable number of years, it is apparent that there are many points of detail which need to be considered if highly reproducible results are to be obtained. Provided care is taken with the selection of appropriate operating conditions, adsorption chromatography has one of the widest ranges of applicability of any LC method for the high-resolution separation of non-ionic species of low molecular weight.

RANGE OF SAMPLE APPLICABILITY

Separations achieved by adsorptive processes are typified by their ability to resolve sample components into their respective classes according to the polar functional groups present in the component molecules, rather than resolving compounds of essentially similar polarity and differing by the extent of aliphatic substitution. This latter style of separation is more often achieved by partition methods.

In addition to separating components of a sample into classes, LSC is particularly effective at resolving mixtures of isomers such as geometrical isomers, *i.e.*, *cis*/*trans* pairs, and positional isomers due to different substitution in an aromatic nucleus. These characteristics are illustrated in Fig.7.1 by a chromatogram of the analysis of technical dinitrotoluene by LSC using a column containing porous silica microspheres. The isomer composition of the sample can readily be seen (qualitatively) as can the separation of the dinitro from the mononitro compounds.

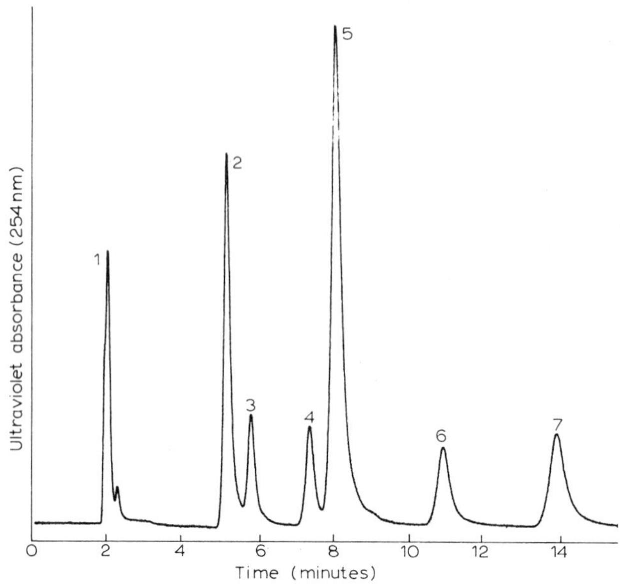

Fig. 7.1. Separation of isomers of dinitrotoluene (DNT). Operating conditions: column, 0.25 m × 2.1 mm I.D.; packing, Zorbax SIL, porous silica; mobile phase, pentane–1% dichloromethane–0.01% methanol; flow-rate, 1.0 ml/min; inlet pressure, 100 bars (1480 p.s.i.); temperature, 25°C. 1 = Mononitrotoluenes; 2 = 2,5-DNT; 3 = 2,6-DNT; 4 = 3,5-DNT; 5 = 2,4-DNT; 6 = 2,3-DNT; 7 = 3,4-DNT.

A wide range of samples may be studied by the LSC method. An oversimplified guide could be that LSC is suitable for substances which are less than 2000 in molecular weight, non-ionic and soluble in at least one organic solvent. From this statement it will be clear that the potential range of applicability is very wide indeed. More precisely, the technique works best with samples of moderate polarity, *i.e.*, molecules with at least one polar functional group. Fig. 7.2 illustrates separations of compounds typical of this polarity class. The figure also demonstrates the difference in selectivity by using alumina in place of silica gel as adsorbent.

Non-polar samples may be analysed by this method using column packings which are highly activated. Details of how this is achieved are given in later sections of this chapter. Weakly ionic species are frequently very strongly retained or elute with poor peak shape when studied on LSC columns. The addition of an acidic or basic solvent to the mobile phase will often reduce dissociation of the sample, depending on its functionality, leading to a significant improvement in the shape of the chromatographed peak.

A novice, when first considering LSC, is tempted to rely on separation data that have been derived from TLC measurements. Although, in principle, both methods might be considered as two ways of performing the same type of separation, considerable caution should be exercised when transferring TLC methods to a modern liquid–solid column chromatographic system.

The reasons for this discrepancy are: firstly, that in most cases a TLC plate is used in a highly activated form whereas a LC column has been deactivated to some extent by the passage of mobile phase through the column prior to injecting the sample. Secondly, with

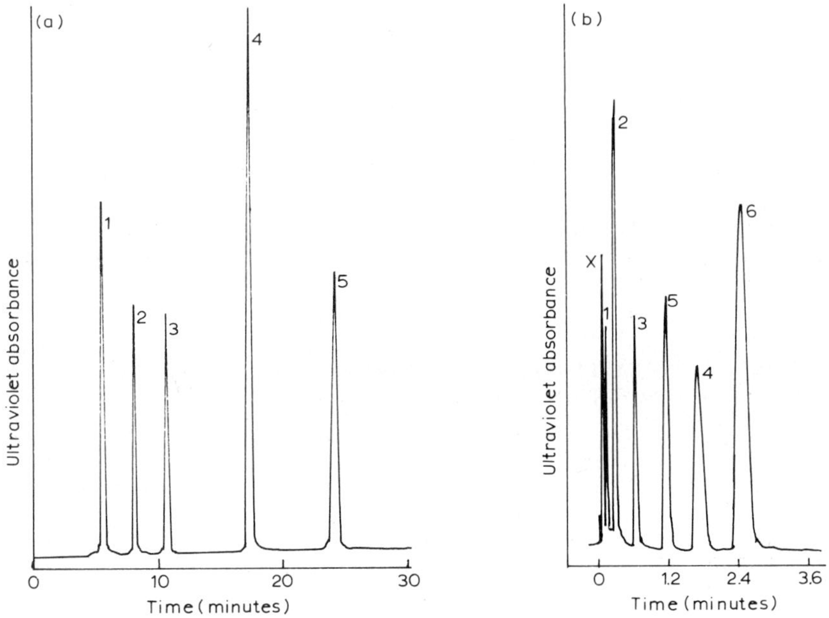

Fig. 7.2. Separation of aromatic compounds on alumina and silica gel adsorbents. Operating conditions: (a) Column, 0.5 m × 2.8 mm I.D.; packing, Spherisorb A5Y; mobile phase, hexane–10% methylene dichloride (water saturated); flow-rate, 0.425 ml/min. (b) Column, 0.15 m × 2.1 mm I.D.; packing, silica gel, 5–10 μm; mobile phase, hexane; flow-rate 6.67 ml/min. X = Impurity; 1 = phenetole; 2 = nitrobenzene; 3 = methyl benzoate; 4 = acetophenone; 5 = carbazole; 6 = 2,4-dinitrobenzene. (Reproduced, with permission, from (a) *Phase Separations Catalogue*, dated January 1975 and (b) R.E. Majors, *Anal. Chem.*, 44 (1972) 1722.)

few exceptions, the type of adsorbent used for TLC varies considerably in terms of particle size distribution, surface area and pore size relative to the LC counterpart. It should also be appreciated that with a TLC plate one is able to observe the position of a "spot" across the entire region from the point of sample introduction to the furthest distance moved by the solvent front. This situation is equivalent to being able to "see" eluting compounds at any point within a LC column, which is clearly not possible in most instances, as the detector must be located at the column outlet.

For a given TLC system, compounds which normally have a high R_F value on a plate will require a weaker solvent (relative to the TLC carrier liquid) to be used in a column system, whereas a compound with a low R_F value will need a stronger eluting solvent to be used as the mobile phase. In the case of multicomponent mixtures, it will be apparent that some form of gradient elution will be necessary if optimum peak shape and speed of analysis are to be achieved.

TYPES OF ADSORPTIVE PACKING

Much of the classical (gravity-fed) liquid–solid column chromatography was carried out with polar adsorbents such as silica gel (sometimes referred to as silicic acid), magnesia, magnesium silicates (*e.g.*, Florisil), alumina, molecular sieves and a range of other mineral-based materials such as bentonite clays. Several non-polar adsorbents have also been employed such as nylon[1], PTFE[2], and charcoal[3]. Unfortunately, many of these materials are fragile and are quite unsuitable as packings for modern chromatographic columns, where, to achieve high efficiency, it is necessary to use finely ground materials which may be subjected to high liquid pressures.

A rather limited number of different chemical types of adsorbent packings have been studied in modern chromatography, at least up to the present time. The most widely used

TABLE 7.1

SOME OF THE COMMERCIALLY AVAILABLE COLUMN PACKING MATERIALS FOR HIGH-PERFORMANCE LIQUID–SOLID (ADSORPTION) CHROMATOGRAPHY

Type	Name	Surface area (m^2/g)	Particle size (μm)	Shape*	Supplier
Silica					
Pellicular	Corasil II	14	37–50	S	Waters
	Vydac adsorbent**	12	30–44	S	Separations
	Pellosil HS	4	37–44	S	Reeve Ange
	Pellosil HC	8	37–44	S	Reeve Ange
	Perisorb A	14	30–40	S	Merck***
	SIL-X-II**	12	30–40	S	Perkin-Elme
Porous	μPorasil	400	10	I	Waters
	Silica A	400	13±5	I	Perkin-Elme
	SIL-X-I**	400	13±5	I	Perkin-Elme
	LiChrosorb§ SI-60	500	5 or 10	I	Merck
	LiChrosorb§ SI-100	400	5 or 10	I	Merck
	Spherisorb S5W, S10W, S20W	200	5, 10 or 20	S	Phase Separ
	Partisil 5, 10, 20	400	5, 10 or 20	I	Reeve Ange
	Zorbax SIL	300	4–6	S	DuPont
	Porasil T	300	15–25	I	Waters
	LiChrospher SI-100	250	10	S	Merck
	LiChrospher SI-500	50	10	S	Merck
	LiChrospher SI-1000	20	10	S	Merck
	LiChrospher SI-4000	6	10	S	Merck
	Micropak SI-5, SI-10	500	5 or 10	I	Varian
Alumina					
Pellicular	Pellumina HS	4	37–44	S	Reeve Ange
	Pellumina HC	8	37–44	S	Reeve Ange
Porous	Woelm Alumina	200+	18–30	I	Woelm
	Spherisorb A5W, A10W, A20W	95	5, 10 or 20	S	Phase Sepa
	Lichrosorb§ Alox T	70	5 or 10	I	Merck
	Micropak Al-5, Al-10	70	5 or 10	I	Varian

*I = Irregular; S = spherical.

**Stated to be chemically deactivated; control of water content in system is less critical.

***E.M. Labs. in the U.S.A.

§Formally marketed under the name Merckosorb.

materials are based on silica, an acidic polar adsorbent, or on alumina, which is generally a basic polar adsorbent but may be chemically modified so as to exhibit acidic or neutral characteristics. Although only these two chemically different types of adsorbent have been widely studied, each has a great number of physical ramifications which offer widely different performances at a wide range of cost. These include materials which differ in particle diameter, in porosity, in being totally porous or porous-layer materials and either irregular or spherical in shape. Each of these properties influence the chromatographic characteristics of the resultant column in that they will decide the ease of packing, pressure drop, column efficiency and sample capacity. Even taking these factors into account, differences in performance of geometrically similar packings, originating from different commercial sources, have been observed. These differences may be attributed in part to the presence of trace elements, particularly residual quantities of heavy metals, in certain packing materials. In Table 7.1 details are given of some of the more widely available forms of adsorbents that are used in modern LC. Unfortunately, the current proliferation of products for modern LC makes it difficult to ensure the data are all inclusive.

A survey of published applications involving the use of these column packing materials is given in Chapter 15.

Many of the products which are employed in liquid–solid (adsorption) chromatography find additional use in steric exclusion work and as supports for stationary phases in partition chromatography. The data contained in Table 7.1 are of some relevance when selecting materials for these latter methods of LC separation.

In addition to the polar adsorbents listed in Table 7.1 there are a number of other materials available which are totally porous and generally possess a fairly broad distribution of particle diameters. Although these materials are not proposed for the highest performance in modern LC, they are available in bulk and at a modest cost. They are typified by those given in Table 7.2.

These materials can be of considerable use in large-scale separations, where the cost of specialised column packings could be prohibitively high, and for cleaning up samples or solvents prior to their use in a high-performance system.

It is also possible to carry out adsorption chromatography using non-polar adsorbents in a manner similar to that of reversed-phase chromatography. Although charcoal is perhaps the most common example of a non-polar adsorbent, it has found little use in modern LC up to the present time. This is probably due to the lack of commercial products which offer sufficiently high purity and good resistance to compression at high pressure.

The great emphasis of adsorption chromatography, however, has been with the development of the polar adsorbents, notably silica and alumina. Of these, silica has been more widely used, as, in general, it offers higher performance in terms of efficiency and linear capacity. The choice between using silica- or alumina-based column packings can be influenced by the acidic nature of silica, which will tend to adsorb basic samples more strongly than would a column packed with basic alumina. Some notable selectivity characteristics of alumina, *e.g.*, its ability to selectively retain certain aromatic hydrocarbon isomers, can be put to advantage. The development of highly efficient column packings based on silica in recent years has made it less necessary to exploit selectivity differences between the various adsorbent types. This situation reflects the inter-relationship of

TABLE 7.2

SOME OF THE LESS EXPENSIVE COLUMN PACKINGS FOR GENERAL USE AS ADSORBENTS IN LSC

Type	Name	Surface area (m^2/g)	Particle size (μm)	Shape*	Supplier
Silica					
Porous	Spherosil XOA-400	350–500	Choice of:	I	Rhone-Progil
	Spherosil XOA-200	125–250	less than 40	S	Rhone-Progil
	Spherosil XOA-075	50–100	or 40–60	S	Rhone-Progil
	Spherosil XOB-030	25–45	or 60–80	S	Rhone-Progil
	Spherosil XOB-015	10–20	or 80–100	S	Rhone-Progil
	Spherosil XOC-005	2–6	or 100–150	S	Rhone-Progil
	Porasil A	350–500	Choice of:	S	Waters
	Porasil B	125–250	37–75	S	Waters
	Porasil C	50–100	or 75–125	S	Waters
	Porasil D	25–45		S	Waters
	Porasil E	10–20		S	Waters
	Porasil F	2–6		S	Waters
	Davison Code 12	800	150+	I	W.R. Grace
	Davison Code 62	350	150+	I	W.R. Grace
	Bio-Sil A	200+	20–44	I	BioRad
	LiChrosorb** SI-60	500	30	I	Merck***
	LiChrosorb** SI-100	400	30	I	Merck
Alumina	Alcoa F-20	200	150+	I	Alcoa
	Bio-Rad AG	200+	less than 75	I	BioRad
	LiChrosorb** Alox T	70	30	I	Merck

*I = Irregular; S = spherical.
**Formally marketed under the name Merckosorb.
***E.M. Labs. in the U.S.A.

efficiency and selectivity that contributes to the resolution of a pair of chromatographic peaks, which was described in Chapter 2. Currently the major proportion of separations achieved by LSC are performed with silica-type column packings.

MECHANISM OF ADSORPTION CHROMATOGRAPHY

In all forms of LSC the material used as column packing has some inherent adsorptive "activity", *i.e.*, the material has the power to concentrate on its surface molecules of sample or solvent drawn from the mobile phase which surrounds the material. The attraction of substances to the surface of the column packing may be considered as a dynamic equilibrium. The diagram shown in Fig. 7.3 illustrates this effect. Before a sample is introduced a state of equilibrium exists whereby molecules of mobile phase are continually being adsorbed on the surface then subsequently desorbed, thus any given molecule will

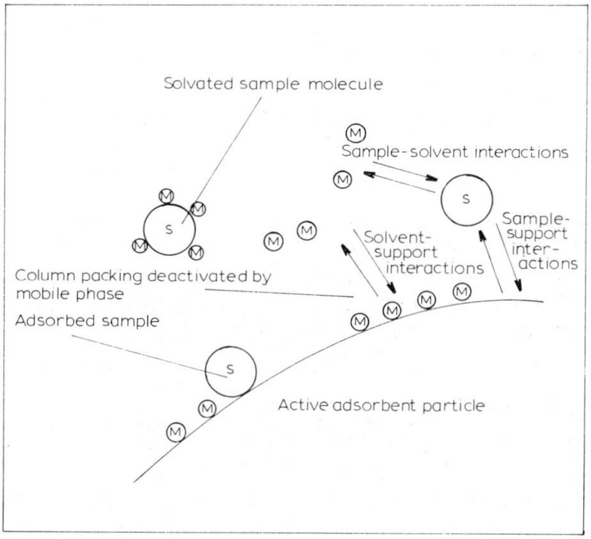

Fig. 7.3. Equilibria at a liquid–solid adsorptive surface.

spend a significant proportion of its "life" in a column adsorbed on the surface of the support. When a sample is introduced, the equilibrium condition between mobile phase and the adsorbed surface is disturbed; molecules of the sample and mobile phase now compete for the adsorptive sites on the surface of the column packing. A strong affinity of the packing for the sample will lead to mobile phase molecules being displaced in favour of sample molecules.

This pictorial situation is complicated by additional interactions between the molecules of the mobile phase which tend to solvate the sample molecules. The overall picture is one where, at equilibrium, the sample molecules are distributed between the surface of the adsorbent and the solvated form in the mobile phase. The molecules of mobile phase are distributed in a likewise manner. This equilibrium state exists at all times even when the mobile phase is not in motion. In many circumstances the equilibrium is not quite as simple as inferred here as any given site on an adsorptive surface may be sufficiently large to accommodate several solvent molecules but only one sample molecule, which is invariably larger in size. Under dynamic conditions, *i.e.*, when the mobile phase is flowing through the column, whether or not the molecules of samples are retained on the column depends on the relative magnitude of adsorptive forces of the packing for the mobile phase or sample molecules, the solvation of the sample, the concentration of all species participating in the equilibrium, *i.e.*, a law of mass action effect, and on the temperature at which the process is being carried out. As regards the latter point, many adsorptive reactions are exothermic, therefore, in principle, raising the temperature will shift the extent of equilibrium in favour of the mobile phase. However, it was noted in the previous chapter that temperature programming is not particularly effective in LC, the reason being principally that the rate of equilibration is often very slow.

For sample retention to occur on introduction into the chromatographic system the choice of mobile phase must be such that molecules of sample are attracted to the surface of the adsorbent at the expense of mobile phase molecules. If the extent of this attraction is overwhelmingly in favour of the sample molecules being adsorbed then the sample would remain on the column packing in the vicinity of where it was injected. For the sample to be able to elute from the column the adsorption of the sample must be represented by an equilibrium distribution with a small, yet significant, proportion of the sample being in the mobile phase. The larger the proportion of molecules in the mobile phase, the more rapid will be the elution of the component.

Since LSC depends on the adsorption of sample molecules on a surface, at the molecular level one can imagine that some selectivity might exist regarding the shape of the adsorptive site and the geometry of the adsorbed species. This effect is realised in practice, as adsorption techniques are particularly effective at resolving samples containing geometrical (*i.e.*, *cis–trans*) and positional isomers.

The generally observed feature that adsorption chromatography, in particular on polar surfaces such as silica gel or alumina, is not effective at resolving homologues may be explained in a pictorial manner from Fig. 7.4, which depicts polar aliphatic molecules such as alcohols adsorbed on a polar surface. Since the point with greatest affinity for the surface is the hydroxyl function of the sample, this will be firmly adsorbed. The polar adsorbent has little affinity for the alkyl chain, which is free to interact with the mobile phase in much the same manner as soap interacts at an oil-to-water interface. If this

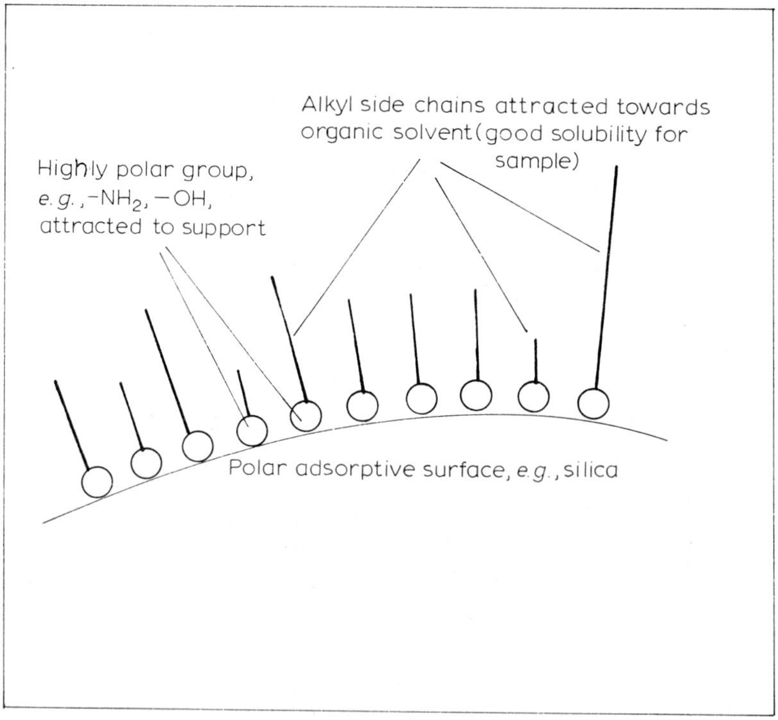

Fig. 7.4. Interaction of compounds belonging to a homologous series with an adsorptive surface (speculative model).

orientation of the sample molecules is accepted, then the fact that homologues generally exhibit a very similar behaviour can be readily visualised, since any increase in length of the alkyl chain will tend to stand up from the surface, thus not influencing to any great extent the interactions at the adsorptive surfaces. Data reported by Scott and Kucera[4] have indicated that the retentive power of silica surfaces is directly related to the concentration of surface hydroxyl functional groups and that siloxyl groups (Si=O) do not contribute to the adsorptive properties of the materials. In this manner the retentive power of silica adsorbent does not appear directly related to the surface area of the support, but only to the extent to which this property influences the concentration of surface hydroxyl (*i.e.*, silanol = SiOH) groups.

CHOICE OF SEPARATING CONDITIONS

In practice the selection of mobile phase, temperature and adsorbent type is made so that ideally the active surface retains each of the components of the sample to a different extent, so that, provided the eluting bands are sufficiently narrow, *i.e.*, the column is efficient, the components will elute separately from the column.

The problem that may arise when seeking to establish such operating conditions is that the range of polarity covered by all the components in a sample is wide, *i.e.*, some components are relatively non-polar whereas others are quite polar. Thus, although one has achieved a different degree of interaction with the support for each sample, the extent of the difference leads to a situation where only some of the components will elute from the column, the rest being strongly retained in the column. In a similar manner, the surface of an adsorbent is not homogenous as some of the adsorptive sites are stronger than others, so that for a given component in the sample some areas of the surface of the packing will be able to retain the component more strongly. This effect leads to a non-linear adsorption isotherm, *i.e.*, the extent of retention is dependent on the mass of sample relative to the surface area of the support. If columns are operated under conditions where this can occur, asymmetric peaks of the types described in Chapter 2 may be obtained. In general, the adsorption isotherm for a totally porous material is linear over a wider range of sample size than for a superficially porous material.

The combination of the effect of the sample components having different polarity and the surface of the chromatographic packing having sites of different adsorptive strength leads to a situation where for many samples there is a risk of some components being completely retained while others are eluted without retention. Also if very strong adsorptive sites are present on the surface of the chromatographic packing, then irreversible adsorption of a proportion or all of a component can occur. This latter problem, together with the possibility of decomposition of samples on the "catalytic" surface of adsorbent packings, has most certainly been known to occur; however, it should be appreciated that these effects are quite rare in all but the most polar of substances, such as peptides and acids. In many other applications there is the distinct impression that most chromatographers have been over-cautious on this particular aspect of LSC, on account of experiences gained with more active adsorptive supports, for example those used in TLC. If one is faced with the problem of irreversible retention or decomposition of a sample, there are several approaches that may be investigated:

(1) Employ a programming technique such as gradient elution where the extent of sample–adsorbent interaction may be decreased by increasing the affinity of the sample for the mobile phase and of the mobile phase for the adsorbent surface.

(2) Deactivate part of the adsorbent surface to remove highly active sites. This will result in a more linear adsorption isotherm, improving sample capacity and peak shape at the expense of some adsorptive strength. This procedure is very important in practice, as will be seen in later sections.

(3) Choose an alternative chromatographic packing which has sufficient capacity for the separation under examination. There are, commercially available, a range of packings for adsorption chromatography differing in surface area, chemical type, and in the form of either totally porous or porous layer materials.

(4) Many samples which chromatograph with difficulty on adsorptive systems are weakly ionic in character. Addition of a small proportion of acid to the mobile phase will effectively suppress the dissociation of a weak acid, leading to improved peak shape and better chromatography. In a similar way, ammonia or a simple amine added to the mobile phase will improve the elution of bases.

(5) Choose an alternative separation method, *e.g.*, liquid–liquid partition.

PRACTICAL ASPECTS OF ADSORPTION CHROMATOGRAPHY

Experience drawn from many publications, especially those of Snyder (*e.g.*, ref. 5), points out that for the best column performance in LSC it is necessary to operate the chromatographic system under conditions where the adsorptive surface is deactivated to some extent. This is most often achieved by addition of a controlled amount of water to the mobile phase or the adsorbent, prior to packing the column, or to both, so as to reduce the most active sites on the adsorptive surface. This leads to improved reproducibility of chromatographic separations and more linear adsorption isotherms which will make retention characteristics of samples less dependent on the sample size and in many instances improve peak shape by reducing peak tailing. Considerable preliminary work and attention to detail are needed to carry out adsorption chromatography with a controlled activity of the column packing material. This is particularly the case when the ideal mobile phase composition is not known and it is necessary to try different solvents as mobile phases. The procedures involved will be described in this and subsequent sections of this chapter. Before doing this, several points of a more general nature should be brought to the attention of the reader.

Chemically modified adsorbents

Several suppliers of adsorptive packings offer products which are described as being chemically treated to obviate the need to control the water content, hence the activity, of the column packing. It is better, however, to consider these materials as ones in which the need to control may have been reduced rather than eliminated, as all adsorbents are sensitive to a greater or lesser extent to the presence of small quantities of highly polar species in the mobile phase.

Maintaining adsorbent activity and mobile phase selection

In a situation where the separation of a completely new sample is to be investigated, it is often found that the initial scouting of possible solvents for use as mobile phases is performed without heed to the activity of the column packing. This approach, although at first sight seeming to be more rapid, will only be successful if it is continually borne in mind that while the adsorbent in the column is reaching an activity level compatible with the composition of the mobile phase, a considerable change in the retention characteristics can occur, thus reducing the possibility of achieving a reproducible separation. Many operators do, however, consider this method as a viable approach, enabling a rough idea of the chromatographic conditions to be obtained very quickly. There are many studies of equilibration rates of columns reported in the literature indicating that in some instances several hundred column volumes of the new mobile phase are required before the column packing attains an equilibrium condition. The series of chromatograms shown in Fig. 7.5 outlines the effect of non-equilibrium between a mobile phase and an adsorbent column packing. The column, packed with Zorbax-SIL porous silica microspheres, was first used with a mobile phase of pentane containing 2% dichloromethane and 0.02% methanol. Later the mobile phase was changed to one containing exactly half the former quantity of polar modifiers. The separations were repeated after 10 column volumes and 100 column volumes of the new mobile phase had passed through

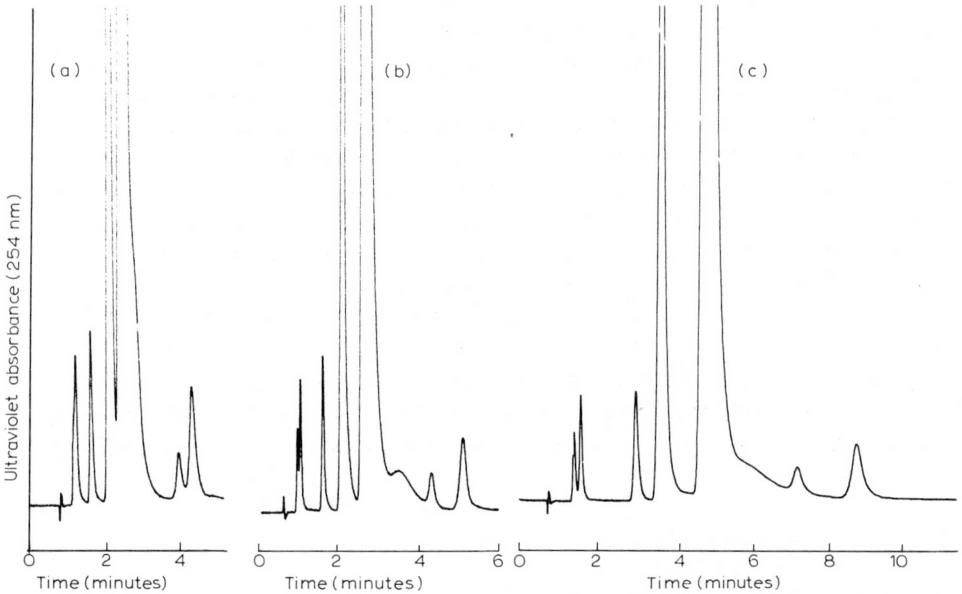

Fig. 7.5. Effect of non-equilibrium conditions on the separation of the isomers of dinitrotoluene. Mobile phase: (a) pentane–2% dichloromethane–0.02% methanol; (b) pentane–1% dichloromethane–0.01% methanol; (c) as for (b). In (b) the sample was injected after ten column volumes of new mobile phase had passed through the column; in (c) it was injected after 100 column volumes of mobile phase had passed through the column. The identity of individual peaks may be made by comparing with Fig. 7.1.

the column: a definite change in sample retention can be observed between the two analyses.

Despite limitations of not controlling the activity of the support, a useful starting point in the selection of the mobile phase is to take two solvents of extreme polarity, for instance hexane and ethanol (or pentane and methanol). The development of the method is started by injecting the sample into the column when only the alcohol is the mobile phase. With adsorptive packings, such as silica and alumina, most samples will elute without retention, *i.e.*, $k'=0$. The next step is to change the mobile phase by using, in turn, mixtures of the alcohol and hydrocarbon solvents, for example 80, 40, 20 and 10% alcohol in hydrocarbon solvent. After at least 20 column volumes of each solvent have passed through the system, the sample is re-injected. This procedure is continued with each solvent mixture, noting the mobile phase composition which causes the components to be just retained and the composition of mobile phase which causes total retention of the sample. If no retention occurs, even when the pure hydrocarbon is used as mobile phase, it will be necessary to choose an alternative separation method, such as reversed-phase chromatography. This scouting exercise, although not operating under completely equilibrated conditions, provides a basic understanding of the polarity range of the mobile phase which will elute the sample from the column. This information makes it possible to decide which solvents in the eluotropic series (given in Table 6.1) have some chance of being used as mobile phases, as they possess a similar eluting strength.

For the experienced chromatographer, the decision of which solvent will yield the best separation of the components of a sample is based largely on intuition, taking into account the known chemistry of the sample and its solubility characteristics. Clearly, intuition comes only from experience and the novice in LC must try to develop his own approach to the subject. On the basis of information derived from the literature, some generalisations regarding selectivity and solvent composition may be proposed, although they are probably not valid for every application. One interesting observation is that greater selectivity is usually obtained between eluting components by using mobile phases formed from mixtures of solvents of differing polarity rather than a single solvent of intermediate polarity, for example a mobile phase formed from a hexane–alcohol mixture might be expected to provide greater selectivity than, say, pure chloroform, which would

TABLE 7.3

EXAMPLES OF THE INFLUENCE OF THE MOBILE PHASE ON THE SELECTIVITY IN ADSORPTION CHROMATOGRAPHY [5]

Mobile phase	Capacity factors, k'		Selectivity, α
	Acetonaphthalene	Dinitronaphthalene	
50% v/v benzene in pentane	5.1	2.5	2.0
23% v/v dichloromethane in pentane	5.5	5.8	1.05
0.05% v/v dimethylsulphoxide in pentane	1.0	3.5	3.5
	Quinoline	Aniline	
Dichloromethane	2.1	1.3	1.6
Benzene	5.4	5.6	1.04
20% v/v diethylamine in pentane	0.4	3.5	8.7

exhibit approximately the same solvent strength. Similarly, the greater the difference in polarity of the solvents forming the mobile phase, the greater the selectivity. Data published by Snyder[5], shown in Table 7.3, illustrates this phenomenon. In practice, the converse of this result is often of equal importance, for instance, when too large a selectivity exists between components, suggesting either excessive analysis times or the need for gradient elution; in this case the use of a single solvent of intermediate polarity may simplify the separation.

Changes in sample selectivity can also be achieved by substituting solvents of different types, *e.g.*, a proton acceptor for a proton donor, an aliphatic for an aromatic solvent, and the use of halogenated solvents in place of esters.

Some efforts have been made to quantify these solvent effects. Since they are considered beyond the scope of this book, however, interested readers are recommended to refer to the data published by Snyder[5].

When using any mobile phase which is essentially immiscible with water it is necessary to control the level of activity of the adsorbent in order to obtain reproducible results. This is achieved by maintaining a small proportion of a polar modifier in each of the solvents used as mobile phases. This low level of polar modifier becomes partially adsorbed on the surface of the column packing, moderating the adsorptive strength, improving the linearity of the adsorption isotherm, and reducing peak tailing.

Two procedures are available for controlling the activity of the column packing. These rely on either the presence of traces of water or lower alcohols in the mobile phase. Since water is essentially immiscible with the solvents under consideration, *e.g.*, hexane, chloroform or ether, special procedures have been adopted to ensure that the water content of these solvents can be controlled precisely. In many applications it is important to operate with the mobile phase partially saturated with water, hence one refers to "hexane (50% water saturated)". A partially water-saturated solvent is prepared by blending appropriate volumes of completely saturated and anhydrous solvents immediately before use.

Preparation of solvents of known water content

Solvents to be used in adsorption chromatography with a known level of water content are prepared in advance according to the following procedure. Each of the water-immiscible solvents to be used should be divided in two parts: one is dried, while the other part is fully saturated with respect to water. Dehydration of most solvents can be effected by passing the solvent through a large column filled with oven-dried (*e.g.*, at 150°C overnight) silica gel. For this purpose inexpensive coarse-grade silica gel may be used. This adsorbent will also tend to remove from the solvents any impurities which might otherwise have been retained on the adsorbent in the high-performance LC column. Finally, the collected solvent is stored over some anhydrous dessicant such as molecular sieve. The portion of solvent which is required in a water-saturated condition is first mixed thoroughly with excess water, *e.g.*, by magnetic stirring overnight. This is followed by passing through a column filled with coarse silica gel, Celite or firebrick which has been loaded with excess water. The collected solvent is stored in contact with excess water until required. It should be appreciated that solvents such as hexane have very little affinity for water and therefore even "water-saturated" hexane contains an extremely low, yet significant, concentration of water. Solvents like ethyl ether, on the other hand, will dissolve very much greater quantities of water.

A common starting point in many reported applications of adsorption chromatography is to employ solvents in the mobile phase that are 50% water saturated. An increase in the percentage of water saturation of the mobile phase will lead to a decrease in the adsorptive capacity of the column and shorter separation times, although it should be remembered that resolution may also suffer as capacity factors are decreased.

The principle advantage of working with solvents saturated to a certain level is that once a column is equilibrated with respect to one solvent at that level, another solvent of the same water content may be introduced as a mobile phase with very little time required for equilibration. This procedure obviates the lengthy time required for an adsorbent to reach equilibrium with the solvent with which it is in contact.

Procedure for changing the level of activity of an adsorbent packing

After packing the chromatographic column, which has most likely been achieved by a slurry method, the adsorbent will exist in a completely deactivated state. Water and any other solvents remaining from the packing procedure must be removed. This may be achieved by passing a definite number of column volumes of dry methanol, acetone, and diethyl ether. At this point the solvent should be changed to one having the appropriate water content, *e.g.*, diethyl ether 50% saturated with water. Passage of this solvent will fairly quickly establish a partial monolayer of water on the surface of the adsorbent covering the most active sites, at the same time drying the solvent passing through the column, *i.e.*, solvent demixing. This should be continued until a state of equilibrium has been achieved, after which time the composition of the mobile phase will be unaltered by passage through the column. The attainment of equilibrium is best monitored by periodic injection of a test compound which is retained to a modest extent (*i.e.*, $k' = 3-10$) and observe the point at which the capacity factor reaches a constant value. This equilibrium state defines a certain activity of the adsorbent column packing material, *i.e.*, a certain level of water content in the packing, a change to another solvent such as dichloromethane or hexane, and subsequent re-equilibration may be made fairly rapidly, provided that the degree of water saturation of the new mobile phase is maintained at the same (50%) level[6]. Diethyl ether is used in this procedure as it has the desirable property of being water immiscible but at the same time is capable of dissolving an appreciable volume of water at room temperature. This relatively high solubility enables sufficient water to be transported into the adsorbent bed with a fairly low volume of mobile phase, conversely anhydrous ether will rapidly dehydrate a column. Other solvents, such as hexane, possess a much reduced affinity for water and would require a much greater volume of solvent to be passed through the column to carry the same quantity of water into the adsorbent. Equilibration using the latter approach has on occasion required passage of several hundred column volumes of mobile phase to pass through the adsorbent bed to attain equilibrium.

Controlling adsorbent activity with alcohols

As an alternative to controlling the activity of the adsorbent with water, some chromatographers prefer to employ anhydrous solvents to which is added a very small proportion of a polar compound such as methanol or isopropanol. In these circumstances there is no problem regarding the limited solubility of the alcohol in the mobile phase, as in most cases it will be completely soluble. The level of alcohol required in the mobile phase to deactivate a silica surface to the same extent as a "50% water-saturated system" is usually in the region of 0.1 to 0.3% by volume, depending on the alcohol. By using this level of alcohol, peak tailing may be significantly reduced and the adsorption isotherms are more linear than for the anhydrous "active" adsorbent.

Detailed studies of the relative merits of modifying adsorbents with partially water-saturated solvents or alcohols[6] suggest that, when practicable, the former method will provide superior results.

Optimisation of mobile phase composition

The methods of solvent selection for optimising a separation in terms of both resolution and speed of analysis have been described fully in Chapter 6. Within the practical restraints imposed by the requirement to maintain a constant adsorbent activity, almost any solvent given in the eluotropic series detailed in Table 6.1 may be employed. A solvent occupying a higher position in the series will cause the sample to be more strongly retained compared with a solvent in a lower position in the table.

REFERENCES

1 H. Beyer and U. Schenk, *J. Chromatogr.*, 61 (1971) 263.
2 D. Hentwen, A. Fournier and J.P. Garel, *Anal. Biochem.*, 53 (1973) 299.
3 J.N. Chapman and H.R. Beard, *Anal. Chem.*, 45 (1973) 2268.
4 R.P.W. Scott and P. Kucera, *J. Chromatogr. Sci.*, 12 (1974) 473.
5 L.R. Snyder, *Anal. Chem.*, 46 (1974) 1384.
6 J.J. Kirkland, *J. Chromatogr.*, 83 (1973) 149.

Chapter 8

Liquid–liquid (partition) chromatography

INTRODUCTION

Classical liquid–liquid partition, on a one-step basis, is performed in a separating funnel where the sample of interest is distributed between two immiscible solvents. The relative concentrations in the two liquid phases are described by the distribution coefficient, which in turn is a function of the solubility of the sample in the two liquids. In partition methods, one is normally striving to selectively extract the required species into one phase while the rest of the sample remains in the other layer. If complete separation of the required species from the sample is needed, then it is normal practice to re-extract each of the separated layers with a fresh portion of the complementary solvent, finally combining all portions of the extraction liquid. This procedure, if required to be performed repeatedly on a given sample, becomes cumbersome, time consuming and can result in significant losses of sample. The method has been mechanised to reduce the extent of manual manipulation, notably by Craig[1], in the form of countercurrent distribution techniques. By this procedure it is possible to perform multiple extractions leading to effective separations of fairly large quantities of complex samples. However, since a considerable time is needed to set up and carry out the separation procedure, the method is not ideally suited to analytical-scale separations.

Liquid–liquid (partition) column chromatography accomplishes similar multi-stage distribution of a sample on a very much smaller scale within the confines of a chromatographic column, where operator manipulations of each distribution stage are eliminated and the number of distribution stages, hence the effectiveness of the separation, is greatly increased.

Partition chromatography is achieved by holding, in a manner to be described later, one liquid phase on the surface of a chromatographic support, *i.e.*, the stationary phase, while the second liquid, *i.e.*, the mobile phase, is passed through the packed column, permitting intimate contact between the two phases. At this stage distribution of the components of the sample can occur.

RANGE OF SAMPLE APPLICABILITY

Liquid partition chromatography may, in the broadest sense, be applicable to any substance which is capable of being distributed between two liquid phases. Since the degree of retention of a sample in a column is primarily a function of the relative, not the absolute, solubility of the sample in the mobile and stationary phases, *i.e.*, the distribution coefficient, it is feasible for compounds which differ widely in absolute solubility to elute from a chromatograph under quite similar conditions.

One of the greatest attractions of liquid–liquid partition is that either of the two liquids may function as the stationary phase, depending on the separation requirements and operating conditions. In practice, liquid partition systems are designated as normal systems when the mobile phase is less polar than the stationary phase, *cf.* adsorption chromatography, and as reversed-phase systems when the mobile phase is the more polar liquid. In most instances a multi-component sample separated by a reversed-phase system will give a completely different order of elution of components compared with that separated by a normal partition system: Often the elution order is completely reversed. The ability to reverse the order of elution of the components in a mixture can greatly simplify analyses where a trace constituent is being sought.

Although in principle partition chromatography is analogous to two-phase distribution, much of modern practice is performed with "pseudo-liquid" stationary phases, where the phase is firmly attached to the chromatographic support material. This arrangement greatly simplifies the practical manipulation of the system, enables gradient elution to be performed with minimal re-equilibration time and provides a separation system that is stable over a long period.

The number of column packings with different chemically bonded stationary phases for both normal and reversed-phase separations is rather limited. However, since each material may be used with mobile phases formed from almost any common solvent, the eluting power of the mobile phase and the range of sample applicability is very wide indeed.

Many beginners in LC find it hard to realise that separations may be achieved in liquid phase systems which show only limited solubility for the sample. Fig.8.1, for instance, shows a reversed-phase separation of polynuclear aromatic hydrocarbons in a system where the mobile phase is aqueous methanol. These compounds are insoluble in water and only sparingly soluble in pure methanol, yet using these two solvents to prepare the mobile phase, good resolution may be obtained between compounds which are very closely related structurally. This result is general to most forms of partition chromatography, *i.e.*, the method is effective in resolving closely related materials[2]. This compares with adsorption chromatography, which is more commonly employed to separate a mixture into classes of compounds.

The characteristic requirement for partition, *i.e.*, a sample must be soluble in more than one solvent, usually pre-supposes that a sample should be non-ionic, since ionic compounds are generally only soluble in water. For the most part, partition LC works best with non-ionic compounds. Very many chemical species have been reported to have been separated by partition chromatography. These include substances such as: phthalate and phosphate plasticisers, hydrocarbons, steroids, organo-chlorine and -phosphorus insecticides, oil-soluble vitamins, and chloro- and nitro-containing compounds.

When potentially ionic substances are studied by partition chromatography, particularly in reversed-phase systems, tailing of the eluting peaks is sometimes observed. In the case of a weakly acidic sample this effect may generally be eliminated by the addition of a dilute acid to the mobile phase. In an analogous manner, a few drops of ammonia per litre of mobile phase will considerably improve the elution of a weak base. Types of samples which are prone to this behaviour include polyphenols, organic acids, *e.g.*, phenoxyacids, amines and substituted amines, for example, alkaloids.

Studies by Eksborg and Schill[3] have demonstrated that the partition technique may be extended to encompass the separation of ionic substances provided the aqueous phase

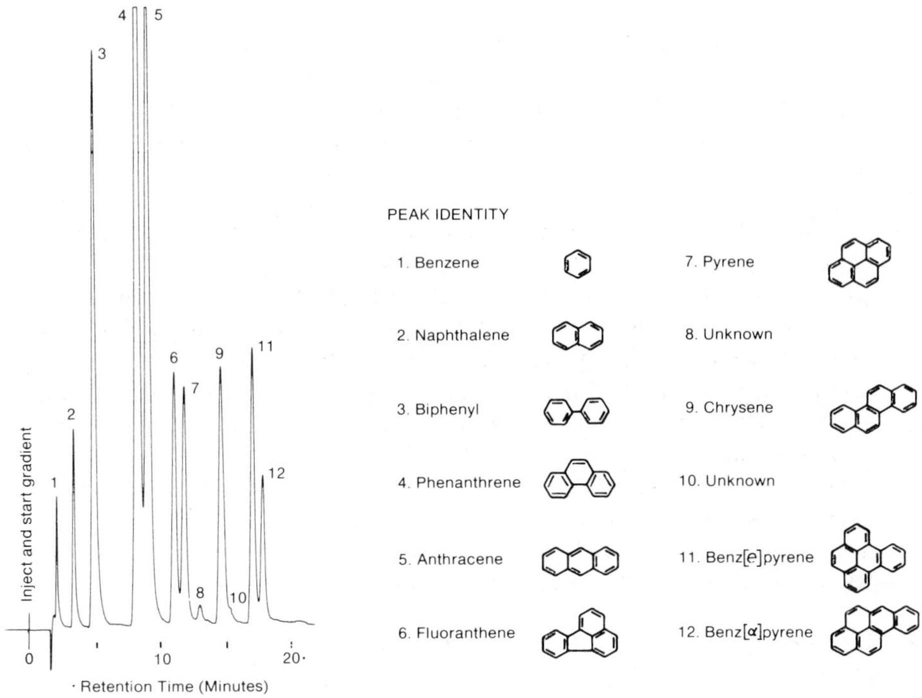

Fig. 8.1. Separation of aromatic hydrocarbons by reversed-phase partition chromatography. Operating conditions: packing, Permaphase ODS; mobile phase, linear gradient from 50% methanol–50% water to 100% methanol; column temperature, 50°C; inlet pressure, 1000 p.s.i.; flow-rate, 1 ml/min; detector. UV photometer. (Reproduced from J.A. Schmit, R.A. Henry, R.C. Williams and J.F. Dieckman, *J. Chromatogr. Sci.*, 9 (1971) 645, with permission.)

contains a counter-ion which combines reversibly with the sample, rendering it soluble in organic solvents. This approach, known as "ion-pair chromatography", is considered in Chapter 9 together with other separation techniques relating to ionic substances.

GENERAL CONSIDERATIONS

The Distribution Law may be stated as follows: If to a system comprising two essentially immiscible liquid layers one adds a third substance which is soluble in both layers, then the substance will, at equilibrium, distribute itself between the two layers irrespective of the total amount of substance present. Thus:

$$K = \frac{a_1}{a_2} \approx \frac{c_1}{c_2} \tag{8.1}$$

where K is the distribution coefficient, a_1 and a_2 are the activities of the substance in the liquid layers 1 and 2, respectively, and c_1 and c_2 are the corresponding concentrations. In dilute solutions, as is usual in analytical LC, the error involved in using concentrations in place of activity is generally negligible.

One of the important consequences of a separation method which relies on the distribution coefficient of a sample is that for many substances the magnitude of the distribution coefficient is independent of the total concentration of the sample in the liquid phases. In these circumstances, since retention of a sample component on a column is a function of its partition coefficient, one may expect no change in the retention characteristics with sample size over a fairly wide range of concentration, thus in a purely partition process symmetrical peaks should be obtained. This situation will exist only if the chromatographic support is essentially inert towards the sample being studied. For optimum results in partition work, no adsorption of the sample on the support should take place.

In general, a distribution coefficient will be independent of concentration if the partition process is not complicated by secondary reactions in one or both liquid layers. There are, however, well documented instances where the distribution coefficient is most definitely concentration dependent. Perhaps one of the best described examples of this situation is the distribution of benzoic acid between benzene and water, where dimerisation in the benzene layer and dissociation of the carboxylic acid function in the aqueous phase lead to a less simple relationship.

Retention in a liquid–liquid chromatographic column, denoted by the capacity factor, k', is related to the distribution coefficient, K, in the following manner:

$$k' = \frac{(M_s)}{(M_m)} = K \cdot \frac{(V_s)}{(V_m)} \tag{8.2}$$

where M_s and M_m are the masses of sample in the stationary and mobile phases, respectively, and V_s and V_m are the volumes of the two phases, respectively. Thus, for increased retention of a component either the distribution coefficient must be selected to give preferential solubility in the stationary phase or the volume of the stationary phase must be increased relative to that of the mobile phase.

Considering the former approach, that of making the distribution of the sample favour the stationary phase, in the classical sense one must carefully select a pair of immiscible liquids which are to serve as mobile and stationary phases. In modern practice, column packings are synthesised with stationary phases firmly bound to the surface of the support. In this way, the stationary phase can be made essentially insoluble in a range of solvents which may then be considered as candidate mobile phases. A range of chemically bonded phases is available commercially, covering the entire range from the very polar, nitrile, amino and hydroxyl phases to the low and non-polar phases such as octadecyl- and aliphatic ether-substituted silanes. Chemically bonded phases can be compatible with almost every solvent, except those which are either strongly acidic or basic. The ultimate stability of a particular material depends on the method of manufacture; restrictions applicable to individual packings are described in later sections of this chapter. The development of bonded phases for partition chromatography has led to a drastic simplification of the method.

The second method by which the sample may be retained more strongly is by increasing the volume of the stationary phase relative to the mobile phase. The approach of simply coating, or bonding, a thicker layer of stationary phase on a chromatographic support, although technically possible, will generally lead to a decrease in the rate of mass transfer with a corresponding reduction in column efficiency. There are two situations, however, where this problem may be minimised. Firstly, by heavily loading the column with a stationary phase of low viscosity, such as water–alcohol mixtures, or secondly, by coating, or bonding, the stationary phase on a support of larger surface area, for example by using particles of smaller diameter.

TYPES OF LIQUID–LIQUID PHASE SYSTEMS

One of the most powerful features of LC is the influence of the composition of the mobile phase on the retention characteristics of the components of the sample being separated. However, in the basic concept of liquid–liquid partition, *i.e.*, two phases formed from essentially immiscible solvents, any change in the composition of the mobile phase would disturb the equilibrium concentration of stationary phase in the mobile phase and subsequently the level of stationary phase held on the support material. This situation could result in complete dissolution of the stationary phase from the column, leading to steadily reducing capacity factors for the sample components until all resolution has been lost. In practice, this problem must be minimised by either taking extensive precautions to avoid dissolution of the stationary phase or, preferably, by using "liquid" coatings which are not removed from the chromatographic support when the mobile phase composition is changed. The various procedures currently used may be sub-divided as follows:

(1) Using a simple liquid stationary phase coating on the surface of the support and ensuring the mobile phase is always completely saturated with stationary phase prior to passage through the separating column.

(2) Using a polymeric substance or one of moderately high molecular weight as the stationary phase which is insoluble in a range of solvents which may be mixed as appropriate to form mobile phases.

(3) Employing speciality column packings where the stationary phase is chemically bonded to the chromatographic support; thus, in principle, any solvent may be used as mobile phase.

(4) A special case of Type 1 – This is obtained by deliberately permitting some mutual solubility of the mobile and stationary phases by addition of a third solvent which is soluble in both phases, in limited quantities to an immiscible pair of solvents. The addition of the third, modifying, solvent to the two immiscible solvents is carried out prior to coating the stationary phase on the chromatographic support.

Partition systems employing two simple liquid phases

Separation systems based on this, essentially classical, approach rely on coating one of the liquids of an immiscible pair on the surface of a suitable chromatographic support. Examples of simple liquid pairs having very low mutual solubility which have proved useful in modern LC are listed in Table 8.1.

The factors governing the choice of support material have been described in Chapter 3. The ideal support material for partition chromatography should possess just sufficient adsorptive activity to retain the stationary phase but be not so strong an adsorbent to leave any residual adsorptive activity on the support which may interfere with the elution and separation characteristics of the samples in subsequent work.

TABLE 8.1

SOME OF THE MORE WIDELY STUDIED LIQUID PAIRS FOR PARTITION CHROMATOGRAPHY

Type of chromatography	Mobile phase	Stationary phase
Normal partition	Aliphatic hydrocarbons, *e.g.*, pentane, hexane, heptane, 2,2,4-trimethylpentane	Water, ethylene glycol, polyethylene glycols, trimethylene glycol, acetonitrile, β,β′-oxydipropionitrile, 1,2,3-tris(2-cyanoethoxy)propane
	Chlorinated solvents, *e.g.*, chloroform, dichloromethane	Water
Reversed phase	Water Acetonitrile	Squalane

A hydrophilic support should be employed for normal partition systems, *i.e.*, where the stationary phase is more polar than the mobile phase; conversely, a hydrophobic support, such as silanised silica, must be used for reversed-phase work. Organic supports, *e.g.*, PTFE, have been described; however, these are only applicable to low-pressure systems. Table 8.2 lists some of the more common commercially available supports which have proved useful in this type of chromatography.

Considerable attention to detail must be given to the coating of the stationary liquid and the subsequent operation of the chromatographic system if reproducible results are to be obtained. An outline of the experimental technique involved is given in Appendix 4. The extent of the practical manipulations is such that many chromatographers have tended to abandon this approach to partition chromatography in favour of the easier to use systems with packing materials possessing chemically bonded phases.

Although the method of using simple liquid coatings is perhaps not enjoying wide popularity, it should be appreciated that, with care, extremely good results may be achieved.

TABLE 8.2

CHROMATOGRAPHIC PACKINGS USED AS SUPPORTS FOR PHYSICALLY LOADED STATIONARY PHASES*

Type	Name	Surface area (m^2/g)	Particle size (μm)	Shape**	Supplier
Pellicular (silica)					
	Corasil I	7	37–50	S	Waters
	Liqua-Chrom	<10	44–53	S	Applied Science
	Zipax	~ 1	25–37	S	DuPont
Porous (diatomaceous earth)					
	Dia-Chrom	10	37–44	I	Applied Science

*All of the materials listed in Table 7.1 may also be considered suitable, although packings with high surface areas may give separations based on a mixed partition–adsorption mechanism.
**I = Irregular; S = spherical.

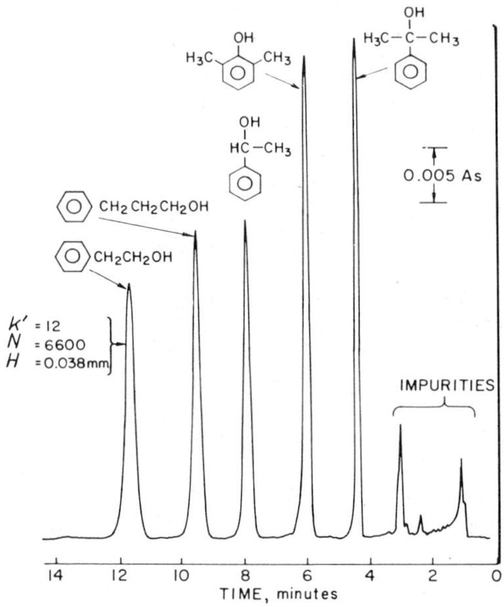

Fig. 8.2. Separation of hydroxylated aromatics by normal partition chromatography using a physically coated stationary phase. Operating conditions: column, 0.25 m × 3.2 mm I.D.; packing, porous silica microspheres, diameter 5–6 μm, pore size 350 Å; stationary phase, β,β′-oxydipropionitrile, approximate loading 30% by weight; mobile phase, hexane, saturated with stationary phase; flow-rate 1 ml/min; inlet pressure, approx. 40 bars (600 p.s.i.); temperature, 27°C. (Reproduced from J.J. Kirkland, *J. Chromatogr. Sci.*, 10 (1972) 593, with permission.)

A particularly modern example of a separation carried out with simple phase systems is shown in Fig. 8.2, where a series of hydroxylated aromatic substances are resolved using a phase system where the stationary phase is β,β′-oxydipropionitrile and the mobile phase is hexane. This separation was reported by Kirkland[4].

Partition systems using polymeric stationary phases

One approach that has overcome some of the practical manipulations necessary when using simple liquid stationary phases is the use of polymeric coatings which are essentially insoluble in a range of solvents. The basis of the method is to increase the molecular weight of the stationary phase sufficiently to reduce its solubility in the mobile phase to an insignificant level, yet at the same time still retain some chromatographic selectivity. In these circumstances it is not necessary to saturate the mobile phase with stationary phase, and thus column equilibrium, the changing of solvents used as mobile phases, and general operation of the column system are far less critical. It should be pointed out that any given polymeric phase will be essentially insoluble in a limited range of solvents, thus although the choice of liquids for the mobile phase is greater than for true liquid–liquid systems, the range is not infinite. Gradient elution can be employed within the recommended limits of the mobile phase composition enabling the partition coefficients of components in the sample to be adjusted during the analysis to effect elution from the column in a realistic time.

Details of some commercially available packings which possess polymeric coatings are given in Table 8.3 together with an indication of their solvent compatibility.

TABLE 8.3

CHROMATOGRAPHIC PACKINGS WITH POLYMERIC STATIONARY PHASES

Name	Coating	General solvent compatibility*	Supplier
Pellidon	Nylon	Hydrocarbons or water plus limited modifier, *e.g.*, alcohol	Reeve Angel
Perisorb PA-6	Polyamide	As above	Merck**
Zipax ANH	Cyanoethyl silicone	As above	DuPont
Zipax PAM	Nylon	As above	DuPont
Zipax HCP	Saturated hydro-carbon polymer	Water plus limited alcoholic modifier	DuPont

*General indication only. Specific details should be obtained from the supplier of the column packing material.
**E.M. Labs. in the U.S.A.

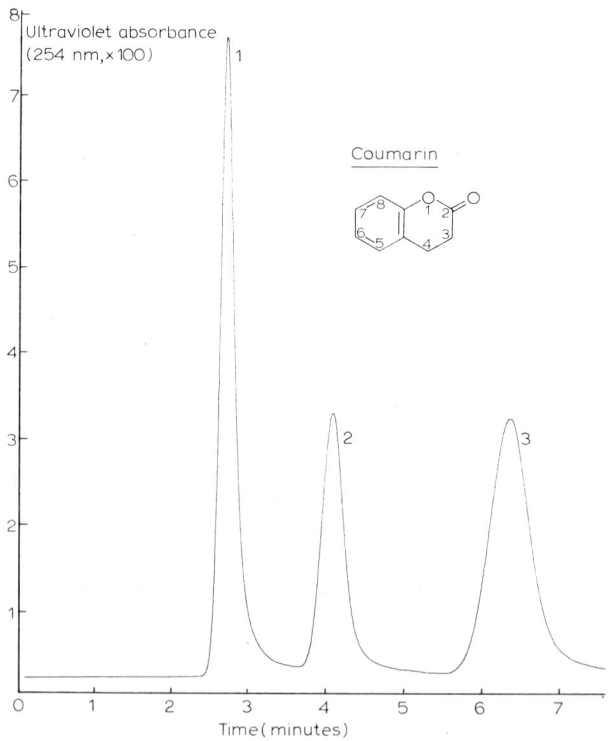

Fig. 8.3. Separation of coumarins on a polymer-coated column. Operating conditions: column, 1 m × 2.1 mm I.D.; packing, Zipax ANH, cyanoethylsilicone; mobile phase, water; flow-rate, 0.5 ml/min; temperature, 25°C. 1 = 6,7-Dihydroxycoumarin 6-glucoside; 2 = 6-hydroxy-7-methoxy-coumarin; 3 = 6,7-dimethoxycoumarin.

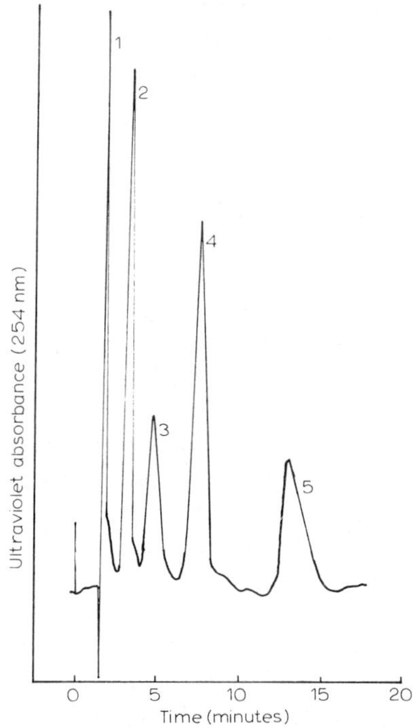

Fig. 8.4. Separation of oil-soluble vitamins by reversed-phase chromatography. Operating conditions: column, 1 m × 2.1 mm I.D.; packing, Zipax HCP, hydrocarbon polymer; mobile phase, methanol–water–phosphoric acid (77:23:0.1); flow-rate, 1 ml/min; inlet pressure, 100 bars (1500 p.s.i.); temperature, ambient. 1 = Solvent response; 2 = chrysene (internal standard); 3 = vitamin A acetate; 4 = vitamin D_2; 5 = vitamin E succinate. (Redrawn from the work of R.C. Williams, J.A. Schmit and R.A. Henry, *J. Chromatogr. Sci.*, 10 (1972) 494, with permission.)

Polymer-coated packing materials have the advantage that they are generally highly selective, particularly when used in the reversed-phase mode, *i.e.*, with mobile phases formed by mixing water and a polar water-miscible solvent. Examples of separations of coumarin derivatives and oil-soluble vitamins are illustrated in Figs. 8.3 and 8.4, respectively. These chromatograms are typical of the style of separation obtained with polymeric phases, that is, highly selective yet rather inefficient. In Fig. 8.4, a point worthy of note is the addition of phosphoric acid to the mobile phase in order to improve peak shape by suppressing the dissociation of the free carboxylic acid function on vitamin E succinate[5]. Polymeric coatings are relatively stable to the presence of acids in the mobile phase, as, unlike the bonded phase packings, adhesion of the stationary phase is not dependent on a chemical linkage.

One limitation with this approach to liquid partition chromatography is that the stationary phase is essentially a solid, although it may be considered as a very viscous liquid, and as such, exhibits poor mass transfer characteristics. Elevation of the temperature of operation leads to some improvement but the range of temperature over which the columns may be operated generally is fairly limited, *i.e.*, from ambient to about 50°C.

In the light of present-day developments in the technology of LC packings, materials with polymeric coatings are best considered as having served a transitionary role as fairly stable, easy-to-use packings – a place which has been taken in recent years by column packings having the stationary phase chemically bonded to the support material. There remains a steady interest in the use of these materials, in part due to their cost being significantly lower than the corresponding bonded phase and also since many phases are resistant to attack by acids and bases which can adversely affect the stability of some types of bonded phases. For similar reasons polymeric coatings continue to be used quite extensively for low-capacity, superficially porous, ion-exchange packings.

Partition chromatography with chemically bonded stationary phases

The introduction of the concept of stationary phases bonded chemically to the surface of the chromatographic support represents a very significant milestone in the development of LC systems. They offer a great separating power of sample mixtures with a high degree of reproducibility while remaining easy-to-use, even by operators with limited experience. In principle a bonded phase could be used in conjunction with any solvent as a mobile phase, thus partition coefficients may be adjusted over a wide range to suit the individual separation requirements.

There have been several methods proposed for the synthesis of these speciality packings, however, not all produce chromatographic packings with optimum performance characteristics. Most of the developments in this area of LC have relied on the use of silica or glass support materials as, at the present time, these tend to give the highest performance in terms of efficiency of packing and resistance to deformation under high pressure. Of the various methods of attaching a stationary phase to a silaceous support, all rely on the presence of silanol groups, *i.e.*, silicon hydroxyl groups, on the surface of the support.

At the present time there is no general agreement as to the concentration of silanol groups on the surface of the support required for optimum bonding of a stationary phase. However, silica surfaces can be modified thermally to reduce the concentration; the temperature of ignition is in the order of 600–900°C. Conversely, acid hydrolysis will increase the silanol concentration[6], the reaction following the form given in eqn.8.3.

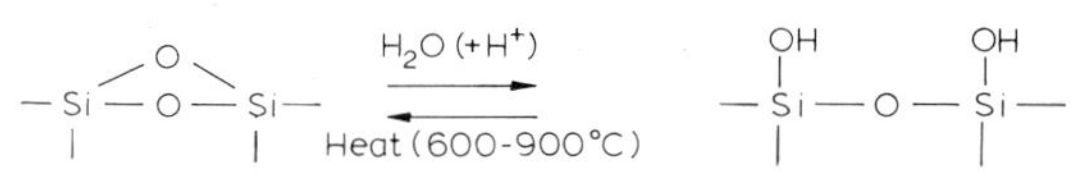

(8.3)

Methods of bonding phases to silanol groups

Formation of a silicate-ester (Si–O–C) linkage

If one considers the generic name for silica gel, *viz*. silicic acid, it is not surprising that it is possible to react the material with an alcohol to form an ester. Eqn.8.4 indicates the general reaction

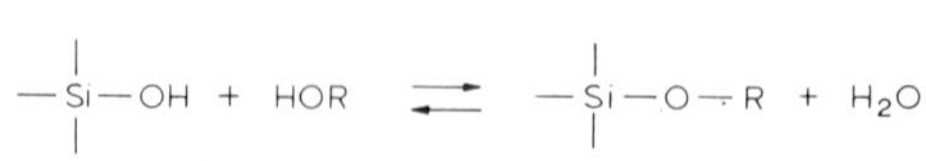

(8.4)

The forward reaction can be made to occur by heating the support in the presence of the appropriate alcohol, either under reflux, for an appreciable time or, in the case of fairly volatile alcohols, in an autoclave.

An alternative method of producing these esters is to initially treat the support with thionyl chloride to produce a "silica chloride", *i.e.*,

$$SOCl_2 + Si{-}OH \rightleftarrows Si{-}Cl + SO_2 + HCl \quad (8.5)$$

This initial approach is attractive, as all products but the one of interest are volatile. Subsequent reaction with the alcohol of interest, ROH, produces the silicate ester with the liberation of further gaseous by-products. Some of the earlier chemically bonded phases for LC were produced in this manner. Unfortunately, these materials are restricted in use to applications with mobile phases which do not contain water or alcohol, as the products are hydrolytically unstable. Many ester-type phases are also thermally labile.

This restriction to not employing hydroxylic mobile phases is too demanding for practical purposes since many solvents contain some, albeit a small proportion of a hydroxylic solvent, for example, most solvents will normally contain a certain amount of water while others, notably chloroform and dichloromethane, have alcohol added in order to improve their storage characteristics.

An additional problem which may occur with any bonded packing material is the existence of residual surface hydroxyl functional groups. These will cause greatest interference when using non-polar solvents, as are required with these silicate-ester phases. Deactivation of supports using, for example, hexamethylenedisilazane, is generally considered to be incomplete, since steric hindrance will not permit every surface proton to be replaced by a trimethylsilyl derivative.

Silicate phases are, however, essentially monomeric, that is, the surface layer is essentially one molecule thick, as the reaction of another alcohol onto an already bonded site is not possible. This configuration leads to materials having good mass transfer characteristics.

Bonded phases relying on a Si–N–C linkage

Bonded phase materials having greater stability, particularly to hydrolytic action, have been described by Brust *et al.*[7]. Here the silica support-to-nitrogen-to-carbon linkage is synthesised rather than the silica support-to-oxygen-to-carbon linkage, as in the case of silicate esters.

These materials are prepared by firstly forming the "silica chloride", as described above, followed by reaction at 80°C with an aromatic diamine, yielding a material with amino functionality. This product can subsequently be modified using any of the reactions normally considered for simple amines to produce a range of packing materials having different functionality. Several possibilities have been reported by Brust *et al.*[7]. They are typified by the reaction of the amine derivative with *p*-nitrobenzylchloride to produce "bristles" of the diamine with a nitrobenzyl head. Thus,

$$-\overset{|}{\underset{|}{Si}}{-}Cl + H_2N{-}CH_2{-}CH_2{-}NH_2 \rightleftarrows -\overset{|}{\underset{|}{Si}}{-}HN{-}CH_2{-}CH_2{-}NH_2 + HCl \quad (8.6)$$

$$-\overset{|}{\underset{|}{Si}}{-}HN{-}CH_2{-}CH_2{-}NH_2 + O_2N{-}C_6H_4{-}CH_2{-}Cl \rightarrow$$

$$\rightarrow -\overset{|}{\underset{|}{Si}}{-}HN{-}CH_2{-}CH_2{-}NH{-}CH_2{-}C_6H_4{-}NO_2 + HCl \quad (8.7)$$

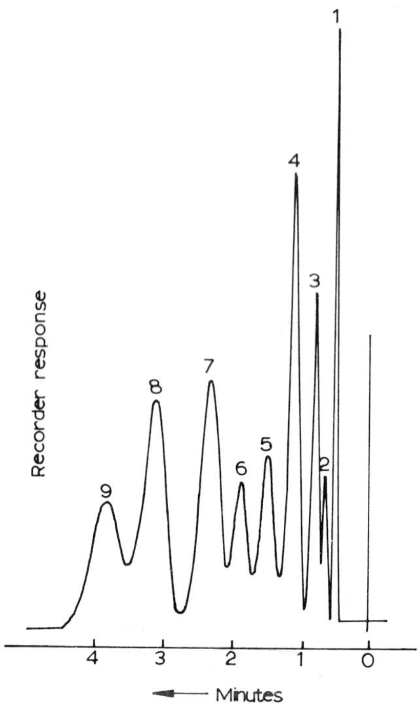

Fig. 8.5. Separation using bonded phases containing a silicon-to-nitrogen bond. Operating conditions: column, 0.5 m × 2 mm I.D.; packing, silica with stationary phase, $Si{\cdot}NH{\cdot}(CH_2)_2{\cdot}NH{\cdot}CH_2{\cdot}Ph{\cdot}NO_2$; particle size, 32–40 μm; mobile phase, *n*-heptane; inlet pressure, 45 bars; linear velocity, 16.5 mm/sec. 1 = *n*-Nonane; 2 = benzene; 3 = 1,5-dimethylnaphthalene; 4 = fluoranthene; 5 = benzopyrene; 6 = anisole; 7 = azobenzene; 8 = nitrobenzene; 9 = *p*-nitrotoluene. (Reproduced from I. Sebestian, O.-E. Brust and I. Halasz, In S.G. Perry (Editor), *Gas Chromatography 1972*, Applied Science Publishers, London, 1973, p. 281, with permission.)

The surface of these materials is considered to be essentially monomeric and thus would be expected to possess good mass transfer characteristics. Studies on the thermal and hydrolytic stabilities of such phases have indicated a marked superiority over the ester type of bonded phases in that they may be used in mobile phases as polar as water within the pH range of 4–8. These materials have as yet not been exploited extensively in the field of modern LC. One of the reported examples[8] of the separating capabilities of these materials is reproduced in Fig. 8.5.

Bonded phases attached via a direct silicon-to-carbon linkage

The most stable chromatographic packings yet devised that possess stationary phases bonded to their surface are those which rely on a silicon-to-carbon linkage. A number of synthesis routes for these materials have been proposed, the commonest procedures involving the use of a Grignard reagent on the "silica chloride" or a several-stage synthesis using chlorosilanes.

The method involving chlorosilanes is outlined in eqn.8.8.

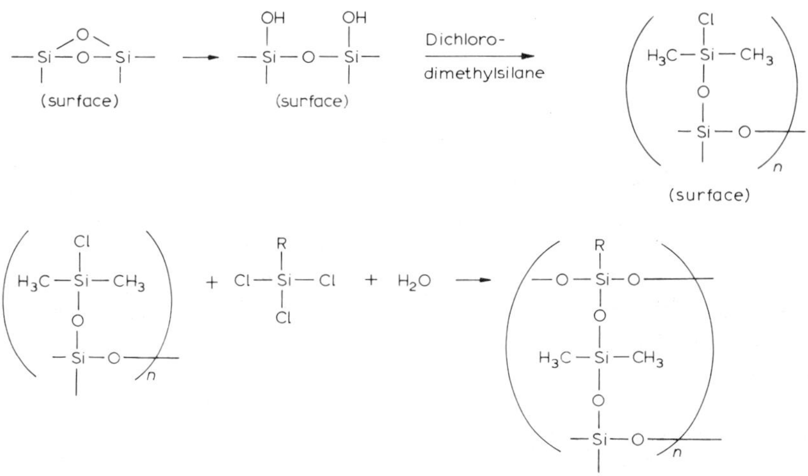

(8.8)

Bonded phases relying on a silicon-to-carbon link generally show very good resistance to hydrolysis in mobile phases in the pH range of 3–8; in some instances even lower pH conditions have been employed with no adverse effects on column stability. Thermal stability depends markedly on the method of manufacture; in all cases data provided by the supplier of the chromatographic packing should be consulted before attempting to work under extreme conditions of pH or temperature.

With these chromatographic packings it is possible to consider varying the mobile phase over the entire polarity range, *i.e.*, from pentane or hexane through to alcohol or water without running the risk of displacing the stationary phase from the column. Experience in use has shown that most of the successful phases equilibrate very rapidly after a change of mobile phase. This opens the possibility of carrying out gradient elution analysis, since a minimum of time will be required to re-establish the initial operating conditions after a chromatographic run.

Details of some of the more readily available chromatographic packings having bonded phases which are available from commercial sources are given in Table 8.4.

In the literature some bonded phases are referred to as modified adsorbents while others are considered as liquid phases firmly anchored to the surface of the support. The theoretical arguments are not important to practical treatment, however, it is useful to appreciate that when polar bonded phases are used, the separation achieved is similar to that obtained by means of adsorption chromatography although very often definite differences in selectivity are observed.

Polar bonded phases, *e.g.*, those with hydroxyl, amino or nitrile functionality, are used in normal partition methods, *i.e.*, with mobile phases prepared from solvents like hexane, chloroform, etc. From the sections related to the selection of mobile phase in Chapter 6 it will be apparent that the chemical composition of mobile phases used in LSC is very similar to that of the solvents used in normal partition methods. It might therefore be reasonable to question the advantages, if any, of this type of partition chromatography. Apart from the possibility of obtaining a change in column selectivity, there are two

TABLE 8.4

COMMERCIAL LC PACKINGS HAVING BONDED STATIONARY PHASES

Surface type	Name	Support type	Particle size (μm)	Supplier
Octadecyl hydrocarbon (C_{18})	Bondapak C_{18}/Corasil	Pellicular	37–50	Waters
	μBondapak C_{18}/Porasil	Porous	10	Waters
	CO:PELL ODS	Pellicular	41	Reeve Angel
	LiChrosorb RP-8	Porous	10	Merck*
	Micropak C-H	Porous	10	Varian
	ODS-SIL-X-I	Porous	13±5	Perkin-Elmer
	ODS-SIL-X-II	Pellicular	30–40	Perkin-Elmer
	Perisorb RP	Pellicular	30–40	Merck
	Permaphase ODS	Pellicular	25–37	DuPont
	Spherisorb ODS	Porous	5 or 10	Phase Separations
	Vydac RP	Pellicular	30–44	Separations Group
	Vydac TP Reversed Phase	Porous	10	Separations Group
	Zorbax ODS	Porous	4–6	DuPont
Other hydrocarbon	Durapak *n*-octane**	Porous	36–75	Waters
	LiChrosorb SI-60, silanised	Porous	5 or 10	Merck
Phenyl	Bondapak Phenyl/Corasil	Pellicular	37–50	Waters
	Phenyl-SIL-X-I	Porous	13±5	Perkin-Elmer
Aliphatic ether	Permaphase ETH	Pellicular	25–37	DuPont
Fluoroether	SIL-X-I-FE	Porous	13±5	Perkin-Elmer
Nitrile	Cyano-SIL-X-I	Porous	13±5	Perkin-Elmer
	Durapak OPN**	Porous	36–75	Waters
	Durapak OPN/Corasil	Pellicular	37–50	Waters
	Micropak C≡N	Porous	10	Varian
	Vydac Polar	Pellicular	30–44	Separations Group
	Vydac TP Polar Bonded Phase	Porous	10	Separations Group
Amino	Amino-SIL-X-I	Porous	13±5	Perkin-Elmer
	Micropak NH_2	Porous	10	Varian
Hydroxyl	Durapak Carbowax 400/Corasil**	Pellicular	37–50	Waters
	Durapak Carbowax 400/Porasil**	Porous	36–75	Waters

further advantages in favour of partition chromatography. Firstly, there is the ability to change solvents rapidly without needing to maintain the activity of the column packing material; no control of the "level of water saturation" is needed in partition work. This feature is reflected in the greater ease of employing gradient elution, over wide ranges of solvent polarity. The second feature of particular importance when dealing with polar samples is that irreversible or partial retention of a component on the column packing is very much reduced, if not eliminated.

Bonded phases of low polarity, typically those with octadecyl hydrocarbon phases, are used exclusively for reversed-phase chromatography where water–alcohol or water–acetonitrile mixtures are used to form the mobile phase. A change in the water to organic solvent ratio changes the capacity factors of sample components markedly. Fig. 8.6 illustrates this effect using, as test mixture, a group of simple aromatic compounds.

Since modern chemically bonded phases are also thermally stable, it is possible to use temperature as a variable. Almost exclusively in reversed-phase work, an increase in column temperature improves peak shape and decreases sample retention; these effects result from a decrease in solvent viscosity, increased solubility, and better mass transfer. Fig. 8.7 shows the effect of a change in column temperature on the separation of several aromatic hydrocarbons; the flow-rate of the mobile phase was held constant throughout.

Bonded phases of intermediate polarity, *e.g.*, an aliphatic ether, can serve as a useful stationary phase in both the normal and reversed-phase modes of operation. Thus, one type of column packing can exhibit a dual role. The pair of chromatograms shown in Fig. 8.8 shows this effect, taking as example a group of herbicides.

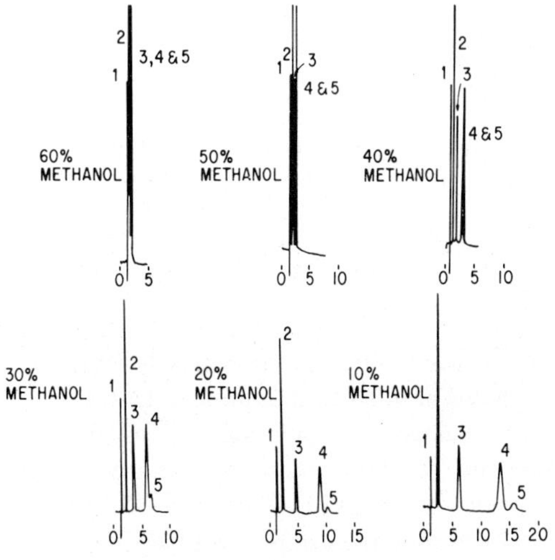

Fig. 8.6. Influence of mobile phase composition in reversed-phase chromatography. Operating conditions: packing, Permaphase ODS; mobile phase, methanol–water; temperature, 60°C; inlet pressure, 1200 p.s.i.; flow-rate, 2 ml/min; detector, UV photometer, 254 nm. 1 = Solvent; 2 = benzene; 3 = monochlorobenzene; 4 = *o*-dichlorobenzene; 5 = iodobenzene. (Reproduced from *DuPont Liquid Chromatographic Methods*, 820M9, dated October 30th, 1970, with permission.)

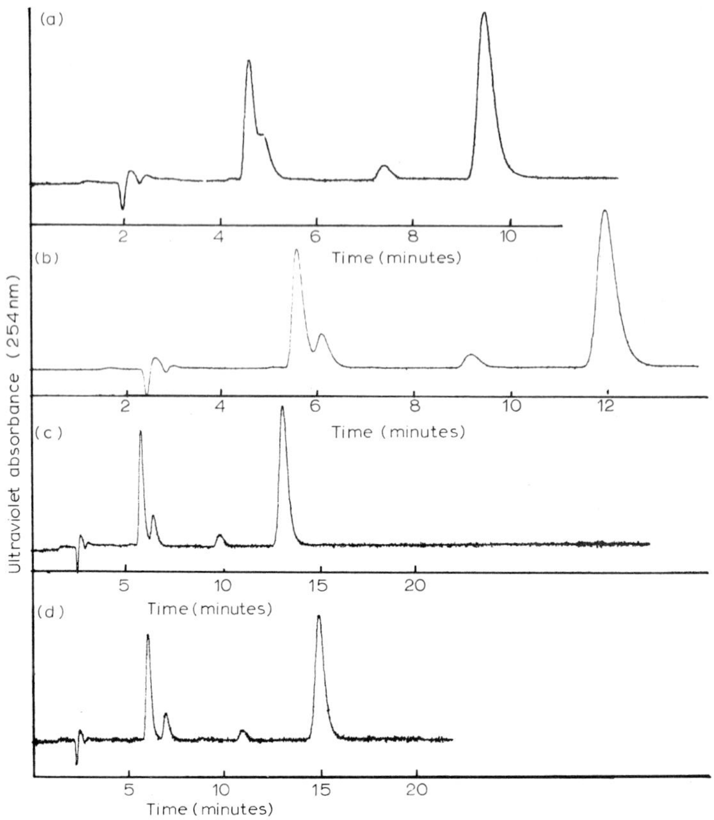

Fig. 8.7. Effect of column temperature in reversed-phase chromatography. Operating conditions: column, 0.25 m × 2.1 mm I.D.; packing, Zorbax ODS, 4–6 μm; mobile phase, methanol–water (78:22); flow-rate, 0.25 ml/min; column temperature, (a) 70°C, (b) 60°C, (c) 50°C, (d) 40°C; inlet pressure, (a) 68 bars (1000 p.s.i.), (b) 75 bars (1100 p.s.i.), (c) 89 bars (1300 p.s.i.), (d) 100 bars (1480 p.s.i.). The first large peak represents impure naphthalene and the second large peak pyrene.

Preparative chromatography is one area of application which benefits greatly from the use of column packings having their stationary phases bonded to the support. In this work, if conventional physically coated stationary phases were employed, the small amount of mutual solubility of the phases which always exists would result in any collected fraction being contaminated with a small amount of stationary phase. If the latter is non-volatile, separation of the collected fraction from this material can prove difficult since the stationary phase will invariably be a good solvent for the sample component. Chemically bonded phases do not bleed under normal circumstances so that, provided re-distilled solvents are employed, isolation of a non-volatile component simply involves evaporation of the collected fraction.

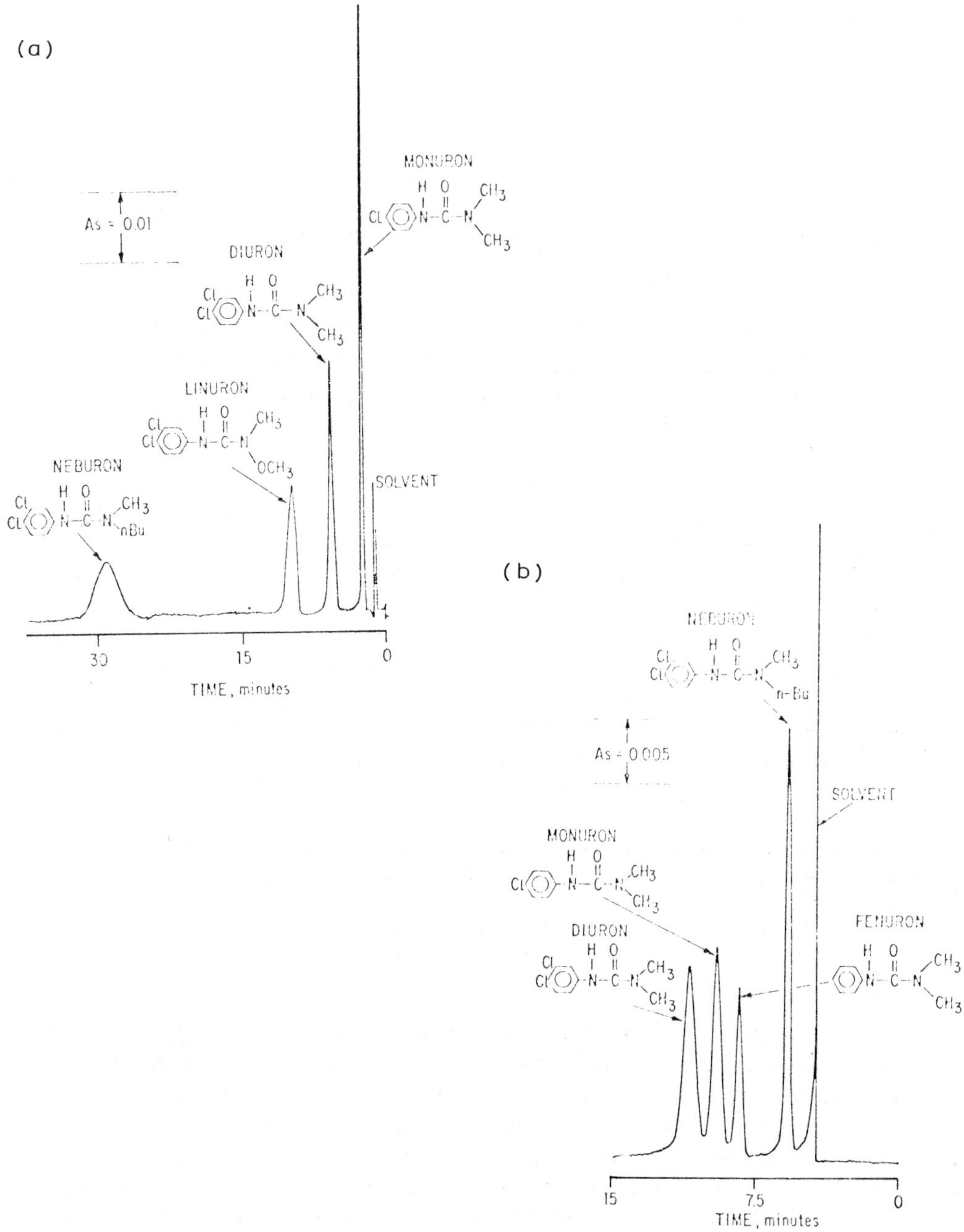

Fig. 8.8. Reversed-phase (a) and normal partition (b) chromatography of urea herbicides on the same bonded stationary phase. Operating conditions: column, 1 m × 2.1 mm I.D.; packing, Permaphase ETH; mobile phase, (a) 35% methanol in water, (b) 1% dioxane in hexane; temperature, (a) 50°C, (b) 27°C; detection, UV absorbance, 254 nm. (Reproduced from J.J. Kirkland, *Anal. Chem.*, 43 (1971) 43A, with permission.)

Partition chromatography using ternary liquid systems

An important extension to the use of two simple liquid phases for partition chromatography has been proposed and developed to a high degree of sophistication by Huber *et al.* Their approach is to select two immiscible solvents that possess widely differing polarities and to add, in limited proportions, a third solvent which is miscible with the other liquids. This action results in the formation of two liquid layers which, at equilibrium, will both contain a finite amount of each solvent differing only in the relative proportions of the solvents present in each layer. For an example, see ref.9.

One of the most widely studied systems has been that formed by the addition of ethanol to the immiscible pair of solvents water and 2,2,4-trimethylpentane (isooctane). Ethanol is readily miscible with the other two solvents and serves to adjust the polarity difference between the two liquid phases formed. As the concentration of ethanol in the system is increased, the difference in the polarities of the two phases decreases until they will have become equal, *i.e.*, when the whole system has become homogeneous. Fig.8.9 illustrates the nature of the phase diagram in a qualitative manner.

With a solvent system of this type it is possible to establish a range of two-phase systems, using three simple liquids, where the difference in polarity may be large or small, depending on the relative proportions of the individual solvents used. There are a number of interesting features of this approach, *e.g.*,

(1) Prior to use in the chromatographic system, the solvents are blended in order to attain as near as possible an equilibrium mixture. At this stage it is possible to reserve a portion of either phase and carry out quantitative solubility tests on prospective samples. Comparison of the solubility data derived from both phases will give the true partition coefficient of the sample in that solvent system and indicate the degree of retention that can be expected in the resultant chromatographic system.

(2) Correlation of chromatographic retention with partition coefficients is feasible and has been demonstrated by Huber *et al.*[10]. This correlation is only possible, however, if no competing processes occur during the chromatographic step, *e.g.*, adsorption on to the support.

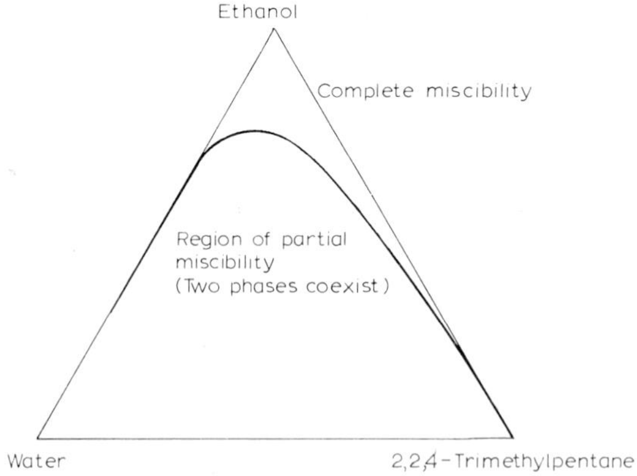

Fig. 8.9. Phase diagram of the ternary system water–ethanol–2,2,4-trimethylpentane at 25°C. (After C.A.M. Meijers, *Thesis*, University of Amsterdam, 1971.)

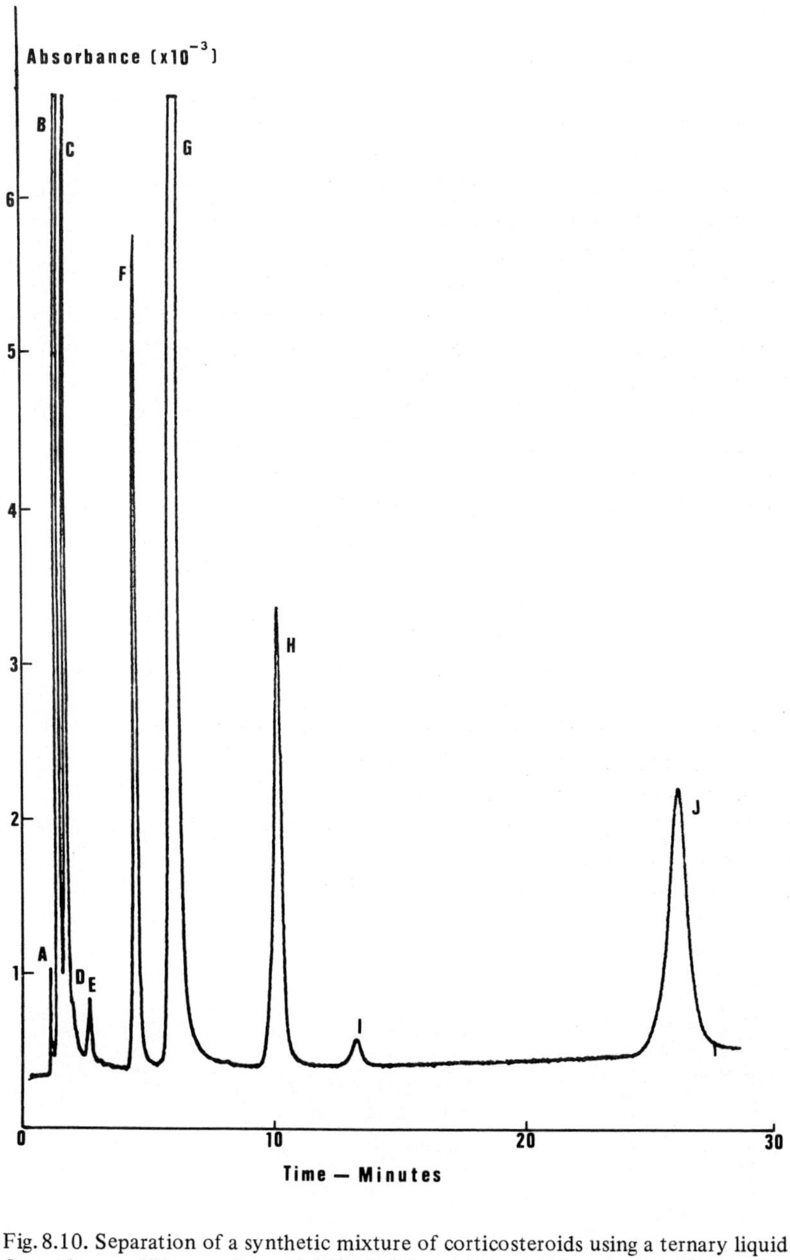

Fig. 8.10. Separation of a synthetic mixture of corticosteroids using a ternary liquid partition system. Operating conditions: column, 0.25 m × 2.1 mm I.D.; packing, Zorbax SIL; the liquid phase is formed by equilibrating dichloromethane–methanol–water (970 ml:10 ml:20 ml), the organic layer being used as the mobile phase; temperature, 25°C; flow-rate, 0.8 ml/min. A = Unknown; B = progesterone; C = 11-deoxycorticosterone; D = unknown; E = unknown; F = 11-deoxycortisol; G = corticosterone; H = cortisone; I = prednisone; J = cortisol. (Reproduced from N.A. Parris, *J. Chromatogr. Sci.*, 12 (1974) 753, with permission.)

TABLE 8.5

SELECTIVITY CHARACTERISTICS OF LIQUID PARTITION CHROMATOGRAPHIC SYSTEMS FOR VARIOUS SUBSTITUTED-UREA HERBICIDES

Phase system		Capacity factor of compound, k'				
Stationary	Mobile	Diuron*	Fenuron*	Linuron*	Monuron*	Neburon*
Oxydipropionitrile on Zipax	Dibutyl ether	1.3	3.9	0.5	2.3	n.d.**
Permaphase ETH	Methanol–water (35:65)	4.3	n.d.	8.3	1.0	25.6
Permaphase ETH	Dioxane–hexane (1:99)	1.5	0.9	n.d.	1.1	0.3
Porous silica microspheres	Dichloromethane (50% water satd.)	23	n.d.	4.0	n.d.	n.d.

*Identification of structures:

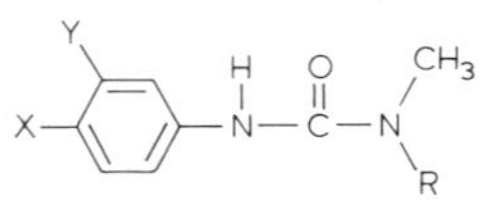

	X	Y	R
Diuron	Cl	Cl	CH_3
Fenuron	H	H	CH_3
Linuron	Cl	Cl	OCH_3
Monuron	Cl	H	CH_3
Neburon	Cl	Cl	n-C_4H_7

**n.d. indicates no data available.

(3) Solubility data from the samples being studied may indicate preferential solubility in the water-rich or hydrocarbon-rich layers. The desirability of having a greater solubility in the stationary phase has been described in Chapter 6. With the solvent system under consideration, the nature of the chromatographic support used in the column, *i.e.*, whether hydrophobic or hydrophilic, will govern which of the two liquid phases will be preferentially retained as the stationary phase.

(4) Having established a chromatographic system, using the methodology described earlier for partition with immiscible pairs of liquids, its retention characteristics may not prove as useful as anticipated. In this situation the stationary phase may be easily removed using an excess of the modifying solvent in the mobile phase, *i.e.*, ethanol in the example cited, to flush out the column. The process may then be repeated using another pair of equilibrated phases.

(5) Columns may be prepared having a high loading of stationary phase. This condition leads to a system offering high capacity and very good selectivity.

The principle disadvantages of this method are the necessity to carefully control the stability of the pair of equilibrated phases in the same way as with the simple immiscible liquid pairs and that it is not possible to employ gradient elution. The great versatility of this approach is particularly attractive when seeking to obtain a highly selective pair of equilibrated phases. Clearly, there are many combinations of other solvents which could be considered in a similar manner, *cf.* for example the work of Hesse and Hovermann[11] and Parris[12]. The separation of corticosteroids shown in Fig.8.10 gives some indication of the high selectivity attainable.

RELATIVE MERITS OF THE VARIOUS FORMS OF PARTITION CHROMATOGRAPHY

The foregoing paragraphs have described a number of approaches which may be used when practising liquid–liquid partition chromatography. Each approach has its relative advantages and these are summarised here to form a survey of the technique.

Partition chromatography has several distinct advantages over related methods, such as adsorption, in that the likelihood of irreversible retention of the sample occurring is considerably reduced and the selectivity possibilities are much greater. At the practical level, the efforts needed to control the activity of an adsorbent for LSC by using partially water-saturated solvents are comparable with the attention to detail necessary when operating partition systems with physically coated stationary phases, be it either the binary or the ternary liquid systems. Consequently, many chromatographers prefer to use bonded stationary phases to simplify their procedures and they increase versatility by using gradient elution techniques.

Changing the order of elution of components of a sample by a considerable change in the selectivity characteristics of a partition system is fairly common. Some idea of the powerful selectivity differences that are available can be obtained from Table 8.5, which summarises retention data, in terms of the capacity factors, for a series of substituted-urea type herbicides. All values are derived from chromatograms published by Kirkland[13,14] By way of comparison, the table includes retention data for similar compounds on an adsorption system. Remembering that the order of elution of components follows the order of increasing capacity factors, it is clear that almost any order of elution could be obtained. This feature of partition systems enables optimum choice of elution order when seeking to determine impurities in the presence of major components, as the minor peaks are more readily detected in a quantitative manner if they elute before the major peak.

Selectivity in partition chromatography can be accomplished by the choice of stationary phase or mobile phase composition or both. The use of packing materials that have stationary phases bonded chemically to the support restricts the selectivity variable for a given column to that attainable by changing the mobile phase composition; however, it should be appreciated that, by using bonded stationary phases, a much wider range of mobile phase types may be used with any given column packing. There is some evidence to suggest that the surface of bonded stationary phases is solvated by the mobile phase, imparting slightly different characteristics; this is considered a consequence of choosing a particular mobile phase rather than an attempt to use a different stationary phase.

The use of simple immiscible liquid pairs for partition chromatography offers a wide choice of column selectivities by coating the column with an alternative stationary phase. Wide differences in retentive power are achieved by using completely different stationary phases, while more subtle changes in column selectivity may be realised by substituting a similar stationary phase, for example ethylene glycol and polyethylene glycols[15]. An increased molecular weight of the stationary phase, although reducing problems associated with column bleed, does lead to an undesirable increase in the resistance to mass transfer. The most selective partition systems based on a physically held liquid are those employing ternary solvent mixtures, where the liquids are all of low molecular weight and relatively non-viscous. These systems are, perhaps, correspondingly those which need the greatest amount of operator care to be stable and reproducible.

The use of packings with stationary phases chemically bonded to the support material offers the greatest ease of use, prolonged column life (some have been used routinely in industrial applications for several years without deterioration), and little chance of being permanently damaged by, for example, the injection of a solvent which is incompatible with the mobile phase. These columns are currently enjoying wide popularity as LC becomes accepted as a routine rather than a specialist method and is performed by less experienced operators.

In terms of predicting what type of methodology will be favoured in the future, current advances in the preparation of packing materials with chemically bonded stationary phases would seem the most promising. There is still much to be learnt about the optimum conditions under which such stationary phases are prepared. Even so, currently columns are being prepared which yield the equivalent of 20,000–40,000 plates per metre and this at the early stages of the intensive development surrounding the combination of very small particles with chemically bonded phases. The future in this area looks very bright indeed. It is apparent that, due to the high degree of selectivity attainable by modifying the mobile phase composition, a very wide range of differently bonded phases will not be necessary. If the proliferation of different stationary phases in a manner similar to the situation in GC can be avoided, it will save much consternation in the future. Many experienced workers in the field of LC are of the opinion that for the vast proportion of partition chromatography probably five or six stationary phases would suffice. A typical choice could be:

An aliphatic non-polar phase, *e.g.*, octadecyl functionality
An aromatic non-polar phase, *e.g.*, phenyl functionality
An electron-rich moderately polar phase, *e.g.*, ether or nitrile functionality
A proton-rich moderately polar phase, *e.g.*, hydroxyl or amine functionality
An electron-rich very polar phase, *e.g.*, polynitrile or ether functionality*
A proton-rich very polar phase, *e.g.*, polyglycols*

REFERENCES

1 L.C. Craig, *J. Chromatogr.*, 83 (1973) 67.
2 R.B. Sleight, *J. Chromatogr.*, 83 (1973) 31.
3 S. Eksborg and G. Schill, *Acta Pharm. Suecica*, 12 (1975) 1.
4 J.J. Kirkland, *J. Chromatogr. Sci.*, 10 (1972) 593.
5 *DuPont Liquid Chromatography Methods Bulletin*, 820M10, March 23, 1972.
6 W.A. Aue and C.R. Hastings, *J. Chromatogr.*, 42 (1969) 319.
7 O-E. Brust, I. Sebestian and I. Halász, *J. Chromatogr.*, 83 (1973) 15.
8 I. Sebestian, O-E. Brust and I. Halász, in S.G. Perry (Editor), *Gas Chromatography 1972*, Applied Science Publishers, London, 1973, p. 281.
9 J.F.K. Huber, in A. Zlatkis (Editor), *Advances in Chromatography 1970*, Chromatography Symposium, Houston, Texas, 1970, p. 348.
10 J.F.K. Huber, C.A.M. Meijers and J.A.R.J. Hulsman, *Anal. Chem.*, 44 (1972) 111.
11 C. Hesse and W. Hovermann, *Chromatographia*, 6 (1973) 345.
12 N.A. Parris, *J. Chromatogr. Sci.*, 12 (1974) 753.
13 J.J. Kirkland, *Anal. Chem.*, 43 (1971) 36A.
14 J.J. Kirkland, *Anal. Chem.*, 40 (1968) 391.
15 J.A. Schmit, in J.J. Kirkland (Editor), *Modern Practice of Liquid Chromatography*, Wiley-Interscience, New York, 1971, p. 375.

*These phases differ in degree from the moderately polar supports, *e.g.*, a secondary amine as compared to a primary amine functionality.

Chapter 9

Ion-exchange chromatography

INTRODUCTION

In many respects ion-exchange chromatography resembles liquid–solid (adsorption) chromatography with the additional characteristic that in the former we are dealing with the adsorption–desorption phenomena associated with ionic, or at least potentially ionic, substances.

In the very simplest picture, one may consider an ion-exchange material having a positive charge (an anion exchanger) which will attract or retain negatively charged species (anions) present in the mobile phase. Similarly, a cation exchanger, having negatively charged sites on its surface, will interact with ions carrying a positive charge. The mechanism of these interactions is explained in more detail in later sections of this chapter.

Since we are dealing with charged species, *i.e.*, ionic compounds, it is not surprising that the technique is most often performed in aqueous media, as, apart from acids, bases and certain speciality solvents like liquid ammonia, such solvent systems have the highest dielectric constant and thus the greatest tendency for compounds to dissociate into ions. Ion-exchange chromatography is ideally suited to the separation of those highly polar substances which, without recourse to derivative formation, cannot be handled by GC. Into this class come amino acids, peptides, sugars, nucleic acids and salts, all of which are of the greatest importance to those working in the life sciences. Although chemically very dissimilar, ion exchange finds considerable use in the analysis of inorganic species, being perhaps in greatest demand for the separation of lanthanide, actinide and noble metals. Using what can be regarded as "classical", *i.e.*, low-pressure ion-exchange chromatography, many of these separations have been possible for several decades albeit in most instances many hours were required to complete a separation. The use of recently developed instrumentation and particles of ion exchanger which are much smaller in diameter than the earlier materials can result in a very considerable improvement in the chromatographic performance and speed of analysis. Perhaps of even greater importance is the development of the new generation of column packings for ion-exchange work that have been specifically designed for modern, high-speed chromatographic methodology. These materials differ considerably from the classical resins in both their physical properties and handling characteristics to such an extent that it is sometimes better to consider them separately.

The use of ion-exchange methods to separate samples of biological origin is interesting in that it is often possible to examine the sample in the same form as it occurs *in vivo*, minimising the risk of rearrangements or isomerism occurring which could complicate and possibly lead to misinterpretation of the end result. With regard to the use of ion exchange for understanding biological systems, the work of Scott and Lee[1], for example, on the examination of physiological fluids that contain literally hundreds of constituents must rate as one of the most challenging.

Throughout the history of ion-exchange chromatography there have been reports of employing organic or semi-organic solvents in the mobile phase. Although some of these separations are based on an ionic mechanism, many have used organic solvents to suppress or enhance the solubility of a component or to achieve elution of hydrophobic compounds from the column. In this latter instance, particularly, it is more likely that many separations are achieved, at least in part, by reversed-phase adsorption of the sample on the surface of the ion-exchange resin rather than by some ionic interaction. This situation is not unexpected when it is considered that most of the earlier column packing materials were based on beads made from styrene–divinylbenzene copolymer, a very hydrophobic material.

More recently developed column packings utilise an inorganic support to impart rigidity to the ion exchanger. In many instances interactions of samples with this support or the coating on its surface contribute to the overall chromatographic selectivity obtained. The behaviour of weakly ionic organic substances on ion-exchange systems is often quite difficult to interpret in view of the various factors which can influence retention. Frequently an empirical approach yields more satisfactory results than one based on the classical concept of ion-exchange behaviour.

The ion-exchange functionality of column packings is most often obtained by incorporating ionic groups, such as sulphonate for cation exchange and quaternary ammonium for anion exchange.

RANGE OF SAMPLE APPLICABILITY

From the introductory section it should be apparent that ion exchange is the method of choice for the separation of ionic species. In general most compounds which are soluble only in water are amenable to analysis by ion-exchange methods either directly, because of inherent polarisation of the molecule, or by complex formation in the aqueous phase. Into this latter category comes the separation of carbohydrates, which form anionic complexes in borate solutions. A possible exception to the general applicability of ion exchange is perhaps the separation of high-molecular-weight substances which are known to be adsorbed strongly on chromatographic packings and column walls.

Applications in the biological sciences

Some of the most important applications of ion exchange are related to the biological sciences, in particular to the separation of amino acids and nucleotides obtained by the hydrolysis of biological samples.

Amino acids have been separated by ion exchange for many years and probably represent one of the earliest applications of LC where the mobile phase was forced through the column under pressure. The separation and detection of amino acids is a rather complex procedure in that gradient elution must be employed to obtain optimum resolution of components while sensitive, selective detection is obtained by using a post-column chemical reaction detector. These detectors, with colorimetric or fluorimetric monitoring, have been described in Chapter 5. Most widely used has been the ninhydrin reaction, whereby amino acids are reacted with ninhydrin reagent to yield an intensely blue colour. The greatest limitation to this approach is that the reaction takes approximately 15 min to go to completion and for quantitative results the column effluent, combined with the reagents, must be held in a reaction coil for this period of time. In more recent work fluorescent derivatives have become more popular as fluorigenic reagents are now available which are capable of reacting with an eluting amino acid in a matter of a few seconds[2]. Developments in both detection and separation aspects of amino acid analysis have brought the speed of analysis down to 1 h, a considerable advance on the separations reported in the late 1950's, which took up to 22 h to complete[3]. For amino acid separations, refined versions of the classical polystyrene-based ion exchangers continue to be used successfully. The increased speed of analysis is largely due to the reduction in the size of the column packing material and having apparatus capable of

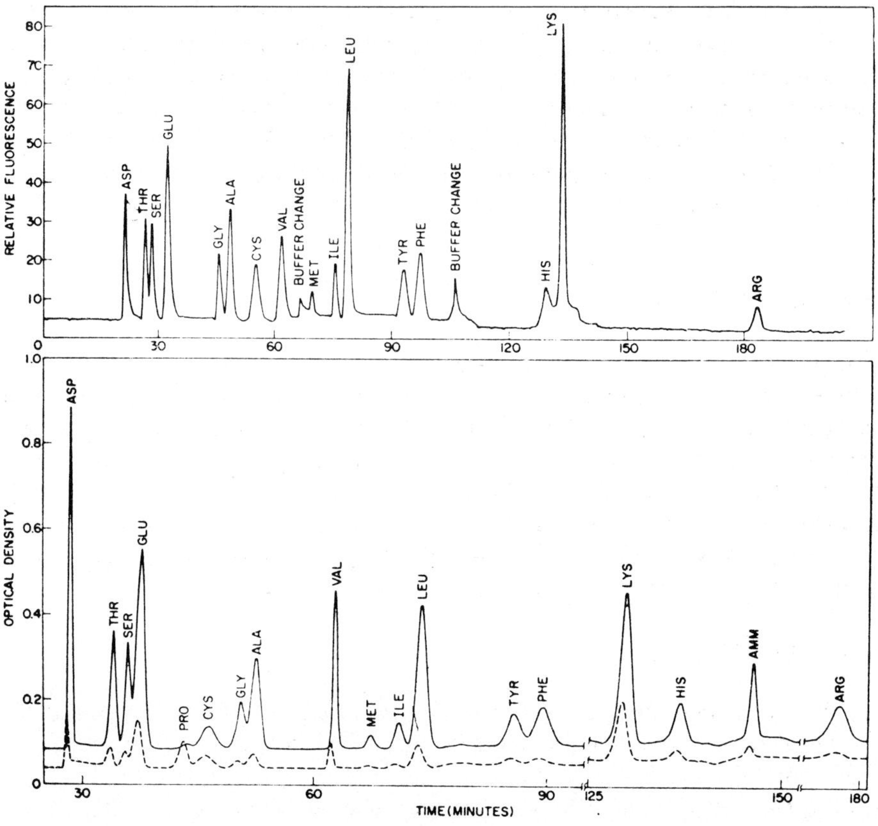

Fig. 9.1. Single-column separation of amino acids using both colorimetric (ninhydrin) and fluorescence detection. Operating conditions: column 0.5 m × 2.6 mm I.D.; packing, Durrum DC-4A resin; temperature, 62°C; mobile phase, sodium citrate buffers in stepwise gradient – (1) 0.2 *M* Na^+, 0.067 *M* citrate, 3% methanol and 0.3 ml/l thiodiglycol at pH 3.27, (2) as buffer (1), but adjusted to pH 3.80, and (3) 0.8 *M* Na^+, 0.2 *M* citrate adjusted to pH 5.90; flow-rate, 10 ml/h; inlet pressure, approximately 53–56 bars (800–850 p.s.i.). ASP = Aspartic acid; THR = threonine; SER = serine; GLU = glutamic acid; GLY = glycine; ALA = alanine; CYS = cysteine; VAL = valine; MET = methionine; ILE = isoleucine; LEU = leucine; TYR = tyrosine; PHE = phenylalanine; HIS = histidine; LYS = lysine; ARG = arginine; AMM = ammonia. (Redrawn from A.G. Georgiadis and J.W. Coffey, *Anal. Biochem.*, 56 (1973) 121, with permission.)

operating at high pressure. Column packing materials in current use have diameters in the order of 10 μm[4]. A typical separation of amino acids is shown in Fig.9.1, where a comparison is made between colorimetric (ninhydrin) and fluorescence detection for this analysis.

Many other important species of biological origin are more amenable to LC by virtue of the greater ease of detection, *e.g.*, separations are capable of being monitored using UV detectors. Areas of application related to nucleotide and purine/pyrimidine bases have been improved considerably in recent years with the advent of more advanced column technology. Fig.9.2 illustrates the present-day capabilities of LC for the separation of nucleotides using in this case a solid core packing with anion-exchange groups bonded chemically to the surface[5].

Cation-exchange chromatography has been employed in a comparable manner to separate organic bases such as purines and pyrimidines of biological importance. The separation reproduced in Fig.9.3 illustrates how compounds which are structurally closely related may be resolved. A considerable quantity of experimental data related to the separation of nucleic acids and related substances has been reported in the literature. A comprehensive survey of this area of application has been compiled by Brown[6].

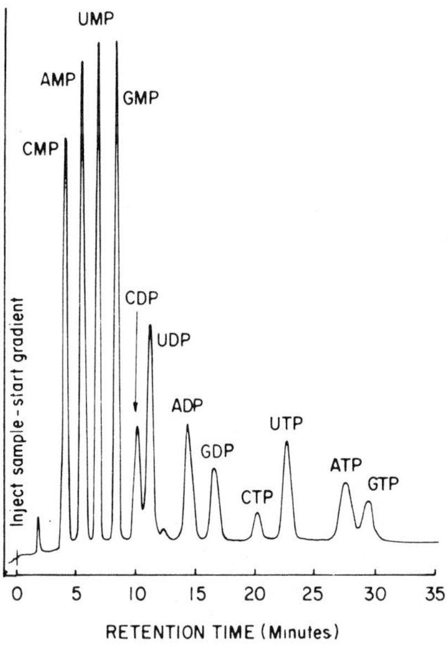

Fig.9.2. Separation of nucleotides by gradient elution ion-exchange chromatography. Operating conditions: column, 1 m × 2.1 mm I.D.; packing, Permaphase AAX; temperature, ambient; mobile phase, gradient from 0.002 *M* potassium phosphate, pH 3.3, to 0.5 *M* potassium phosphate at a gradient rate of 3%/min; inlet pressure, 67 bars (1000 p.s.i.); flow-rate, 1 ml/min; detector, UV absorbance. CMP = Cytidine-5′-monophosphate; AMP = adenosine-5′-monophosphate; UMP = uridine-5′-monophosphate; GMP = guanosine-5′-monophosphate; CDP = cytidine-5′-diphosphate; UDP = uridine-5′-diphosphate; ADP = adenosine-5′-diphosphate; GDP = guanosine-5′-diphosphate; CTP = cytidine-5′-triphosphate; UTP = uridine-5′-triphosphate; ATP = adenosine-5′-triphosphate; GTP = guanosine-5′-triphosphate. (Reproduced from R.A. Henry, J.A. Schmit and R.C. Williams, *J. Chromatogr. Sci.*, 11 (1973) 358, with permission.)

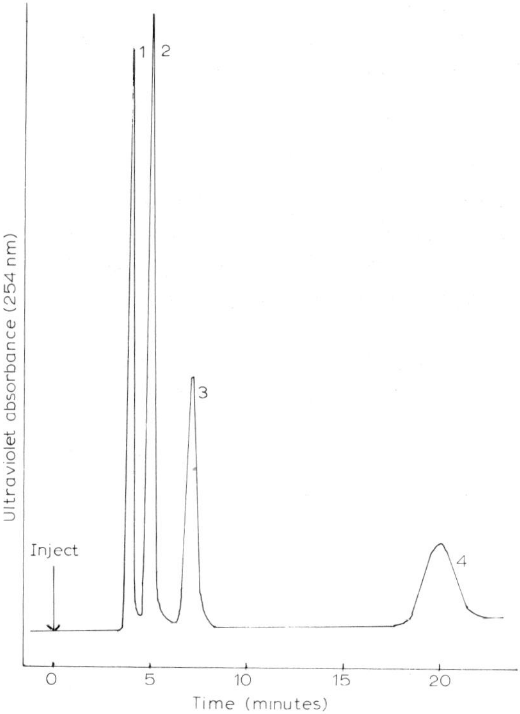

Fig. 9.3. Separation of several purine bases by cation-exchange chromatography. Operating conditions: column, 1 m × 2.1 mm I.D.; packing, Zipax SCX; temperature, 50°C; mobile phase, 0.01 *M* nitric acid + 0.04 *M* sodium perchlorate; inlet pressure, 40 bars (600 p.s.i.); flow-rate, 1 ml/min. 1 = Inosine; 2 = hypoxanthine; 3 = adenine; 4 = adenosine.

Much interest has been shown in the possible clinical uses of ion-exchange chromatography as a means of screening biological fluids in order to highlight abnormalities. This area of application poses a major challenge in that most body fluids contain literally hundreds of different compounds that should be separated. The significance of variations in the analysis of body fluids from individual subjects is a separate and equally complex subject. Recent work in this area has dealt with the detection of the metabolites of biogenic amines[7] and the screening of urine for possible abnormalities. Fig. 9.4 shows a series of separations reported by Scott and Lee[1] illustrating the increased speed of separation achieved by using coupled columns, one containing microparticulate ion-exchange resin of the polystyrene type and the second packed with a modern pellicular material. It should be noted that the separations shown were completed in 80 h, which was considered a significant reduction in time as previous studies involved a 240-h separation – such is the complexity of high-resolution analyses of biological fluids.

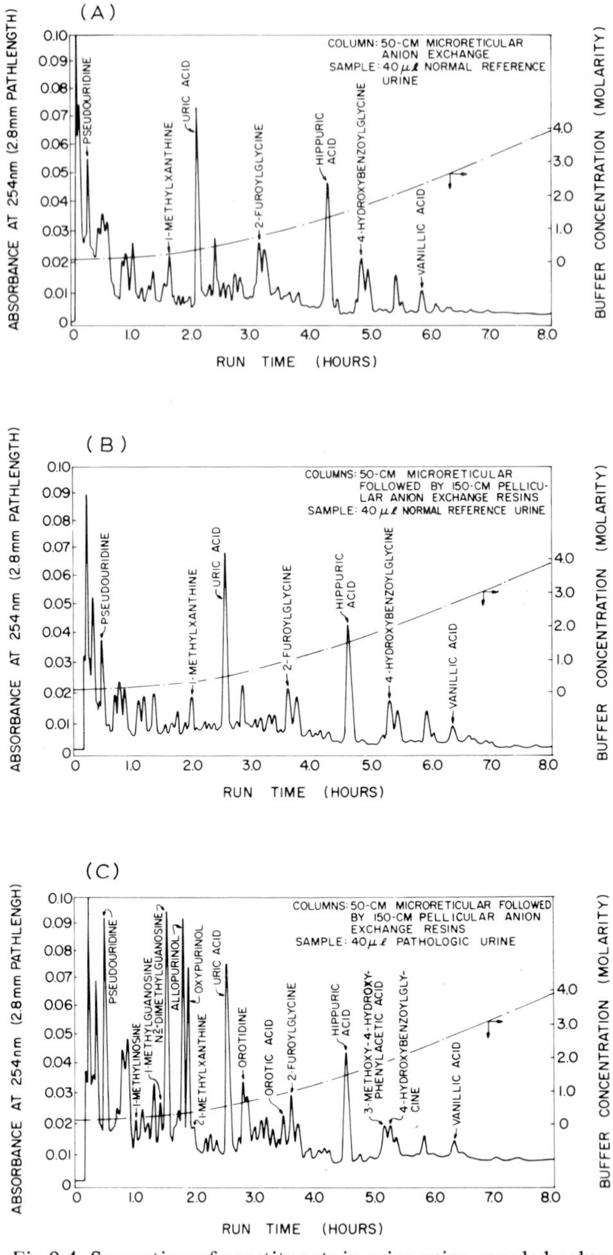

Fig. 9.4. Separation of constituents in urine using coupled columns. Comparison of the separation of the UV-absorbing constituents of urine on a short, 50-cm, column (A) of microreticular anion-exchange resin (Aminex A-27, 12–15 μm diameter) and on sequential columns of microreticular (50 cm) and pellicular (Pellionex AS) (150 cm) resins (B and C). Eluent, acetate buffer (pH 4.4); average flow-rate, 12.0 ml/h; temperature, increasing from ambient to 60° and 40°, respectively, for the two columns at 1 h. Samples: (A and B) 40 μl normal reference urine; (C) 40 μl pathologic urine. (Reproduced with permission from C.D. Scott and N.E. Lee, *J. Chromatogr.*, 83 (1973) 383.)

Other applications of ion exchange

The more established, polystyrene-based, resins possess a high exchange capacity, in the order of 5 mequiv. per gram of packing. A high exchange capacity can be a distinct advantage in applications where large sample sizes are required, however in many instances the high capacity leads to very strong retention of the sample on the column. Elution of the sample components under these conditions is only possible by the action of relatively concentrated buffer or salt solutions, typically in the order of 1–5 *M*. High concentrations of salts may be quite acceptable in glass columns, but such solutions can be particularly aggressive towards apparatus made of stainless steel and also cause problems if salts are allowed to crystallise in fine-bore tubing.

The more recently developed ion-exchange packings, in general, possess low exchange capacities (approximately two orders lower than the microparticulate polystyrene-based materials) and are correspondingly much less retentive. As a consequence, eluting solvents require considerably lower concentrations of buffers and salts, 10–100 m*M* being typical, to elute the sample components. Separations using modern packing materials can often be achieved quite rapidly, usually in less than 20 min.

Many polar compounds of pharmaceutical interest may be easily chromatographed in

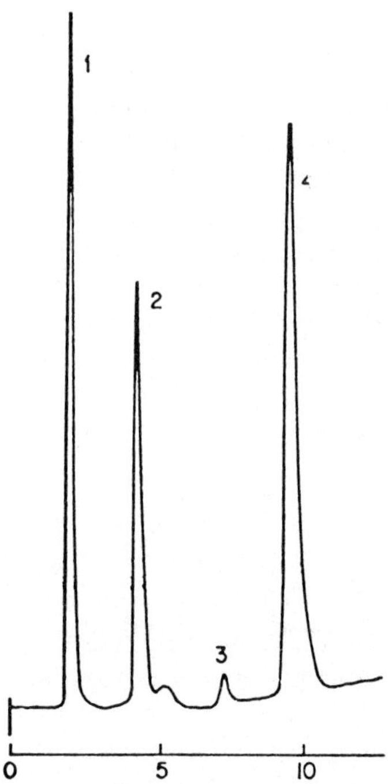

Fig. 9.5. Separation of several compounds of the vitamin B group by ion-exchange chromatography. Operating conditions: column, 1 m × 2.1 mm I.D.; packing, Permaphase AAX; mobile phase, gradient elution from 0.004 *M* sodium phosphate, pH 4.4, to 0.2 *M* sodium phosphate at a gradient rate (non-linear) of 10%/min; inlet pressure, 100 bars (1500 p.s.i.); detector, UV absorbance, 254 nm. 1 = Nicotinic acid; 2 = riboflavin monophosphate; 3 = impurity; 4 = folic acid. (Reproduced from R.C. Williams, R.A. Henry and J.A. Schmit, *J. Chromatogr. Sci.*, 11 (1973) 618, with permission.)

the manner described. Some of the well documented applications include: barbiturates, sulphonamides, analgesics, and antibiotics such as tetracyclines and cephalosporins. Chapter 15 contains a considerable number of references to literature sources where further details of these applications may be found.

Fig.9.5 illustrates the type of results commonly obtained with the high-speed, ion-exchange methods in the separation of several compounds of the vitamin B group. The low concentration buffer in the mobile phase and the speed of analysis are quite different from those obtained with the conventional, polystyrene-type, ion exchangers.

In addition to the more obvious separation of acids and bases, ion exchange has found widespread application in the food industry for the analysis of food-colouring agents and artificial sweeteners such as saccharin.

MECHANISM OF ION-EXCHANGE SEPARATIONS

Classical ideas on ion-exchange equilibria

In the opening paragraphs of this chapter it was stated that ion-exchange resins may be considered as either anionic or cationic, depending on the nature of the functional groups which are attached to the supporting matrix. In the discussion of the mechanism of separation, only one form will be discussed in depth, that of the cation-exchange system; the ideas put forward apply equally to an anion-exchange system, excepting the polarity of the species considered will be opposite, *i.e.*, for anion read cation, etc.

In a chromatographic column filled with a cation exchanger, the column packing, if it were in isolation, would possess a net negative charge. In practice, electroneutrality is maintained by the resin being in the "salt form", *i.e.*, each negative site on the resin has a cation held as an ion pair. These cations are commonly hydrogen or sodium ions, thus a resin is said to be in the "hydrogen form" or "sodium form", respectively. Conversion of a resin from one form to another is accomplished by passing an ionic solution through the column containing an excess of the desired cation. Thus to convert a resin from the hydrogen form into the sodium form one passes a solution of a sodium salt through the column. The following exchange reaction takes place at the surface of the resin

$$Na^+ + [(\text{resin})^- H^+] \underset{H^+}{\overset{Na^+}{\rightleftharpoons}} H^+ + [(\text{resin})^- Na^+] \tag{9.1}$$

It is important to note that this reaction is in reality a dynamic equilibrium following the laws of mass action. If an excess of hydrogen ions were present, *i.e.*, the solution was acidic, then the reverse reaction would take place.

The extent to which the exchange reaction occurs depends on the concentrations (strictly the activities) of the species present. The equilibrium constant defines this condition quantitatively (eqn.9.2).

$$K = \frac{[(\text{resin})^- Na^+]\ [H^+]}{[(\text{resin})^- H^+]\ [Na^+]} \tag{9.2}$$

A greater value of K indicates that more of the resin will be converted into the sodium form. In practice, ion-exchange chromatography is performed with the mobile phase "buffered" at a definite pH so the situation may be considered as ions in solution (counter ions) in dynamic equilibrium with similar ions forming ion pairs with the charged sites on the resin. This equilibrium condition is dependent on operational variables such as pH, temperature, and the ionic strength of the mobile phase.

If in place of the sodium ions a sample is applied which also forms cations on being introduced into the column, then a similar equilibrium distribution of this compound will be established, in competition with the distribution of the counter ions. In an analogous manner to the conversion of a resin from one ionic form to another, as described above, if no other selectivity factors were involved, the ionic species in greatest concentration would interact most strongly with the resin and thus be preferentially retained. Thus, if it were the counter ion which was in the highest concentration, one would anticipate that the sample would be eluted rapidly. Similarly, if one decreases the concentration of the counter ions in the mobile phase, increased retention of the sample would be obtained. This effect is observed in practice; thus a complex mixture of components which is retained may be eluted progressively by the application of gradient elution, where the concentration of the counter ions is increased during the course of the separation.

Although the ion-exchange resin is commonly a strongly acidic or basic material, *i.e.*, is fully ionised at all pH values, the same comment does not apply to samples, as these may be only weakly or moderately dissociated. The degree of interaction of sample with an ion-exchange resin is a function of the concentration of the ions it produces in solution, strictly the ratio of ionised to neutral molecules and not the total concentration of "ionised + non-ionised" sample. It follows that a change in the pH of the mobile phase which will cause a partially dissociated molecule to ionise further will lead to increased retention, whereas a change in pH which suppresses the dissociation will cause the sample to elute more rapidly from the column. The same feature also applies to weak ion-exchange resins, where the surface groups only ionise under favourable pH conditions, a typical example being weak cation exchangers having a carboxylic acid functionality.

Secondary interactions contributing to the separation

Points on the mechanism of separation outlined so far have assumed that separations in ion-exchange chromatography are due solely to ionic interactions between the sample and the column packings. However, in many applications, what might appear to be separations based on ion exchange, is, in fact, complicated by secondary interactions which are essentially non-ionic in nature. These interactions arise from adsorption or hydrogen bonding of the sample to the non-ionic part of the matrix of the column packing, or simply the limited solubility of the sample in the mobile phase. These effects are particularly noticeable when studying organic substances as distinct from the more completely ionised inorganic anions and cations. The mechanism of ion exchange is further complicated by the selectivity introduced by the charge and radius of the ions competing for the charged sites on the resin, where generally the larger the radius of the ion, the stronger will be its affinity towards the resin. This effect is described in greater detail in the section of this chapter dealing with the selection of the mobile phase.

The overall mechanism by which separations are accomplished in "ion-exchange" chromatography is thus a very complex subject owing to the many complementary processes which can affect the retention of samples. It is difficult to predict what operating conditions will provide the best separation, since the relative magnitude of the processes contributing to an overall separation vary with sample type. In the case of organic substances, it is sometimes possible to separate anions on a cation-exchange column and *vice versa*, completely contradicting the simple ideas of ion exchange. Not surprisingly, many chromatographers tend to adopt an empirical rather than scientific approach when examining samples by ion-exchange methods.

STRUCTURE OF COLUMN PACKINGS FOR ION-EXCHANGE CHROMATOGRAPHY

The most important naturally occurring materials to show ion-exchange properties are the zeolite class of alumino-silicates. These materials possess a characteristic open framework structure having the general composition of $M_{x/n}(AlO_2)_x \cdot (SiO_2)_y \cdot zH_2O$, where the charge on the cation, M, is n and z is the degree of hydration. The extent of the hydration and the relative proportions of Al to Si vary from one example to another. Although these materials were originally found to occur naturally, some have been synthesised to yield useful ion exchangers and molecular sieves. This latter property is dealt with in more detail in the chapter describing steric exclusion chromatography.

Other types of inorganic ion exchangers have been developed commercially. Examples are those relying on a microcrystallised gel structure of zirconium oxide (anion or cation exchanger, depending on the pH of the mobile phase), zirconium phosphate (cation exchanger) and ammonium phosphomolybdate (cation exchanger especially selective for alkali metal ions).

Inorganic ion exchangers do not enjoy particularly wide popularity as there are organic-gel based materials available which offer superior performance in respect of column efficiency and the number of different types of selectivity.

The widespread usage of ion-exchange methods based on open-column methods and the long-standing interest in amino acid analysis have resulted in a large number of organic ion-exchange resins being commercially available. The number of variations possible is too large to be discussed in this text, but suffice it to say, many of these "classical" materials are not suitable for high-resolution work. The principal reason for their general inapplicability is that, with few exceptions, the particle size is too large to enable packed columns to give highly efficient separations at high mobile phase velocities. This feature arises as a result of the fact that the materials are totally porous, enabling the formation of "stagnant pools of mobile phase" which limit the rate of mass transfer. Earlier chapters have explained that this form of mobile phase mass transfer can be reduced if the particle diameter is reduced to less than 10 μm. This condition has been approached in several commercial products, notably those offered by Durrum and the "Aminex" resins from Bio-Rad. Both of these groups of materials were originally introduced for high-speed amino acid analyses but have subsequently proved of value in other applications. A reduction in the overall diameter of the ion-exchange beads leads to a marked decrease in the column permeability, consequently high pressures must be employed if high-speed analyses are required.

Most conventional ion exchangers utilise the styrene–divinylbenzene type of copolymer as the supporting matrix, where the divinylbenzene is incorporated to cross-link the resin to give it rigidity. For use in modern LC the physical strength of the matrix of a packing material must be sufficient to withstand high pressures without compression and for this reason alone hardly any of the resins widely used in low-pressure systems can be used. Most conventional ion exchangers also swell or shrink if the pH or ionic strength of the mobile phase is varied, thus creating serious dimensional changes. These features, which are intolerable in a high-resolution system, can be reduced to some extent by heavily cross-linking the copolymer (usually with more than 8% divinylbenzene) in an attempt to make the structure more rigid. This action, however, often decreases the permeability of the column and leads to exclusion of large molecules from the resin. As an alternative, one can produce a packing material where the principal structure of the particle is inorganic, *e.g.*, a silica or glass bead, and the ion-exchange functionality is incorporated as a thin surface coating of resin or by the ionic groups being chemically bonded directly to the surface of the support in a manner analogous to the stationary phases for partition chromatography. The latter approach is also advantageous from the point of view of stationary phase mass transfer, for chemically bonded functional groups can be incorporated as a mono-molecular layer, which leads to improved rates of mass transfer. This situation contrasts with that using ion exchangers based on totally organic beads or pellicular (surface layer) packings, where the polymeric material restricts the rate of mass transfer.

The capacity of ion-exchange materials is expressed in terms of the number of equivalents of exchangeable ions available per gram of packing material. Values for the various types of ion exchangers differ considerably. For instance, a conventional resin for open-column chromatography typically has a capacity in the range of 3–7 mequiv. per gram of dry resin. High-resolution copolymer resin-type beads with particle diameters in the region of 10 μm have similar exchange capacities of approximately 3–5 mequiv./g, depending on the particular material. This level of exchange capacity is generally considered as high. In practice, this high capacity will lead to strong retention of ionic samples unless the mobile phase contains appreciable (1–5 M) concentrations of buffer or ionic modifiers. A particular merit of these resins is that large samples may be introduced when necessary, for instance, when needing to overcome limited detector sensitivity. When one considers the pellicular or controlled surface porosity types of ion-exchange material, the capacity is very considerably lower, in the region of 10 μequiv. per gram of packing. This feature of the surface layer materials restricts their use to systems equipped with very sensitive detectors, as the size of the sample that can be separated must be kept small. A general guide to sample size in ion-exchange work is to limit the maximum size of the sample to less than 5% of the total exchange capacity of the column packing.

Clearly, the design of the ion-exchange chromatographic packing material has an important influence on its resultant performance characteristics. The principal types – totally porous polymeric resin, pellicular, controlled surface porosity and chemically bonded – are depicted diagrammatically in Fig.9.6.

The relative advantages and disadvantages of these different materials may be summarised as follows.

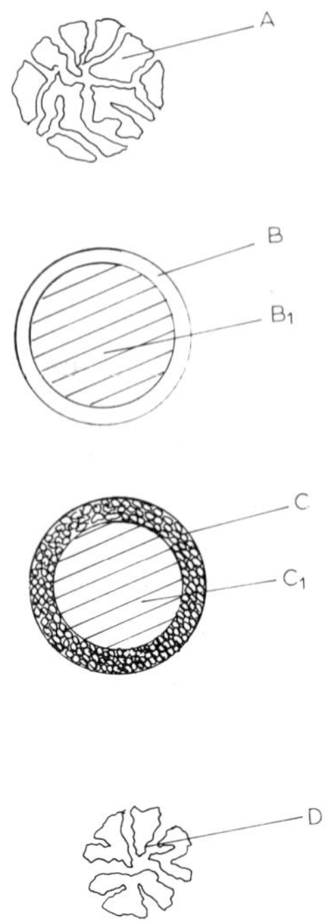

Fig. 9.6. General structure of different ion exchangers. (A) Styrene–divinylbenzene copolymer, porous, with ionic functional groups chemically attached; particle diameters 8 μm upwards; material liable to swell. (B) Thin (1-μm) layer of resin similar to above, coated on to inorganic support; (B1) impervious glass bead; particle diameter typically 25–50 μm. (C) Microsphere layer (porous) on which ionic functional groups are coated or chemically bonded; (C1) Inner impervious silica or glass bead, diameter about 30 μm. (D) Porous silica microparticle with ionic functionality bonded to surface. Particle size typically 10 μm.

Porous polymer resins

Porous polymer resins offer an exchange capacity two to three orders higher than the surface layer type of packings and find use in systems employing less sensitive detectors and in methods involving complex sample mixtures. Porous resins are susceptible to compression under high pressure unless highly cross-linked polymers are used in the manufacture of the supporting matrix. The latter approach, however, leads to materials having a lower permeability, which leads to the exclusion of large molecules from the resin. Poor mass transfer in a totally porous resin can limit overall column efficiency. The efficiency can be improved, however, by working with small resin beads with a mean diameter in the order of 5–10 μm. A disadvantage of these materials is that a change in the concentration or the nature of the counter-ion in solution can cause swelling or shrinking, which is unacceptable in a high-resolution system.

Pellicular or controlled surface porosity ion exchangers

Ion exchangers which possess a thin surface layer of resin on an impervious support lack exchange capacity and lay a restriction on the size of sample that may be applied to a column. Mass transfer is, however, much improved compared with that using porous resin materials and is best when the individual ion-exchange functional groups are bonded to the inorganic surface via a "spacer" group, usually a short aliphatic chain. Ion exchangers which are bonded to the inorganic surface may be used in semi-aqueous or non-aqueous media, in marked contrast to the simpler, polymer coatings, which dissolve quite readily in many organic solvents. The inorganic matrix of these materials is not susceptible to swelling or shrinking as the nature of the mobile phase is changed. However, the thin layer of ion-exchanger on the beads can swell to some extent, but this effect is limited in extent by the inorganic "backbone" which maintains the structure of the column bed. Since the ion-exchange processes are limited to the thin surface layer, the equilibration time following a change of mobile phase is quite short, generally less than 30 min, depending on the mobile phase velocity. This speed is in marked contrast to the many hours that are frequently required to equilibrate high-capacity, totally porous resins.

Ion exchangers bonded to small, totally porous inorganic supports

Perhaps the most interesting materials for really high-performance ion exchange will result from the recent introduction of packings formed from very small (10 μm diameter or less), totally porous, silica particles. These packings, due to their high surface area, enable a modest proportion of ionic functional groups to be incorporated. Provided the internal pores of the silica support are comparatively wide to permit ready access of both sample and counter ions, rapid mass transfer should be possible, leading to high column efficiencies. The high surface area should in turn permit an exchange capacity much higher than that obtainable with surface layer packings based on solid glass beads.

Sometimes ion exchangers are classified by being described as either microreticular or macroreticular materials. This distinction, originating from more traditional forms of ion-exchange chromatography, differentiates by the dimensions of the internal pores. Microreticular resins have internal pores of comparatively small diameter, which allow solvent molecules, *viz.* water and small ions, to penetrate the polymer matrix yet exclude larger molecules. Most common polymer-based ion exchangers are of the microreticular type. Macroreticular, or macroporous as the name suggests, implies that the pore structure is sufficiently large to allow penetration of larger molecules. Resins of the latter type have been used mostly in low-pressure chromatography for separations performed in semi-aqueous media. If the inorganic support of a bonded ion exchanger has pores of a sufficiently large diameter, then it can be anticipated that it will be suitable for the separation of ionic species of moderately high molecular weight. Currently materials of this type are only just becoming available, however the future high-resolution ion exchangers could well be based on such materials.

TABLE 9.1

SOME COLUMN PACKINGS FOR MODERN ION-EXCHANGE CHROMATOGRAPHY

Type	Name	Approx. exchange capacity	Particle size (μm)	Supplier
Pellicular strong anion	Ion-X-SA	n.d.**	30–40	Perkin-Elm
	"Pellicular Anion"	10	~40	Varian
	AE-Pellionex SAX	10	44–53	Reeve Ange
	AS-Pellionex SAX	10	44–53	Reeve Ange
	Perisorb AN	30	30–40	Merck***
	Permaphase AAX*	100	25–37	DuPont
	Permaphase ABX*	60	25–37	DuPont
	Vydac Anion Exchange*	100	30–44	Separations
	Zipax SAX	12	25–37	DuPont
Pellicular weak anion	AL-Pellionex WAX	n.d.	44–53	Reeve Ang
	Zipax WAX	n.d.	25–37	DuPont
Pellicular strong cation	Ion-X-SC	n.d.	30–40	Perkin-Elm
	"Pellicular Cation"	10	~40	Varian
	HC-Pellionex SCX	60	44–53	Reeve Ang
	HS-Pellionex SCX	8–10	44–53	Reeve Ang
	Perisorb KAT	50	30–40	Merck***
	Vydac Cation Exchanger*	100	30–44	Separation
	Zipax SCX	3.2	25–37	DuPont
Porous anion (polymer support)	Aminex A-14	3400	17–23	Bio-Rad
	Aminex A-25	3200	15.5–19.5	Bio-Rad
	Aminex A-27	3200	12–15	Bio-Rad
	Aminex A-28	3200	7–11	Bio-Rad
	DA-X8A	4000	6–10	Durrum
	DA-X4	2000	15–25	Durrum
	DA-X2	2000	15–25	Durrum
Porous cation (polymer support)	Aminex A-4	5000	16–24	Bio-Rad
	Aminex A-5	5000	11–15	Bio-Rad
	Aminex A-6	5000	15.5–19.5	Bio-Rad
	Aminex A-7	5000	7–11	Bio-Rad
	AA-15	5000	16–28	Beckman
	PA-28	5000	16	Beckman
	PA-35	5000	13	Beckman
	DC-1A	5000	15–21	Durrum
	DC-2A	5000	9–15	Durrum
	DC-4A	5000	6–10	Durrum
	AN-90	5200	16–28	Hamilton
	B-80	5200	10–20	Hamilton
	H-70	5200	18–30	Hamilton
Porous anion (inorganic support)	Partisil-10-SAX*	n.d	10	Reeve Ang
	Vydac TP Anion Exchange	n.d.	10	Separatior
Porous cation (inorganic support)	Partisil-10-SCX	n.d.	10	Reeve Ang
	Vydac TP Cation Exchange	n.d.	10	Separatior

*Chemically bonded phases.

** n.d. indicates no data available.

***E.M. Labs. in the U.S.A.

COMMERCIALLY AVAILABLE ION-EXCHANGE MATERIALS

So far, the geometry of ion-exchangers has been described. In the next section the nature of the functional groups participating in the exchange processes are considered. Materials for ion exchange are conveniently classified in terms of the capabilities of the resin, thus they are generally referred to as weak or strong cation exchangers or weak or strong anion exchangers. The principal difference between strong and weak ion exchangers is that the functional groups on the former type are fully ionised at almost any pH and will therefore always exhibit some exchange characteristics. The weaker resins, on the other hand, possess functional groups which dissociate only under certain pH conditions and can be made more or less retentive by adjusting the pH of the mobile phase. This latter facility is useful when attempting to analyse samples which are themselves strongly ionic and which would otherwise be very strongly retained on strong ion exchangers. The possible combinations of the geometry of the packing material and the nature of the functional groups can give rise to a very large number of chromatographic packings. Some of the more common ion-exchange materials that are available commercially are listed in Table 9.1. References to technical papers describing their use can be found in Chapter 15.

A number of the column packing materials which are included in Table 9.1 have been introduced only fairly recently. Consequently there are very little, if any, data reported which can be used as a guide to the properties of a particular material. In these circumstances, until such a time when more data will be available, the best guide is to consult the manufacturers who in many instances will answer queries with regard to the use of a particular material. Attempting to find ion exchangers from different commercial sources yet offering the same selectivity characteristics can be a tedious exercise as most commercial data only indicate the nature of the functional groups which participate in the ion-exchange processes, the capacity in micro- or milli-equivalents per gram, and the general nature and size of the support material. Although such data are of great value, when the separation of organic species and of substances of large molecular weight is desired, other factors like the exclusion from a resin of large or highly solvated species should be considered; also adsorptive or hydrogen-bonding interactions can occur between the sample components being chromatographed and the matrix of the packing material at sites other than the exchangeable functional group. These features of a column packing are difficult to describe quantitatively as they vary with the nature of the mobile phase and also fine details of the method of preparation of the packing material might need to be described which may well form part of a proprietary process.

PRACTICAL ASPECTS OF ION-EXCHANGE CHROMATOGRAPHY

The methodology used in developing separations based on ion exchange can, in principle, form a very logical sequence in all separations where the mechanism of separation is solely due to ionic interactions. In practice, complications occur as secondary effects due to matrix–sample interactions often affect the elution characteristics. This situation, although frustrating when wishing to deduce a separation mechanism, should not be ignored as many highly successful separations have been reported which would appear to be possible only by the combined mechanism. The practical considerations outlined in this section deal essentially with the understanding of ion-exchange processes free of complications.

General sample applicability of the method

Perhaps it goes without question that samples which are amenable to separation by ion-exchange methods must be capable of forming charged species in solution. This situation may arise naturally in the case of strong electrolytes, typically inorganic salts and those formed from strong organic acids and bases. Alternatively, the ionic character may be induced by dissociation or protonation of an otherwise weakly ionic substance using high or low pH conditions. Thus, in the case of weak acids, alkaline solution will enhance dissociation to form an anion and similarly in acid conditions a weak base would be protonated. Some substances are amphoteric in that they may be protonated in acid solution and dissociated in alkaline solution, in a manner to that shown in eqn.9.3, which takes as example the ionisation of an amide link common to many heterocyclic molecules.

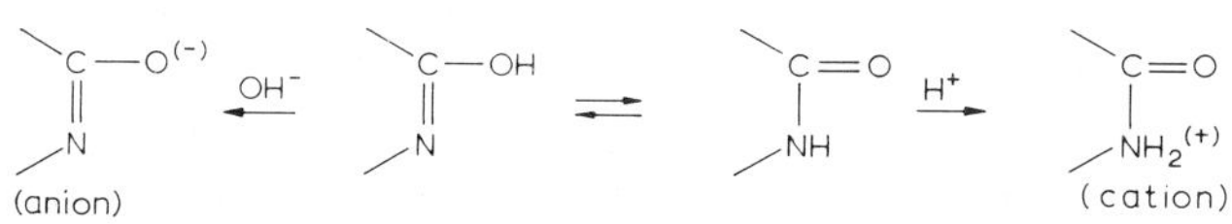

(9.3)

Other compounds, although ionic, form internal salts, known as Zwitterions, the most important example being the amino acids. Their ionic character is outlined in eqn.9.4.

$$\underset{\text{(anion)}}{NH_2{-}\overset{R}{\overset{|}{C}}H{-}COO^{(-)}} \xleftarrow{OH^-} NH_3^{(+)}{-}\overset{R}{\overset{|}{C}}H{-}COO^{(-)} \rightleftarrows$$

$$\rightleftarrows NH_2{-}\overset{R}{\overset{|}{C}}H{-}COOH \xrightarrow{H^+} \underset{\text{(cation)}}{NH_3^+{-}\overset{R}{\overset{|}{C}}H{-}COOH} \qquad (9.4)$$

Clearly with both of the last examples the choice of mobile phase pH will govern whether the sample will behave as a cation or an anion. The pH value at which a Zwitterion is neutral is known as its isoelectric point and its value is dependent on the structure of the molecule. Although in most applications the value of the isoelectric point will not be known, the example is taken to illustrate the variations in elution behaviour possible by a change in the pH of the mobile phase.

An important method of preparing ionic species from neutral species, rendering them amenable to ion exchange, is to form ionic complexes. The most widely studied system using this approach must be the formation of borate complexes with *cis*-1,2- and -1,3-diols, particularly the application of this reaction to carbohydrate analysis[8]. Borates react with diols to form complex anions, according to the reaction shown in eqn.9.5, which renders them amenable to analysis by anion-exchange chromatography.

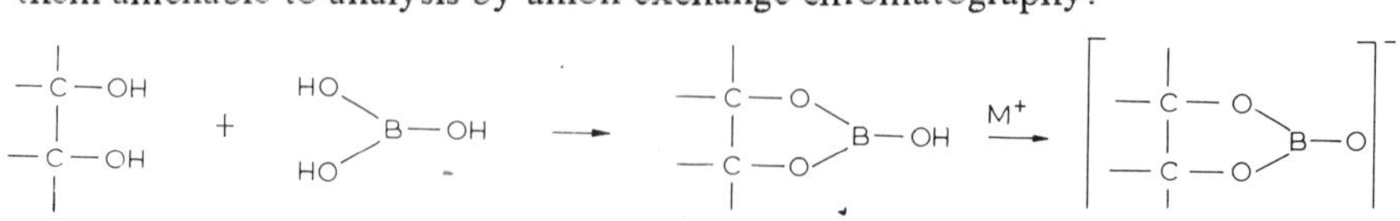

(9.5)

Packing columns with ion-exchange materials

The general methodology of packing chromatographic columns has been described in Chapter 3, however, packing ion-exchange resins of the porous polymer type can involve some special considerations. These resins tend to shrink or swell depending on the nature of the liquid with which they are in contact. The net result on a packed column is a marked change in permeability, leading, in extreme cases, to either the mobile phase being able to flow easily through voids in the column bed or to a plugged column. Unless there is very good reason to the contrary, *i.e.*, a previously published method, the inexperienced chromatographer is strongly advised to consider using the more recently developed packings, which have a rigid, inorganic, supporting matrix. These last-mentioned materials may be handled in a manner analogous to the corresponding materials for adsorption or partition chromatography, *i.e.*, particles greater than 20 μm may be dry packed into columns whereas smaller particles should be slurry packed. Swelling of modern packings is seldom a problem, thus a change of solvent can be effected without difficulty.

In the event that porous polymer type packings have to be used, the most satisfactory method of preparing the column is by using a slurry technique. The most satisfactory solvent in which to slurry the chromatographic resin is the mobile phase which is to be used in the separation procedure once the system has been set up. This situation is complicated by two factors, firstly, it pre-supposes a knowledge of which mobile phase will be required and, secondly, particles of the ion exchanger will tend to settle in this liquid medium, leading to differences in permeability throughout the column unless all particles of the ion exchanger are of the same size.

The former complication can be overcome in instances where the analysis involved is well established or by accepting that the first column to be studied of a particular material may be degraded in terms of efficiency as different mobile phases will have to be passed through in an attempt to establish the optimum mobile phase for the separation under consideration. Once this has been found, a second column can be prepared with this liquid, giving hopefully optimum efficiency and selectivity. The foregoing remarks may be taken to point out the extreme case of working with porous polymeric resins. In many instances, *e.g.*, in gradient elution work, a change of mobile phase may not significantly disturb the physical characteristics of the packed bed.

With regard to inhomogeneities in the packed bed due to partial separation of particles on the basis of size, sedimentation effects can readily occur during the introduction of a slurry if the density of the packing material is not identical to that of the liquid phase. The only completely satisfactory method of eliminating this effect is to use packing materials which have been very closely fractionated according to their particle diameter, for example ion-exchange materials are commercially available with particle diameters in the range of 7–11 and 12–15 μm. The use of high-pressure slurry packing techniques as described in Chapter 3 will also minimise the sedimentation effect to a certain extent, however, due to the inherent lack of rigidity of polymer beads as distinct from rigid inorganic structures, extreme pressures cannot be employed.

Factors influencing selection of mobile phase

In the earlier section of this chapter dealing with the mechanism of ion-exchange processes, the effect of pH and ionic strength of the mobile phase were discussed in general terms, indicating the selectivity which an exchanger will possess for different cations and anions. In practice, the buffer solution used in an ion-exchange separation should be selected on the basis of three factors, *viz.*

(a) pH – With samples or weak ion exchangers that are only partially ionised, the pH of the mobile phase will regulate the degree of ionisation, hence the concentration of ions in solution. An increase in the number of ionic sites on a resin and/or the ratio of sample ions to neutral molecule in solution will lead to stronger retention of the sample.

(b) Concentration (ionic strength) – An increase in the concentration of counter ions in solution relative to ions from the sample will result in the counterions, *i.e.*, ions from the buffer solution, being preferentially held on the resin. A decrease in ionic strength (concentration of the buffer) leads to stronger retention of the sample. Many of the separations achieved with superficially porous ion exchangers have required mobile phases containing only very dilute buffers, *i.e.*, in the range of 1–100 m*M*. In contrast, the high-capacity porous polymer resins most often require a buffer concentration of approximately 0.1–10 *M*.

(c) Selectivity of the counter ion – This effect results from the ability of ion-exchange resins to discriminate between ions of similar charge but differing in their geometry. Factors influencing the selectivity include the magnitude of the ionic charge, the radius, the degree of solvation of the ion and interactions with the support matrix. Data indicating the relative affinity or ion selectivity of ion exchangers are available for the more classical forms of resins that have been studied widely in low-pressure systems for many years. Unfortunately, tables giving details of the ion selectivity of resins seldom, if ever, take into account other practical considerations of mobile phase selection, for instance, corrosion aspects and compatibility with the method of detection. Halides, reducing agents and strongly UV absorbing ions will normally need to be avoided. The most popular anions used in modern LC systems are: phosphates, borates, nitrates, perchlorates and, to a lesser extent, sulphates, acetates and citrates. (Caution: Some organic anions can remove the protective oxide surface from certain grades of stainless steel leading to corrosion.) In cation-exchange chromatography nearly all reported separations employ one of the following cations: sodium, potassium, ammonium, or hydrogen. Accurate prediction of ion-selectivity effects is not always possible when working with modern pellicular or bonded ion exchangers as secondary interactions can contribute considerably to the overall separation. With the more conventional porous polymer resins, selectivity characteristics are often supplied in manufacturers' literature. (See, for example, catalogues supplied by Bio-Rad Laboratories.)

Optimisation of mobile phase

The variables under consideration are the pH, the concentration and the nature of the counter ions.

One semi-empirical approach for deciding the operating conditions for a separation based on ion exchange is to employ a dilute buffer solution as the initial mobile phase. If no other data are available, the pH and the concentration of the buffer are decided by experiment. For many pellicular ion exchangers a useful starting concentration is 10 m*M*; buffers in the pH range of 3–10 can be formed by mixing phosphoric acid and sodium hydroxide solutions, monitoring the neutralisation with a pH meter. Citrates (low pH) and borates (high pH) can be considered as alternative buffers for systems which are incompatible with phosphates. Outside this pH range dilute acid or alkali can be used. In the initial stages of developing a method the "buffer only" mobile phase is selected for optimum retention of the sample components. (Note: After changing the pH of the system, check that the liquid entering and leaving the column has the same pH.)

Having decided a mobile phase pH which retains all components of the sample, the ionic strength is progressively increased, ideally with a counter-ion having a high affinity for the resin, until the sample is displaced. Nitrates and perchlorates are particularly effective at displacing samples from anion-exchange columns; perchlorates have an additional advantage in absorbing less in the UV region of the spectrum giving more stable

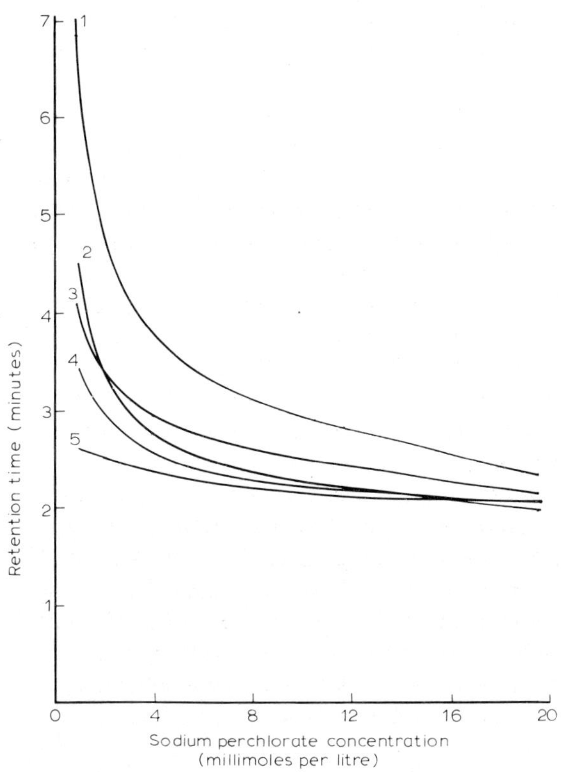

Fig. 9.7. Influence of counter-ion concentration on the retention time of barbiturates on a strong anion exchanger. Operating conditions: column, 1 m × 2.1 mm I.D.; packing, Zipax SAX, strong anion exchanger; mobile phase, 10 m*M* sodium borate, pH 9.2 + sodium perchlorate; flow-rate, 1.0 ml/min; inlet pressure, 100 bars (1470 p.s.i.g.); temperature, 25°C. 1 = Secobarbital; 2 = phenobarbital; 3 = amobarbital; 4 = isobutyl allylbarbital; 5 = barbital.

baselines to chromatograms run under gradient elution conditions. Fig.9.7 illustrates graphically the dependence of retention of samples on the concentration of counter ion for a number of barbiturate drugs.

A more preferable approach, leading to greater selectivity, is to operate the column–mobile phase system at a pH which is closer to the pK value (dissociation constant) of the acid or base being studied, assuming this value is known. In this range of hydrogen ion concentration the components of the sample will be only partially ionised, the extent to which each component ionises being related to the dissociation characteristics of the individual component. As retention of a sample on an ion exchanger is a function of the ratio of the concentration of ions relative to the neutral molecules it yields in solution, this value and hence the retention characteristics will differ with the pK of the sample components. Careful selection of the pH of the mobile phase will enable an element of selectivity to be introduced into the chromatographic system without necessarily having to change the type of column employed.

According to Smith *et al.*[9] a mobile phase buffered at approximately 1.5 pH units above the pK value of a base will provide a useful starting point for the selection of mobile phase for cation-exchange chromatography. At this pH, less than 10% of the sample component will be in the ionic form, thus small changes in the pH of the mobile phase will lead to a significant change in the concentration of ions in solution, changing the retention characteristics considerably. In many instances, organic samples will be insoluble in water when present as their non-ionised forms. Addition of a water-miscible organic solvent such as alcohol is then necessary to ensure complete solution of the sample. In this respect, attention should be paid to the stability characteristics of the ion-exchange material being used as some packings can deteriorate rapidly in the presence of organic solvents.

From the preceding paragraphs it will be noted that the displacement of components from a column by the addition of a neutral salt having a high affinity for the column packing is most often achieved using gradient elution. During the early stages of the introduction of a neutral salt by means of a gradient it is sometimes observed that the column packing will totally adsorb the added counter-ions until it has reached saturation. Thus, if for example, the salt is introduced in a manner whereby the concentration entering the column is increasing linearly with respect to time, the early part of the gradient profile leaving the column will be eliminated until, at some point after the start of the gradient, the concentration of added salt will suddenly increase in a manner not expected from the rate and shape of the concentration profile at the column inlet. Beyond this "breakthrough" point the change in concentration of the neutral salt leaving the column will follow that entering the column. The effect of this adsorption of ions from a neutral salt is analogous to demixing or dehomogenisation of multicomponent mobile phases that is observed in adsorption chromatography. When monitoring the separation with a photometric detector, this effect is manifest as a spurious peak occurring at a retention volume slightly larger than that anticipated for the breakthrough of the mobile phase containing the neutral salt. The spurious peak may be minimised, if not eliminated, by operating the system with a small proportion of the neutral salt present in the mobile phase at all times; the concentration should be sufficiently low not to lead to premature elution of the retained components.

Contamination of ion-exchange packings with materials that become irreversibly adsorbed can sometimes cause practical difficulties, particularly when working with samples taken from biological origin. The use of a disposable guard column packed, as described in Chapter 4, with an ion exchanger identical to the main separating column can reduce the frequency with which columns need to be renewed.

ION-PAIR PARTITION CHROMATOGRAPHY

This method may be considered a hybrid of ion-exchange and liquid–liquid partition chromatography. Its development and subsequent application has resulted largely from the research performed at the University of Uppsala by Schill and coworkers. In essence, ionic compounds that would normally be soluble only in aqueous phases are rendered more soluble in organic solvents by the formation of hydrophobic "ion pairs" with an aqueous counter ion. The ion pair is then able to partition between the organic and aqueous layers in much the same manner as a neutral molecule. The approach is attractive as a method of analysis, for the nature of the counter ion may be selected to provide optimum chromatographic selectivity characteristics or, by using a counter ion which possesses a high UV absorbance, provide an otherwise UV-transparent sample with a chromophoric group enabling high-sensitivity detection.

The primary interaction in ion-pair extraction may be written as follows

$$A^{+}_{aq} + B^{-}_{aq} \rightleftarrows AB_{org} \qquad (9.6)$$

where A^+ is the cationic species originating from the sample and B^- the counter-ion which will render the ion pair hydrophobic. As in any other reversible process, the extent of the forward reaction may be expressed quantitatively using the distribution coefficient, K.

$$K = \frac{[AB_{org}]}{[A^{+}_{aq}]\,[B^{-}_{aq}]} \qquad (9.7)$$

The distribution of the sample (represented by A^+ and AB in eqns.9.6 and 9.7) will be governed by the concentration of the counter ion, B^-.

In principle ion-pair partition methods may be performed in either the normal or reversed-phase modes. The most widely reported systems are those which rely on an organic mobile and an aqueous stationary phase. It has proved particularly important in ion-pair work to select a chromatographic support which is as far as possible inert, *i.e.*, will not influence the separation by any adsorption effects. In this respect, ion-pair methods are more critical since, by virtue of their very polar nature, they will tend to be adsorbed from the solution on to the surface of the support. Successful chromatographic systems have been devised with supports of the substituted cellulose type, *e.g.*, ethanolised and silicone treated. Inorganic supports, which are normally preferred because of superior stability at high pressures, must be carefully deactivated to eliminate adsorption effects.

Many counter ions have been found suitable for use in ion-pair chromatography. They can be considered in three main groups, *viz.* (a) anions which render bases, *i.e.*, cations soluble in organic phases, (b) cations which combine with water-soluble anions to yield

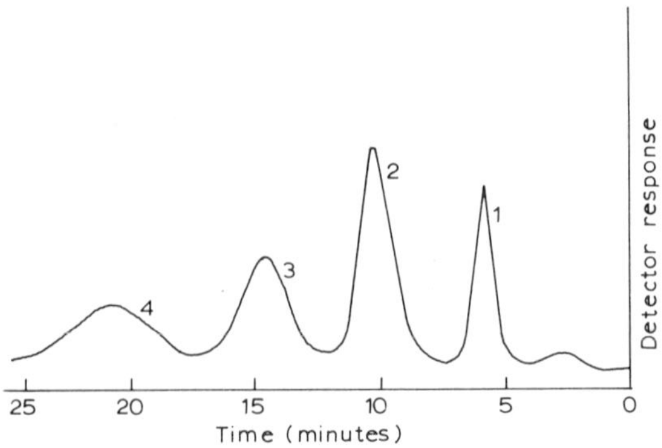

Fig. 9.8. Ion-pair chromatography: separation of amino phenols. Operating conditions: column, 0.3 m × 4 mm I.D.; packing, silicone-treated cellulose; stationary phase, bis-(2-ethylhexyl)-phosphoric acid in chloroform; mobile phase, citrate buffer, pH 3.8. 1 = Epinephrine; 2 = synephrine; 3 = norphenephrine; 4 = *p*-hydroxynorephedrine. (Reproduced from S. Eksborg, P.-O. Lagerström, R. Modin and G. Schill, *J. Chromatogr.*, 83 (1973) 99, with permission.)

an organic solvent-soluble product, and (c) cations or anions with high detector response characteristics, *e.g.*, high UV absorption, which when coupled to the sample in the form of an ion pair will enhance the detection of the components.

Details of these different applications with related experimental procedures can be found in the reported works from the University of Uppsala. (See, for example, refs. 10 and 11.) An example of the first group of counter ions is the use of tetrabutylammonium ions for the extraction of anions, both organic and inorganic, into organic solvents such as chloroform. In an analogous manner, the second group is typified by organic bases and indeed metal ions being rendered soluble in organic solvents containing bis-(2-ethylhexyl)-phosphoric acid. A separation achieved by ion-pair chromatography in the reversed-phase mode is illustrated in Fig. 9.8, where four aminophenols are resolved with a system offering a high degree of selectivity[10].

The third use of ion-pair formation, that of enhancing the detection of a sample by using a highly absorbing counter ion, presents one of the most potentially useful aspects of this technique. This is particularly so in the case of UV detectors. There are some very sensitive, yet moderately priced detectors available, *e.g.*, those using the 254-nm emission line from a low-pressure mercury lamp, that suffer from the inability to detect compounds which do not absorb at that wavelength. Eksborg *et al.*[10] have demonstrated the feasibility of increasing the detectability of anionic samples in ion-pair chromatography by using N,N-dimethylprotriptyline as the counter ion, which results in ion pairs having a high UV absorbance at 254 nm. Molar absorptivity values in the order of $4 \cdot 10^3$ have been claimed.

REFERENCES

1 C.D. Scott and N.E. Lee, *J. Chromatogr.*, 83 (1973) 383.
2 S. Udenfriend, S. Stein, P. Böhlen, W. Dairman, W. Leimgruber and M. Weigele, *Science*, 178 (1972) 871.

3 D.H. Spackman, W.H. Stein and S. Moore, *Anal. Chem.*, 30 (1958) 1190.
4 A.G. Georgiadis and J.W. Coffey, *Anal. Biochem.*, 56 (1973) 121.
5 R.A. Henry, J.A. Schmit and R.C. Williams, *J. Chromatogr. Sci.*, 11 (1973) 358.
6 P.R. Brown, *High-Pressure Liquid Chromatography; Biochemical and Biomedical Applications*, Academic Press, New York, 1972.
7 B.A. Persson and B.L. Karger, *J. Chromatogr. Sci.*, 12 (1974) 521.
8 J.I. Ohms, J. Zec, J.V. Benson and J.A. Patterson, *Anal. Biochem.*, 20 (1967) 51.
9 J.B. Smith, J.A. Mollica, H.K. Govan and I.M. Nunes, *Intern. Lab.*, 2 (1972) 15.
10 S. Eksborg, P.-O. Lagerström, R. Modin and G. Schill, *J. Chromatogr.*, 83 (1973) 99.
11 S. Eksborg, *Acta Pharm. Suecica*, 12 (1975) 19.

Chapter 10

Steric exclusion chromatography

INTRODUCTION

As the name implies, steric exclusion chromatography is the separation of sample components according to differences in their size or shape. This method of separation is unique among LC procedures as it relies entirely on the physical restriction of molecules moving through a packed column rather than on any interactive effect, *e.g.*, adsorption or partition into a stationary phase.

The steric exclusion method can be performed in both aqueous and non-aqueous media. Historically, these practical variations have developed independently and as a consequence are often distinguished by the use of different names. Thus we have gel filtration, which is the separation of samples in aqueous solution using, as the chromatographic packings, porous hydrophilic gels such as agarose and cross-linked dextrans. Gel permeation chromatography (GPC) describes similar separation processes performed in non-aqueous media with hydrophobic gels, the most popular of which are based on polystyrene. The two techniques are sometimes referred to collectively as gel chromatography.

The mode of separation, which is discussed in more detail in the next section, relies on the restricted diffusion imposed by the physical dimensions of the internal pore structure of the chromatographic packings. The examples of column packing already mentioned are organic based and consequently some of these, notably the cross-linked dextrans, can suffer deformation if subjected to pressure in systems where accelerated liquid flow is being sought. Alternative packings based on a rigid inorganic structure, *e.g.*, silica or glass, have been employed to obviate this problem, thus the term "gel chromatography" is strictly inapplicable; the name "steric exclusion" is, on the other hand, all embracing.

RANGE OF APPLICABILITY OF THE METHOD

Steric exclusion methods find wide application in the separation of samples of both high and low molecular weight. The molecular sizes of samples which may be studied in both aqueous and non-aqueous systems extend from a molecular weight of $<10^2$ to approximately 10^7. With such a wide range of applicability, it is easy to see how this technique has proved invaluable to polymer chemists for establishing the molecular weight distribution of a sample and to those illucidating complex biological systems. There are numerous reports in the literature of separations of high-molecular-weight polymers, proteins and enzymes based on steric exclusion methods that would have been very difficult to achieve by other techniques.

The use of steric exclusion chromatography with samples of low, *i.e.*, less than 2000, molecular weight presents an interesting approach to separation problems in that optimisation of the mobile phase composition is not required in many applications. If the sample is readily soluble in the mobile phase, elution of the components within a given volume of effluent is virtually assured. The method may be employed to routinely separate simple mixtures consisting of components of almost any molecular size. This generalisation does presuppose that adequate chromatographic resolution is attainable within a short period of time.

MECHANISM OF SEPARATION

The simplest way to describe the mechanism of separations based on steric exclusion is to adopt a pictorial approach. There are two basic requirements for exclusion chromatographic systems, firstly the solvent used as mobile phase should readily dissolve the sample, secondly the chromatographic packing which is employed should be inert towards the sample, *i.e.*, there should be no adsorption effects, and have a totally porous structure, the diameter of the pores being in the same order of size as the diameter of the sample molecules to be separated. In practice, the sample will usually contain a mixture of components of differing size and the column packing will possess a range of pore diameters.

Prior to introduction of the sample, the situation in the chromatographic column is a packed bed of porous material with the mobile phase flowing through the column. Within the pore structure of the chromatographic packing is held the same liquid which due to poor mass transfer is essentially static, *i.e.*, "stagnant pools of mobile phase", as shown in Fig. 10.1(A). If, for example, a two-component mixture is introduced having one component

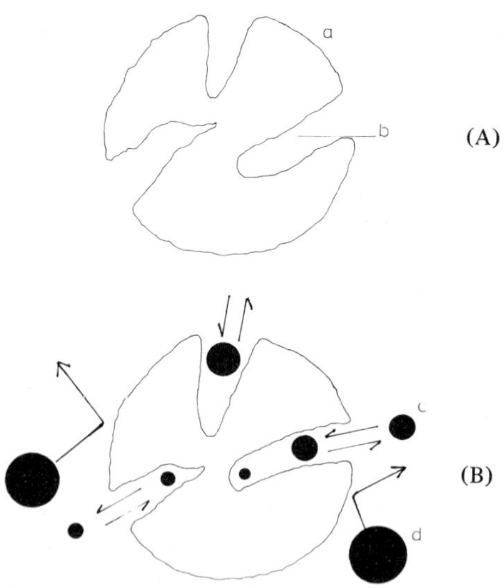

Fig.10.1. Pictorial concept of steric exclusion chromatography. (A) Open pore structure containing "stagnant pools of mobile phase". (a) Freely moving mobile phase; (b) slowly moving/stationary mobile phase. (B) Separation of sample components by their ability to permeate pore structure of column packing material. (c) Smaller molecules permeating the interstitial pores; (d) large excluded molecules.

of high and the other of low molecular size, the larger-diameter species will be unable to penetrate the pores of the column packing material. The volume available to the large molecules during their passage through the column is less than the total volume of liquid held in the column and, if the molecules are totally excluded from the pores in the packing they will elute in the void volume (interstitial volume) of the column. The small molecules of the sample are free to diffuse into the stagnant pools of mobile phase held in the internal pores of the column packing material; consequently, elution of these smaller species from the column will require a larger volume of mobile phase (Fig.10.1(B)).

Several interesting features arise from this concept. Firstly, as the mobile phase and stagnant pools are the same chemical entity, the method could be considered as liquid–liquid partition between two volumes of the same solvent. In this situation the partition coefficient would be unity. In a steric exclusion column, this situation does exist for the smallest molecules, *i.e.*, those which have access to all regions of the column in which mobile phase is effectively stagnant. Sample molecules which satisfy this condition are referred to as completely permeating the column. All molecules small enough to achieve this effect will all possess the same, and necessarily the largest, elution volume from the column being employed.

Secondly, sample molecules of larger size, although still being equally soluble in the mobile and "stagnant" liquid phases are partially or totally excluded from the pores in the chromatographic packing. This physical impediment creates a situation where the partition coefficient appears to be less than unity since the solubility in the "moving" mobile phase is not restricted. This results in these molecules eluting earlier than those which could totally permeate the structure of the chromatographic column; the greater the degree of exclusion from the inner pores of the packing, the earlier will the sample elute.

It has become accepted practice to continue to use the term distribution coefficient for this quantity. In steric exclusion studies this coefficient may be defined as

$$K = \frac{V_{i\,\mathrm{acc}}}{V_i}$$

where V_i is the total pore volume and $V_{i\,\mathrm{acc}}$ is the pore volume accessible to a given size of molecular species.

The retention volume for the same species may be expressed as

$$V_R = V_0 + K \cdot V_i$$

where V_0 is equal to the void or interstitial volume.

The volume contained within a steric exclusion column may be considered as comprising three distinctly different parts, *viz.* the volume occupied by the solid portion of the packing material, the "moving" mobile phase (V_0), which for the columns containing spherical particles represents about 35% of the total volume, and the pore volume (V_i) that is filled with "stagnant" mobile phase, having the same chemical composition as the mobile phase. The relative volumes occupied by the solid matrix and the stagnant mobile phase are a characteristic governing the performance of each individual chromatographic packing material. In a purely exclusion process components can only elute from the column with retention volumes falling between V_0 and ($V_0 + V_i$), thus for maximum separation of sample components it is of considerable practical importance to use columns with a large pore volume and the lowest void volume. As the overall volume of mobile phase in which samples may elute from the column is fixed, it follows that for maximum resolution of components it is most necessary to reduce the broadening of the eluting bands. This is achieved, as in other LC systems, by strict attention to dead volume effects in the injector, column couplings, and detector flow cell.

For the steric exclusion method to be unhindered by the influence of side effects such as adsorption, it has already been stated that the column packing should be inert and the mobile phase should be a good solvent for the sample. A feature of practical importance arising from this situation is that one can be certain that all the sample components injected into a column will elute within a definite volume without needing to optimise or change the mobile phase composition, *e.g.*, by employing gradient elution. No peaks will be retained for a period longer than the time taken to elute components having a distribution coefficient of unity. This characteristic is particularly useful when performing a series of analyses using the same column, as the point after which no further components will elute may be decided with some degree of accuracy without prior knowledge of the nature of the sample.

This simple pictorial explanation of the mechanism of separation in steric exclusion chromatography is adequate for most purposes. In practice, certain anomalies do occur such that a compound elutes in a volume different from that expected, or indeed in a volume larger than the theoretical maximum elution volume for the column system. In most instances these effects can be attributed to some form of interaction between the sample and the matrix of the column packing or to the sample being solvated to an extent different from that anticipated.

In other liquid phase separation methods considerable efforts have been made to improve mass transfer by minimising "stagnant pools of mobile phase". Although generally regarded as deleterious for high-speed LC, diffusion processes occurring within the pore structure of the column are responsible for the separations obtained in steric exclusion chromatography; thus elimination of the possibilities of formation of stagnant pools of mobile phase would eliminate the possibilities of achieving a separation. It follows that in general chromatographic peaks will be broader than those obtained from separations based on other LC methods when performed in an analogous manner. In earlier chapters the influence of extra column band broadening was considered to be most critical with components of low retention, *i.e.*, with capacity factors lower than 2, in steric exclusion work the capacity factors are invariably below $k' = 1$ and consequently considerable attention to detail in the construction of the apparatus is necessary if good resolution of sample components is to be achieved.

COLUMN PACKINGS FOR STERIC EXCLUSION CHROMATOGRAPHY

General considerations

Many different materials have been proposed as column packings for exclusion chromatography, the choice of physical and chemical characteristics being considerable.

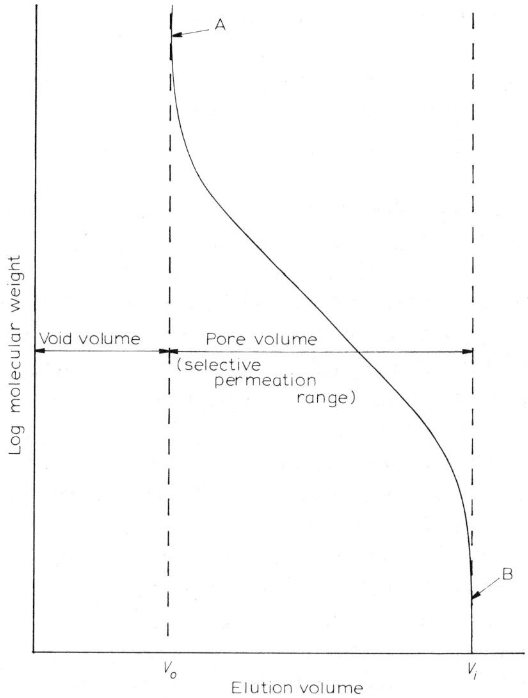

Fig.10.2. Typical plot of molecular weight against elution volume for a steric exclusion chromatographic column. A = Exclusion limit; B = total permeation.

Some explanation of the properties necessary for good performance need to be described to enable the merits of the many commercially available packings to be discussed. In the previous section it was mentioned that the useful working range of column effluent occurs between volume V_0 and $(V_0 + V_i)$. The curve shown in Fig.10.2, which is representative of many exclusion types of columns, indicates the general relationship that exists between the elution volume and the molecular size of the sample component*.

With reference to Fig.10.2, most of the important performance characteristics of a steric exclusion column may be described.

(1) Exclusion limit is the size of sample molecules below which they are able to permeate the gel, *i.e.*, molecules any larger are totally excluded and elute within the void volume of the column.

(2) Total permeation of a column occurs with all molecules smaller than those expected to elute within the total liquid volume of the column $(V_0 + V_i)$. It is only possible to achieve different retention characteristics for samples larger than this value.

(3) The lower the gradient of the centre, selective permeation, portion of the curve, the larger will be the volume of column effluent needed to elute compounds of fairly similar molecular size. This will lead to the best resolution of components over a given range.

*This curve is commonly referred to as a calibration curve and is most often constructed on a log–linear basis.

(4) Interpretation of the size of sample components is made easier if the centre portion of the curve is linear over a wide range.

(5) The difference in the size of molecular species at the exclusion limit and the total permeation limit indicates the range of sample sizes for which the column is applicable.

(6) The maximum value of the capacity factor of a sample eluting from a steric exclusion column is determined by the ratio of the pore volume to the void volume. Since the effectiveness of a column depends only on the pore volume, a packing material offering the highest capacity factors will be the preferred type.

These features of steric exclusion columns indicate their chromatographic performance in terms of ability to resolve sample mixtures and the types of samples which are capable of being separated. If such characteristics are ignored, it is possible to attempt to separate a mixture of components which although differing in size might all be of such a size that they are all excluded or totally permeate the column packing – hence no separation would be possible.

In many applications using steric exclusion chromatography it is accepted practice to link columns together in series. This arrangement is necessary to overcome two practical limitations of existing exclusion columns. Firstly, although all the sample may elute from the column within the volume described as the "selective permeation" range of the calibration curve, the resolution between successive components may be inadequate, *i.e.*, the column does not possess a high enough efficiency. In these circumstances columns of similar exclusion characteristics are connected in series to increase the separating power of the system. An alternative situation requiring coupled columns is where part of the sample is excluded or totally permeates the column. Coupling columns having exclusion properties for larger or smaller molecules, respectively, will obviate this problem. A number of columns, each of different exclusion characteristics, can be coupled to yield a system capable of permeation over a very wide range of molecular size. This approach is attractive in that a given sample may neither totally permeate, nor be totally excluded from the column system as a whole. It is important to realise, however, that if a sample passes through one or several columns in which the components of the sample do not permeate the packing in a selective manner, then these columns merely add to the total void volume of the chromatographic system – leading to excessive band broadening, loss of resolution, and increased separation times.

An experienced operator will often be able to recognise the conditions of total exclusion or total permeation of a sample by the general shape of the chromatographic trace. The steric exclusion curve depicted in Fig. 10.3 illustrates the somewhat larger, sharper "false" peaks at either end of the chromatogram resulting from the column having too narrow a permeation range for the sample being studied. Indubitably, the best way of deciding if the permeation range is sufficient is to couple an additional column to extend the permeation range and re-examine the sample.

Total exclusion or permeation of a sample from a column can be positively identified from the chromatogram by reference to the calibration curve for the system, if one is available or has been constructed. It is important to realise that steric exclusion is based on molecular size and not molecular weight, thus great care should be taken when attempting to correlate exclusion limits for chemically different samples from a calibration curve constructed with the aid of reference materials, of say polystyrene, which may have quite

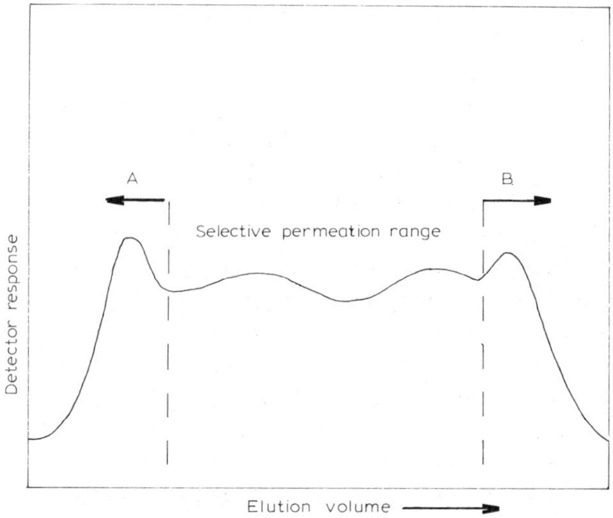

Fig.10.3. Chromatographic artifacts due to the use of a column packing with too narrow an exclusion range. A = Material totally excluded from the gel (molecular weight too high for the column used); B = material completely permeating the gel (molecular weight too low for the column used).

different size–weight characteristics. This latter problem has been overcome by Grubisic *et al.*[1], who have developed a procedure for molecular weight determinations from steric exclusion data by the additional correlation with the intrinsic viscosity of the column effluent; the relationship and method are described more fully in later sections dealing with the determination of molecular weights.

In addition to these properties of the chromatographic column system, there are the physical properties of the packings which govern the choice of operating conditions. Amongst these are included the physical strength of the packing, solvent compatibility, and particle size.

Physical properties

Soft gels

A great deal of steric exclusion chromatography has been performed using column packings that are relatively soft and are swollen by the action of the carrier solvent. Because of this characteristic, it is important to swell the packing material in the solvent to be employed as the mobile phase, ideally by an overnight soak in excess solvent, to ensure that no further change in volume occurs after the column has been packed. Some of the most widely used materials of this type are cross-linked dextrans and agarose gels, which are primarily intended for use with aqueous carriers. These swollen "gels" are easily compressed, thus it is not always possible to apply a pressurised liquid flow to the system in order to accelerate the speed of analysis. This is particularly the case with gels having high exclusion limits, *i.e.*, in the molecular weight range of 10^5–10^7, where even

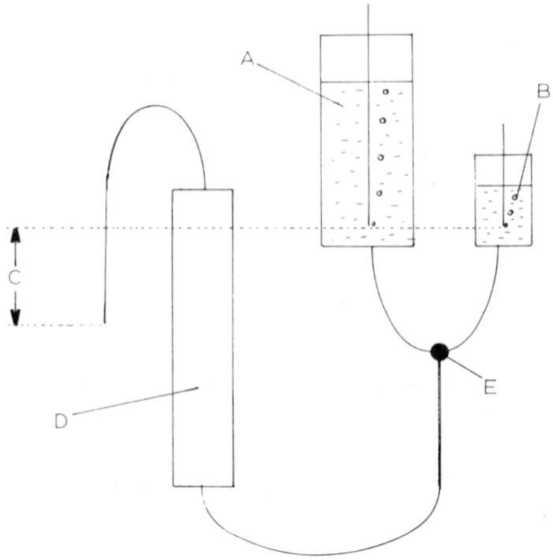

Fig.10.4. Siphon for control of liquid flow in columns containing compressible gels. A = Marriotte flask with eluent; B = Mariotte flask with sample; C = operating pressure; D = soft gel packing; E = two-way valve. (Reproduced from L. Fischer, *An Introduction to Gel Chromatography*, North-Holland, Amsterdam, 1971, p. 249, with permission.)

the pressure exerted by a head of liquid above the column packing is sufficient to compress the gel bed, leading to a change in the permeability and the pore dimensions. In these cases, liquid flow is achieved by employing a special siphon arrangement on the column outlet, as shown in Fig. 10.4. These gels probably represent the most extreme example of fragile column packings that have been used in LC. Despite this obvious operating problem, these materials have proved of considerable value for the separation of high-molecular-weight, water-soluble species of biological interest, *e.g.*, proteins and enzymes. The aims of modern LC require that we attempt to replace these softer gels with more rigid materials which offer similar separation capabilities. Currently, this situation is far from being realised in practice and for completeness technical descriptions of a number of commercially available "soft" gels are presented in Table 10.1. If columns packed with soft gels must be used in modern high-pressure apparatus, it is good practice to fit a pre-column filled with some rigid support ahead of the sample introduction point. This pre-column will moderate any possible flow fluctuations from the pump.

Soft packings for steric exclusion work generally possess a wide distribution of pore dimensions within a given type of packing. From this it follows that the molecular weight range of samples that may be studied with a given packing material is quite large, *i.e.*, the slope of the selective permeation (centre) part of the calibration curve in Fig. 10.2 is fairly steep. It should be remembered that soft gels swell when in contact with a compatible solvent; consequently the pore dimensions are increased. Subsequently, when the column is packed, the resultant pore volume, V_i, is large relative to the interstitial (void) volume, V_0, *i.e.*, yielding a column of fairly high capacity.

An important practical problem that can occur with some soft gels, notably the cross-linked dextrans, is that they can be attacked by bacteria or fungi which will degrade the chromatographic performance and may also infect the collected fractions. Columns are most prone to damage in this manner when they are left unattended rather than in opera-

TABLE 10.1

SOFT-TYPE GELS FOR EXCLUSION CHROMATOGRAPHY

Each type described represents a series of packings, each having different pore diameters, hence exclusion characteristics. The data given represent the mol. wt. range for the whole series.

Name	Type	Useful mol.wt. range*	Method**	Supplier
Bio-Beads S	Styrene–divinylbenzene copolymer	up to 1.4×10^4	GPC	Bio-Rad
Bio-Gel A	Agarose	2.0×10^4–1.5×10^8	GFC	Bio-Rad
Bio-Gel P	Polyacrylamide	10^2–4.0×10^5	GFC	Bio-Rad
Merckogel*** OR	Polyvinyl acetate	up to 10^6	GPC	Merck
Sephadex G	Dextran (cross-linked)	up to 8.0×10^5	GFC	Pharmacia
Sephadex LH-20	Derivatised dextran	10^2–4.0×10^3	GPC	Pharmacia
Sepharose	Agarose	3.0×10^5–2.5×10^7	GFC	Pharmacia

*Approximate range determined with polystyrene standards for GPC and with globular proteins for GFC.
**Abbreviations: GFC = gel filtration chromatography (aqueous solvents); GPC = gel permeation chromatography (organic solvents).
***Merckogel is sold under the name EM Gel in the U.S.A.

tion. A number of chemicals have been used to minimise this problem; these include chloroform, cresol, formalin, and sodium azide. Each of these chemicals are effective but can suffer from several disadvantages, for instance, cresols and formalin will cause proteins to precipitate irreversibly, sodium azide possesses a rather high UV absorbance at wavelengths commonly used in LC, and chloroform is capable of being sorbed by any plastic components in the system. (The use of plastics is quite common in certain low-pressure, aqueous systems.) Sodium azide in concentrations in the region of 0.02% has been found to give satisfactory protection to chromatographic columns of the type described. Further details of other chemicals suitable for the sterilisation of chromatographic columns have been described by Fischer[2].

Semi-rigid packings

Clearly, it is not an easy matter to sharply divide column packings into the categories "soft" and "semi-rigid". This latter term is, however, normally reserved for materials which may be employed in chromatographic systems designed to operate at elevated pressures, as distinct from gravity-fed solvent supply systems.

A fairly wide range of packings are available in this category, most of which are intended for use in organic rather than aqueous media. Packings based on polystyrene, *i.e.*, styrene–divinylbenzene copolymers similar to the base matrix of many earlier types of ion-exchange resins, have enjoyed considerable success in this area of work as highly cross-linked varieties are capable of withstanding liquid pressures of several hundred bars. The use of other materials, such as polyvinyl acetate, has been much less widely studied.

In general, semi-rigid packings exhibit a high column permeability, thus it is possible to operate at a relatively high linear velocity, *e.g.*, 10 mm/sec, of the mobile phase without fear of seriously deforming the column packing. Materials are available with different exclusion limits so that an optimum grade may be selected for a given application. Chromatographic columns based on polystyrene are generally compatible with most

TABLE 10.2

SEMI-RIGID GELS FOR EXCLUSION CHROMATOGRAPHY

Name	Type	Useful mol.wt. range*	Method	Supplier
Aquapak	Styrene divinylbenzene rendered water compatible		GFC	Waters
Ar Gel**	Styrene divinylbenzene	up to 2.0×10^7	GPC	Applied Research Labs.
Hydrogel	As for Aquapak		GFC	Waters
Poragel A	Styrene divinylbenzene	10^2–2.0×10^4	GPC	Waters
Styragel***	Styrene divinylbenzene	10^2–5.0×10^8	GPC	Waters

*The molecular weight is based on measurements with polystyrene.
**Ar Gel is not available in the U.S.A.
***Styragel is also available in a small-particle version called μStyragel.

organic solvents, excepting the lower alcohols and solvents which may contain water; the most popular solvents used are tetrahydrofuran and chloroform, because of their excellent solvent characteristics. Resins formed from polyvinyl acetate, on the other hand, are compatible with alcohols and in this respect may be considered as complementary to the polystyrenes. There are only a limited number of semi-rigid column packings for steric exclusion chromatography which may be used with aqueous solvents. The principle feature of all packings which limits their solvent compatibility is that the mobile phase must "wet" the surface of the column packing. If this does not occur, interfacial surface tension prevents molecules from permeating into the inner pore structure of the matrix. As most of the semi-rigid packings described here are of the polystyrene type, the reason for their non-compatibility with water should be evident. Introduction of ionic sites on the resin will yield a water-compatible packing material, *viz*. an ion exchanger. Since for true steric exclusion behaviour it is important to eliminate any sample–support interactions, this procedure can only be considered for completely non-ionic water-soluble samples. The converse situation, *i.e.*, steric effects occurring during an analysis based on an ion-exchange mechanism, is one of the reasons why the behaviour of sample components in some ion-exchange separations is sometimes hard to predict.

Table 10.2 gives details of a number of semi-rigid packings for steric exclusion chromatography that are currently available commercially. Like the soft gels, these materials can be produced with a broad spectrum of pore dimensions leading to a fairly wide usable size range for each grade of packing. These more rigid resins, being highly cross-linked, do not swell to the same extent as the soft resins in the presence of the carrier liquids. In these circumstances, the maximum capacity, *i.e.*, V_i/V_0, is significantly lower. The nature of each of the packing materials varies considerably with the method of manufacture, particle size, etc., but in general the semi-rigid, polystyrene resins offer some of the most efficient columns available for steric exclusion chromatography at the present time.

Rigid packings

These chromatographic materials are invariably based on inorganic supports such as glass or silica that have been manufactured with precisely controlled internal pore dimensions. Generally, highly efficient columns may be readily prepared using these packings in much the same manner as chromatographic supports for adsorption and partition chromatography, *i.e.*, particles with a diameter larger than about 30 μm are dry packed, slurry packing being employed for the material of smaller diameter. These columns have a high permeability so that, in general, a high rate of liquid flow is achieved with a modest inlet pressure, yet the bed of packing is sufficiently rigid to be able to withstand high pressures where the particle diameter or the desired mobile phase velocity makes it mandatory.

The distribution of pore size within the inorganic packings is often much narrower than in the organic resins. This feature leads to a rather narrow molecular size range for selective permeation, while at the same time offering high resolution within a given, fairly narrow, size range.

Practically all solvents are capable of wetting a glass or silica surface, thus it could be assumed that inorganic steric exclusion packings offer the greatest versatility with regard to solvent compatibility. Although wetting of the surface and the possibility of performing some type of separation based on exclusion is quite feasible, adsorption on the surface of the support can prove of considerable difficulty. In aqueous solutions a number of proteins have been found to be strongly adsorbed, eliminating the possibility of a separation based solely on an exclusion mechanism.

In some work with aqueous solvents, the addition of non-ionic detergents to the mobile phase has improved the situation. Alternatively adsorption of ionic substances can be reduced by judicious selection of the mobile phase pH and ionic strength. It should be remembered, however, that glasses and silicas will dissolve in alkaline media, especially at elevated temperatures.

When exclusion is performed with organic solvents deactivation of the column packing, for example by silanisation, can reduce the problem of adsorption, but as yet there does not appear to be a perfectly satisfactory method. A paper by several Russian authors, Eltekov *et al.*[3], described the use of γ-aminopropyltriethylsiloxane as a deactivating agent for silica surfaces that would subsequently enable silica-based packings to be used for the chromatography of proteins. Little appears to have been published subsequently to substantiate what would appear to be an interesting approach.

While working with samples of biological origin it is sometimes advantageous to be able to operate under sterile conditions. Porous glasses are particularly attractive in these circumstances as they may be sterilised by either thermal or chemical methods without fear of degrading the column packing material.

A summary of the technical details of inorganic chromatographic packings suitable for steric exclusion studies is given in Table 10.3. It is considered most probable that this facet of LC column packing technology will advance considerably in the near future, the most likely advances being made by the introduction of very small, *e.g.*, 5-μm-diameter, totally porous packings with a range of pore dimensions which cover the entire spectrum of molecular size. It is conceivable that the use of products with less problems of adsorption

TABLE 10.3

RIGID PACKINGS FOR EXCLUSION CHROMATOGRAPHY

All silica or glass packings tend to adsorb sensitive samples, particularly packings of small-pore diameter, *i.e.*, high surface area. Most suppliers offer chemically deactivated varieties which reduce, but rarely eliminate adsorption.

Name	Type	Exclusion limit (mol. wt.)	Supplier
Bio-Glass	Porous glass	5.0×10^6	Bio-Rad
CPG-10	Controlled porosity glass	2.0×10^7	Electro-Nucleonics
LiChrospher	Porous silica spheres	$>2.0 \times 10^6$	Merck*
Porasil	Porous silica spheres	$>4.0 \times 10^6$	Waters
μPorasil	Porous silica	$>4.0 \times 10^6$	Waters
Spherosil	As for Porasil	$>4.0 \times 10^6$	Rhone-Progil
Vit-X	Porous glass	1.2×10^6	Perkin-Elmer

*E.M. Labs. in the U.S.A.

will be possible, either from the use of different starting materials to produce the support or by chemically deactivating the surface. An illustration of possible high-speed separations is given by Fig. 10.5, which shows the resolution of several polystyrene molecular weight standards using a column packed with porous silica microspheres of 5 μm diameter.

CHOICE OF MOBILE PHASES FOR STERIC EXCLUSION CHROMATOGRAPHY

In a number of places in this chapter it has already been indicated that the composition of the mobile phase is not selected to optimise interactions with the surface of the chromatographic support or a stationary phase thereon; rather, if true steric exclusion is sought it is most important to eliminate such interactions. The mobile phase in steric

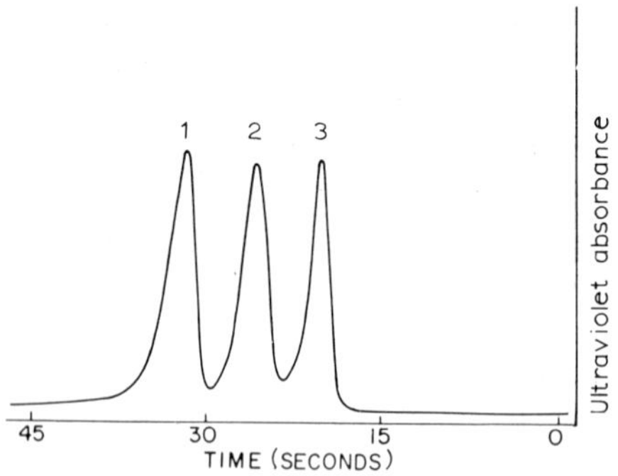

Fig.10.5. Fast size separation of polystyrene standards. Operating conditions: column, 0.25 m × 2.1 mm I.D.; packing, porous silica microspheres, diameter 5–6 μm, pore diameter, 350 Å; mobile phase, tetrahydrofuran; inlet pressure, 109 bars (1625 p.s.i.); flow-rate, 1 ml/min. (1) Molecular weight 2030; (2) molecular weight 51,000; (3) molecular weight 411,000. (Reproduced from J.J. Kirkland, *J. Chromatogr. Sci.*, 10 (1972) 593, with permission.)

exclusion chromatography is selected on the basis of the following requirements:

(1) It must be a good solvent for the sample, either at ambient temperature or at the temperature at which the separation is to be performed. It is important to note that for a number of applications of steric exclusion to the characterisation of high polymers, sufficient solubility of the sample in the carrier liquid is only achieved at elevated temperatures. The viscosity of polymer solutions is also reduced significantly by working above ambient temperature, provided the sample and column packing material are capable of withstanding the temperature.

(2) The mobile phase should not be reactive towards the column packing material, yet must be capable of "wetting" its surface, so that samples will permeate freely within the pores of the packing by diffusion processes alone. In the case of "soft" gels, the resin will be swollen appreciably by the carrier liquid, the degree of swelling being a function of the packing–solvent combination. This situation leads to very effective steric exclusion chromatography, but the swollen gel is rather fragile, severely limiting the liquid flow-rate and pressures that may be employed.

(3) Depending on the nature of the mobile phase and the samples being studied, a certain degree of solvation of the sample will occur. It is fairly common, particularly in the case of the small molecular species, that a change in the carrier solvent will lead to an apparent change in the elution volume suggesting a different molecular size. This is most often attributable to different degrees of solvation by the mobile phases concerned and also to changes in the pore dimensions due to the column packing swelling to a different extent.

A similar phenomenon occurs with samples which tend to associate, *e.g.*, dimerise, in some solvents while remaining monomeric in others.

(4) Another very important consideration of the mobile phase is that it should be compatible with the detectors likely to be employed. It is unfortunate that some of the best solvents for steric exclusion chromatography of organic polymers must rate as some of the poorest for detector compatibility. For instance, toluene, tetrahydrofuran and halogenated benzenes are widely used because of their excellent solvent characteristics, the latter particularly in high-temperature work. All three solvents are extremely flammable, of dubious toxicity and present difficulties when working with UV photometric detectors. Tetrahydrofuran, if pure, is transparent at wavelengths longer than 230 nm, however, in practice strongly UV absorbing stabilisers are added by most suppliers to improve the storage characteristics and minimise the formation of explosive peroxides. Typical stabilisers are hydroquinone and butylated hydroxytoluene. The use of tetrahydrofuran with these stabilisers present is not recommended from the point of view of detection by either a UV or moving wire detector. Similarly any slight leakage of solvent from the system will, on evaporation, leave a crystalline residue which can lead to minor, yet irritating, problems such as crystals forming over windows of the detector cells and blockage of fine capillary tubes.

Removal of the stabilisers from tetrahydrofuran is best effected by careful distillation*.

---

*This is by no means a task for inexperienced personnel as considerable care should be taken to avoid the formation of explosive peroxides. Having distilled the solvent, it should be stored in the dark, ideally under nitrogen, and checked routinely for the presence of peroxides.

TABLE 10.4

SOLVENT COMPATIBILITY OF PACKINGS FOR STERIC EXCLUSION CHROMATOGRAPHY

Column type	Compatible solvents
Agarose	Water and salt solutions in pH range 4–9
Derivatised dextran	Water and organic solvents pH above 2
Dextran	Water above pH 2
Glass	All solvents (avoid strong alkalis). Adsorption effects can be reduced by chemically treating surface, but then incompatible with water
Polyacrylamide	Water and salt solutions in pH range 1–10
Polystyrene (styrene divinylbenzene)	Organic solvents only. Avoid water and lower alcohols
Polyvinyl acetate	Organic solvents only
Silica	As for glass

A guide to the choice of solvents which have been shown to give acceptable results with the various types of column packings for steric exclusion is given in Table 10.4. As an empirical guide, the complications due to adsorptive effects will be minimised if the solvent employed as mobile phase resembles the structure of the column packings, for example, toluene is a very effective solvent for separations involving packings based on styrene–divinylbenzene. Similarly, water, being a hydroxylic solvent, is the most suitable solvent for work with soft gels based on a carbohydrate structure, *e.g.*, dextrans, which possess many hydroxyl groups.

GENERAL SCOPE OF STERIC EXCLUSION CHROMATOGRAPHY

Relative merits of the method

Unlike other separation methods in LC, the exclusion technique is applicable to samples of any molecular species, whether it be large or small, provided simply that it will dissolve in true solution in a suitable solvent. These very general restrictions indicate the considerable potential that exists for the exclusion method. Table 10.5 details some of the substances that have been studied, clearly showing the wide applicability of the technique. Some of the solvents indicated in Table 10.5, specifically cresol and di- and trichlorobenzene, are used invariably at elevated temperature, often above 100°C, to ensure adequate solubility of the polymer sample. Operation under these conditions should only be considered when absolutely necessary, since most equipment, particularly the modular, self-assembled, type is normally quite unsuitable. Common problems which occur include precipitation of polymers in cooler parts of the apparatus leading to blocked capillary tubing, unstable detectors, potential hazards from fire and harmful solvent vapours, and extended equilibration times. Several commercial instruments are available which are designed to minimise these problems and should be seriously considered if this type of application is to be performed routinely. In most applications, less noxious solvents, such as tetrahydrofuran and chloroform, serve adequately as mobile phases for exclusion chromatography.

TABLE 10.5

SOME SUBSTANCES THAT HAVE BEEN STUDIED BY EXCLUSION CHROMATOGRAPHY (GPC AND GFC)

(Reprinted from *J. Chem. Educ.*, 47 (1970) A461, A505, with permission.)
Abbreviations: X = studied by GPC in this solvent; U = usually soluble in this solvent; N = usually insoluble in this solvent.

Substances fractionated by GPC	*o*-Dichloro-benzene	Benzene or toluene	Methylene chloride	Tetrahydro-furan	Chloro-form	Dimethyl-formamide	*m*-Cresol	1,2,4-Trichloro-benzene	Water
Acenaphthylene–MMA copolymer				X		X			N
Acenaphthylene–styrene acrylics				X		X			N
Acrylic styrene–butadiene				X		X			N
Acrylates		X		X		X			N
Acrylonitrile–butadiene rubber	U	X						X	N
Alkyd resins		X		X					
Antioxidants for polymers				X					
Asphalts★,★★	X	X	X	X	X	X	X	X	N
Polybutene-1	X	X	N	X★	N			X	N
Butyl rubber	X	X		X				X	N
Carbowaxes	X			X		X		X	
Cellulose acetate				X		X			
Cellulose nitrate				X		X			N
Butadiene, *cis*-polymer	X	X		X				X	N
Coal tar pitch★★	X	X	X	X	X	X	X	X	N
Dextrans									X
Dialkyl phthalates	X	X		X				X	N
Dimethyl polysiloxanes	X	X	X	X	U			X	N
Drying oils				X					
Epichlorohydrin								X	
Epoxy resins, uncured		X		X					
Ethyl acrylate polymers	X	X		X		X			N
Ethylene–vinyl acetate copolymer	X							X	N
Ethylene–propylene copolymer	X							X	N
Fatty acids and derivatives	X	X		X				X	
Furfuryl alcohol	X			X				X	
Glycerides	X			X				X	

(*Continued on p. 206*)

TABLE 10.5 *(continued)*

Substances fractionated by GPC	*o*-Dichloro-benzene	Benzene or toluene	Methylene chloride	Tetrahydro-furan	Chloro-form	Dimethyl-formamide	*m*-Cresol	1,2,4-Trichloro-benzene	Water
Isocyanates				X		X			
Lexan (see Polycarbonates)									
Lignin sulphonates									X
Lipids			X	X					
Lubricating oils	X	X	X	X	X	X		X	N
Melamines	X			X				X	N
Methacrylates		X		X		X			N
Methyl methacrylate–styrene copolymer	X	X		X		X		U	N
Mineral oil	X	X		X				X	N
Neoprene (see Rubber, neoprene)									
Non-ionic surfactants				X					
Nylons (4, 6, 66, etc.)							X		N
Phenolic resins				X					N
Phenol formaldehyde				X			X		X
Plasticizers, various	X	X	X	X		X	X	X	
Polyalkylene glycols	X	X	X	X		X	X	X	
Polybutadiene	X	X		X				X	N
Polycaprolactam							X		
Polycarbonates	X		X	X				X	N
Polyelectrolytes				X					X
Polyesters, non-linear and unsaturated	X			X			X	X	N
Polyethers		X				X			N
Polyethylene, branched		N	N	N	N	N		X	N
Polyethylene, linear		N	N	N	N	N		X	N
Polyethylene oxide						X			
Polyethylene terephthalate				N			X		N
Polyisobutylene		X							N
Polyisobutylene copolymers		X							N
Polyisoprene		X							N
Polyols				X		X			
Polynuclear aromatics	X	X	X	X		X		X	N

Polyphenylene oxide								X	N
Polypropylene								X	N
Polystyrene	X	X	X	X		X	X	X	N
Polysulphonates	X	X		X				X	
Polysulphones				X					
Polyurethanes			X			X			N
Polyvinyl acetate	X			X		X	X		
Polyvinyl acetate copolymers	X			X				X	
Polyvinyl alcohol									X
Polyvinyl butyral				X		X			
Polyvinyl chloride		X		X				X	N
Polyvinyl fluoride						X			
Polyvinyl methyl ether				X		X			
Propylene–(butene-1) copolymers	X	X						X	N
Rubber, acrylonitrile–butadiene		X						X	N
Rubber, butyl	X	X		X				X	N
Rubber, natural	X	X						U	N
Rubber, neoprene		X							N
Rubber, styrene–butadiene	X	X		X				X	N
Silicones	X	X		X	X			X	
Styrene–acrylonitrile copolymer				X		X			
Styrene–isoprene copolymer				X					
Trifluorostyrene				X					
Urethane prepolymers				X		X			
UV stabilizers for polymers				X					
Waxes (hydrocarbon)		X	X	X		X	X	X	
Vinyl chloride–vinyl acetate–maleic acid terpolymer						X			

*Less than 20,000 molecular weight only.

**Only partially soluble in all solvents.

With samples with molecular weights greater than 2000, steric exclusion becomes the method of choice for the separation of one species from another on a relatively non-specific basis, *i.e.*, independent of the chemical nature of the sample. In specialised applications other techniques are complementary, *e.g.*, isoelectric focussing and affinity chromatography for biological samples such as proteins and enzymes. High-speed centrifugation is also applicable for the separation of mixtures of large molecules. Steric exclusion has the advantage of requiring relatively simple apparatus and not subjecting the sample to any treatment more severe than dissolving and flowing through a column packed with an inert, porous packing.

Samples having molecular weights less than about 2000 are still directly amenable to exclusion methods, but in this case retentive chromatographic procedures such as ion exchange, adsorption and partition often give a higher resolution of components within a shorter time period, albeit not giving a separation based on size discrimination. With regard to the speed of separation, this situation could change significantly within the next few years as highly refined packing materials are introduced. In the lower molecular weight range the "sizes" of the samples are often modified considerably by the influences of solvation and molecular association in some solvents, leading to a separation which might be somewhat unexpected on the basis of the molecular weight of the components.

Applications

Determination of the molecular weight distribution of polymers

Since the steric exclusion method relies on the discrimination of sample components due to differences in their molecular size, it forms the basis of an indirect method for determining the molecular weight of samples. Perhaps of greater importance is that the method serves as a means to determine the distribution of molecular size or weights in a heterogeneous or polydispersed sample. The name gel permeation chromatography, which as described earlier is one of several exclusion methods, has become synonymous with molecular weight determination of polymers.

The all-important feature necessary when establishing such a method is to ensure that the exclusion column system is properly calibrated. Fig. 10.2 illustrated the general relationship between elution volume and the logarithm of the molecular weight. This relationship is valid only for compounds of similar chemical type, for only then is there a constant relationship between molecular weight and size. Direct calibration of exclusion columns using, for example, a series of essentially monodisperse polystyrene standards of known molecular weight is, without doubt, the ideal method when wishing to characterise polydispersed polystyrene samples. A range of fairly well characterised polymer samples are available commercially, a list of the principal suppliers will be found in Appendix 5.

Unfortunately, in many applications no reference standards exist which are of the same chemical type as the sample being studied. Complications created by this situation may be overcome, however, by applying a universal calibration procedure which is applicable to all chemical types. In this method, use is made of the observation that elution volumes of polymer samples in steric exclusion systems may be correlated with their hydrodynamic volumes. This latter value is determined by viscosity measurements. It is found that the hydrodynamic volume, which is calculated by multiplying the molecular weight of a substance by its intrinsic viscosity, is a universal parameter, independent of the chemical nature and the shape of the molecule. A single calibration curve is thus obtained when the logarithms of the hydrodynamic volumes are plotted against elution volumes observed on steric exclusion columns. The reason for this universality of the plot is that the value obtained for the intrinsic viscosity measurement takes into account geometrical effects and solution effects such as solvation of the sample.

Having carefully calibrated a series of columns for their molecular weight characteristics, it is clearly of importance to maintain the calibration over a long period. This is best achieved by keeping the temperature constant and not changing the mobile phase to a different liquid which might swell or shrink the gel. The use of an internal standard is one method of overcoming minor changes in the calibration characteristics of an exclusion column or operating technique. One proposed method[4] that may be applied to UV-transparent samples is to add a very low concentration of known polystyrene standards to the sample solution prior to analysis. By using two detectors connected in series, a UV absorbance and the normal differential refractive index, a dual trace is obtained, *viz.* the UV output corresponding to the added polystyrene standards and the refractive index trace due to the sample. The relative sensitivities of the two detectors are such that only a very small proportion of polystyrene needs to be added and this is not detected by the refractometer. Separations with a ratio of sample concentration to standard of 1:0.003 have been demonstrated.

When polydispersed samples are studied by steric exclusion chromatography, the recorded trace obtained is most often a broad distribution rather than a series of discrete peaks as obtained in other forms of LC. Simplified diagrams illustrating typical results are shown in Fig.10.6. The result from a very simple polymerised sample would resemble curve A. In practice, curves of type B are more generally observed, *i.e.*, a broad distribution profile indicating the presence of a wide range of molecular species. Occasionally, results in the form of curve C are observed where one can clearly distinguish the polydisperse nature of the sample.

If qualitative differences between batches of essentially the same polymeric sample are to be monitored, differential exclusion chromatography can be used to advantage[5]. In this method the mobile phase is a dilute solution of the reference grade sample in, say, tetrahydrofuran or chloroform. This solution is pumped through the entire chromatographic system including columns and detectors as if it were the normal mobile phase. Samples to be compared against the reference material are dissolved in the same solvent, at the same concentration as the reference and then injected into the chromatograph. When the test sample and reference material are identical, a straight baseline is obtained, however, minor differences in composition are clearly discernable. This method enables the differences in samples to be observed and is therefore of considerable value in quality assurance testing. When performing this and other exclusion methods it is vital that the solvent used to dissolve the sample is exactly the same as that used in the mobile phase. A common error, particularly when using tetrahydrofuran, is to dissolve the sample in "stabiliser-free" solvent while operating with stabilised solvent as the mobile phase. The action leads to a "vacancy" effect, where a negative peak is observed at the position where the stabiliser would have eluted if present in the sample. Similar effects can be observed when using

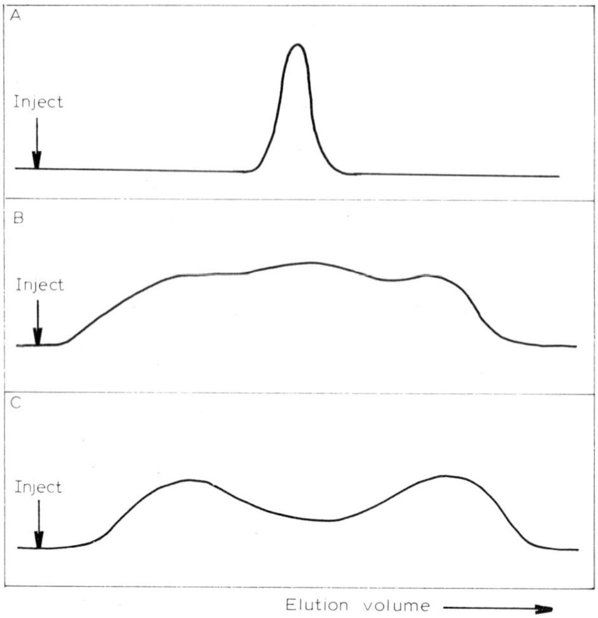

Fig.10.6. Some typical exclusion (GPC) curves of polymer samples. (A) Simple monodisperse polymer; (B) sample containing broad distribution of molecular sizes; (C) sample clearly possessing fractions of different molecular size.

refractometric detection at high sensitivity due to the solvent injected being saturated with air while the mobile phase has been thoroughly degassed.

A major task is involved in quantitatively determining the molecular weight distribution of a sample from such a steric exclusion chromatogram. Several points in the interpretation of the chromatogram need considerable attention to detail if the end result is to be meaningful. Firstly, the detector response must be corrected to take into account any selectivity in its response characteristics towards different components in the sample. This step is simplified if a non-selective detector is employed; to this end a great deal of use has been made of differential refractive index detectors, which, for samples of comparable chemical type, are reasonably non-selective.

The assignment of a particular detector response, *i.e.*, a given fraction of the sample, to a certain molecular weight is only possible in systems which are capable of repeating elution characteristics of a sample to a high degree of precision. Since the elution volume is related to the logarithm of the molecular size, a small error in the volume measurement will cause a large error in the calculated molecular weight. Many of the pumping systems used in early work were unable to deliver the mobile phase with the required high precision of flow. For this reason, it has become established practice to fit a liquid flow monitor after the detection system. The most popular device has been the simple liquid siphon in which the column effluent collects and which, when full, automatically discharges the liquid into a waste container. The action of discharging the liquid is sensed as the meniscus passes a photocell/lamp assembly and records the event as a spike on the chromatographic trace. Correlation of the eluted peaks on a chromatogram with elution volume can subsequently be made by summing the number of spikes marked on the chart from the point of injection to the peak, where the distance between successive spikes represents the liquid volume held in the siphon, typically 1 or 5 ml. The reproducibility of these siphon counting devices is in the order of 1% and although better than some of the earlier pumping systems represents one of the limiting features of this approach. Recently, the development of sophisticated pumping systems with the possibility of feed-back control of the output of the pump has resulted in systems in which the flow-rate through the column system may be precisely controlled, making interpretation of molecular weight of a fraction based on elution time, *i.e.*, length along the recorder chart, more reliable.

Before accurate interpretation of the distribution of molecular weight can be made, it is important to correct the chromatograms for dispersion or peak broadening which does not arise from permeation processes. This effect is best explained by example; if a genuine monodisperse substance was passed through a column system, the substance would elute as a peak, *i.e.*, it would have a finite width; if one relates this observation to a calibration curve of elution volume *versus* molecular weight, it would suggest that the substance contained a distribution of molecular components. This band broadening is due to inefficiencies in the chromatographic system, typically arising in the column connections and the injection and detection systems.

Correction of data for dispersion before correlating elution volumes to molecular weights with the aid of a calibration curve is a particularly involved mathematical process. If the components in the sample are incompletely resolved, as illustrated in the chromatograms in Fig. 10.6, considerable additional manipulation of the data is required. For the most accurate assignment of a molecular weight distribution of a sample the amount of calculation involved becomes formidable if manual processing of the data is considered. Solution of this problem has been achieved, however, by employing Fourier transform and computer techniques, the details of which are considered beyond the scope of this text. More detailed procedures for the complete characterisation of the molecular weight distribution of a polymer sample can be found in the publications of Tung (*e.g.*, ref. 6) and Hess and Kratz[7].

Separation of species of high molecular weight

Another very important usage of the exclusion technique is the separation of large molecular species on either the analytical or preparative scale. This is particularly the case in the isolation of substances from biological fluids. Separations in this field are either concerned with the removal of very much smaller molecules, *e.g.*, salt from a larger species, or the separation of a number of complex molecules from one another. Separations of these types have almost exclusively been achieved using soft gel types of chromatographic packings and aqueous solvents, *i.e.*, gel filtration. Although some applications are indicated in this text, without doubt the most comprehensive guide to specific applications can be obtained from suppliers of chromatographic packings and equipment. For the use of soft, water-compatible gels in the case of separations involving samples of biological origin, the technical information available from Pharmacia Fine Chemicals (Uppsala, Sweden) is to be recommended.

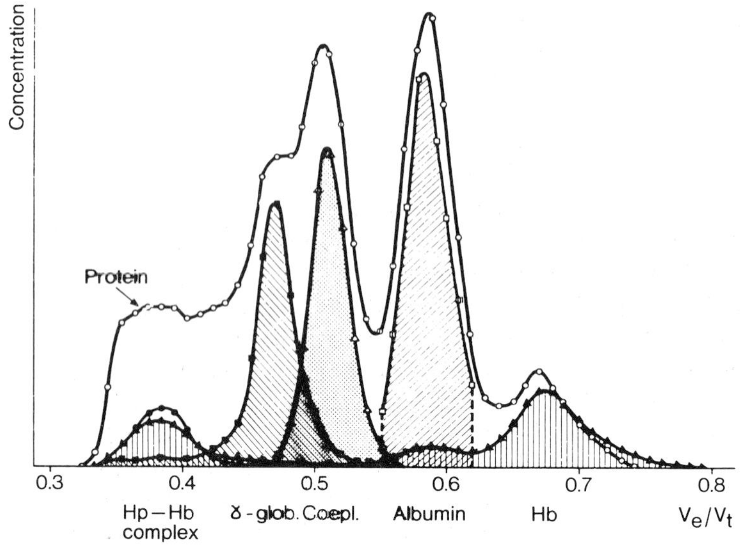

Fig.10.7. Separation of human serum proteins and haemoglobin by exclusion chromatography. Operating conditions: column, 0.735 m × 42 mm I.D.; packing, Sephadex G-200; mobile phase, 0.1 *M* Tris–HCl buffer and 1 *M* sodium chloride; flow-rate, 10–20 ml/h. (Reproduced from J. Killander, *Biochim. Biophys. Acta*, 93 (1964) 1, with permission.)

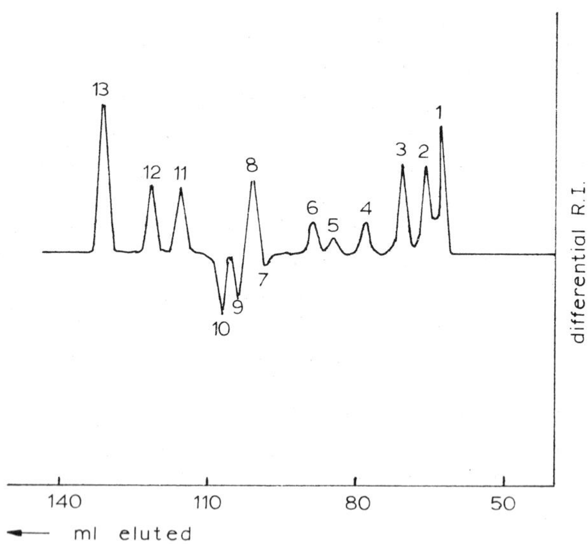

Fig.10.8. Separation of species of low molecular weight by exclusion chromatography. Operating conditions: column, approx. 3.7 m × 8 mm I.D. (actually, 12 ft. × 0.376 in. O.D.); packing, Styragel, 40 Å porosity; mobile phase, tetrahydrofuran; detection, refractive index. 1 = P-4000; 2 = *n*-octadecyl ether; 3 = *n*-dodecyl ether; 4 = *n*-octyl ether; 5 = *n*-dodecane; 6 = *n*-heptanol; 7 = *n*-heptane; 8 = 1,4-dichlorobutane; 9 = *n*-pentane; 10 = ethyl ether; 11 = ethyl iodide; 12 = methyl iodide; 13 = carbon disulfide. (Reproduced with permission from J.G. Hendrickson, *Anal. Chem.*, 40 (1968) 49, copyright by the American Chemical Society.)

As an illustration of the separation of high-molecular weight materials by this method Fig. 10.7 shows the fractionation of human serum proteins and haemoglobin using the soft, cross-linked dextran gel Sephadex G-200.

Separation of samples of low molecular weight

Although strictly an extension of the previous section, the value in use when studying small molecules changes sufficiently to merit this application being dealt with separately.

With highly efficient columns, which can be produced using, for example, the semi-rigid, polystyrene type of packings, it is possible to obtain a sufficiently high resolving power that will allow the complete separation of samples which differ little in molecular size. The use of modern, pressurised LC systems with long, highly efficient columns, *i.e.*, with a large number of theoretical plates, can provide a very practical separation system. From the properties of steric exclusion chromatography it is certain that whatever sample is introduced will be eluted from the column within a definite volume. It follows that one is able to inject any sample which is soluble in the mobile phase and be sure that it will be eluted without needing to first optimise the mobile phase conditions or to use gradient elution. This is in contrast to retentive forms of chromatography, where it is often difficult to decide when or if all the sample has eluted from the column. Hendrickson[8] has demonstrated that, given a highly efficient column and adequate time, results could be obtained which could be considered as those obtained with a type of "liquid phase size spectrometer".

The concept is particularly attractive in that quite large samples may be introduced, for it should be remembered that in steric exclusion the sample remains dissolved in one liquid, the mobile phase, which is selected to be a good solvent for the sample. Precipitation within a column system is therefore very unlikely unless the column temperature is appreciably lower than the temperature of the samples; because of this the capacity of the column is higher than that of those relying on other separation methods.

One of a number of very interesting chromatograms reported[8] which illustrates the capabilities of steric exclusion in the separation of small molecules is reproduced in Fig. 10.8.

It should be appreciated that this chromatogram was first published in 1968 and as considerable advances in column technology and speeds of analysis have been made in the intervening years it is almost certain that superior results could be achieved at the present time.

High-molecular-weight fractions present in a sample which, if strongly retained, can lead to deterioration of retentive chromatographic column systems, will elute first as they are completely excluded from the pore matrix of a steric exclusion packing whose pore dimensions have been selected for optimum resolution of small molecules.

Application of exclusion chromatography as a clean-up technique

Freedom from contamination in this manner makes the method ideal for use as a clean-up method prior to applying a more selective chromatographic method to one or several fractions taken from the effluent of the exclusion column. This approach has been utilised for the examination of natural products, notably pyrethrins[9] and components of fruit

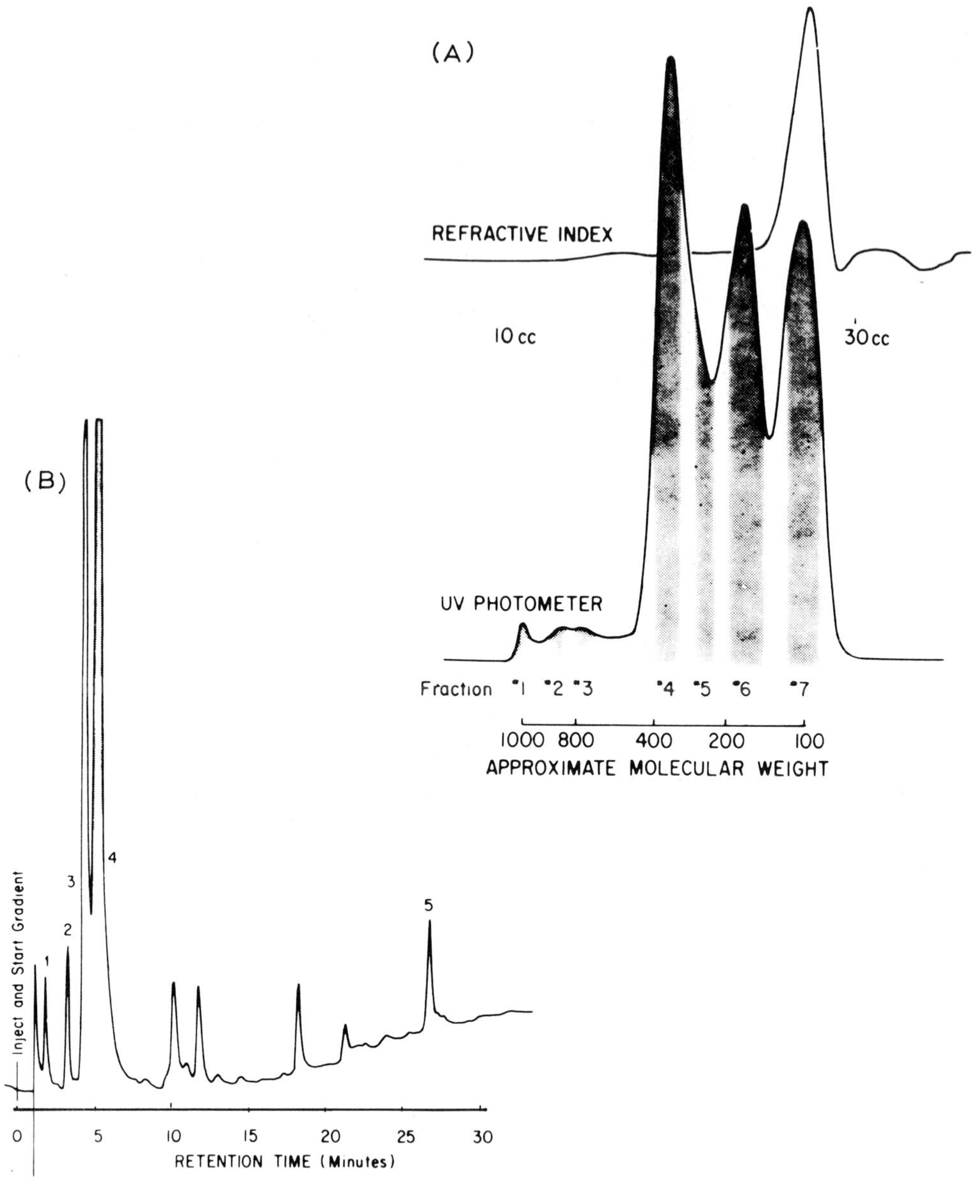

Fig.10.9. Analysis of fruit juice extracts by combined exclusion and partition chromatography. (A) Steric exclusion of Valencia orange oil; (B) partition chromatography of Valencia orange oil, fraction 6. Operating conditions: (A) Column, 1 m × 7.9 mm I.D.; packing, Bio-Beads SX-2; mobile phase, chloroform; flow-rate, 0.8 ml/min; temperature, ambient; detection, UV absorbance, 254 nm. (B) column, 1 m × 2.1 mm I.D.; packing, Permaphase ODS; mobile phase, linear gradient from 5% methanol in water to 100% methanol, at 3% change/min; flow-rate, 1.5 ml/min; temperature, 50°C; detection, UV absorbance, 254 nm. 1 = Pinene; 2 = limonene; 3 = neral; 4 = geranial; 5 = codinene. (Reproduced from J.A. Schmit, R.C. Williams and R.A. Henry, *J. Agr. Food Chem.*, 21 (1973) 551, with permission.)

juices[10]. Fig. 10.9 illustrates this approach in firstly separating the components of orange juice on the basis of molecular size (curve A); the fraction containing components of molecular size 100–200 was subsequently analysed by high-resolution partition chromatography, which enabled closely related components of similar size to be resolved (curve B).

The potential of using steric exclusion chromatography as a clean-up method prior to a more sophisticated analysis has, perhaps, been somewhat underestimated up to the present time. The high sample capacity and the ability to elute the total sample with a definite volume of column effluent must surely appeal to those wishing to re-use chromatographic columns for multiple samples, which may in themselves be very crude. Of the relatively few papers which have been published describing this approach, some of the most detailed have been by Stalling *et al.*, who in one paper[11] described the chromatographic procedures involved in the clean-up of fish lipids for pesticide residue analysis.

REFERENCES

1 Z. Grubisic, P. Remp and H. Benoit, *J. Polym. Sci., Part B*, 5 (1967) 753.
2 L. Fischer, *An Introduction to Gel Chromatography*, North-Holland, Amsterdam, 1971, p. 232.
3 Y.A. Elketov, A.V. Kiselev, T.D. Khokhlova and Y.S. Nikitin, *Chromatographia*, 6 (1973) 187.
4 R.C. Williams, J.A. Schmit and H.L. Suchan, *J. Polym. Sci., Part B*, 9 (1971) 413.
5 J-Y. Chuang and J.F. Johnson, *J. Appl. Polym. Sci.*, 17 (1973) 2123.
6 L.H. Tung, *J. Appl. Polym. Sci.*, 10 (1966) 375.
7 M. Hess and R.F. Kratz, *J. Polym. Sci., Part A*, 2 (1966) 731.
8 J.G. Hendrickson, *Anal. Chem.*, 40 (1968) 49.
9 *DuPont Liquid Chromatography Methods Bulletin*, 820M12, Sept. 1972.
10 J.A. Schmit, R.C. Williams and R.A. Henry, *J. Agr. Food Chem.*, 21 (1973) 551.
11 D.L. Stalling, R.C. Tindle and J.L. Johnson, *J. Ass. Offic. Anal. Chem.*, 55 (1972) 32.

USES OF LIQUID CHROMATOGRAPHIC PROCEDURES

Chapter 11

Qualitative analysis

INTRODUCTION

The main purpose of qualitative analysis is to establish the identity of an, as yet, uncharacterised sample. This requirement may occur for one of several reasons:

(a) The sample may have resulted from a new synthesis programme, or the new synthesis, by a different route, of an established material. It is important to establish if the product is the same as the one being sought and if impurities, particularly undesirable impurities, may have been introduced.

(b) Isolation of compounds from complex naturally occurring products, *e.g.*, alkaloids from plant material.

(c) Confirmation of the identity of potentially hazardous or restricted chemicals. For example, in the enforcement of legislative procedures concerned with pollution and forensic science.

All who are experienced in analytical chemical methods will be all too familiar with the problems that can arise when more attention is given to quantitative rather than to qualitative assessment of a sample. Typical examples are remarks like:

(a) "The sample definitely contained three impurities". – These were shown much later to be impurities in the solvent used to dissolve the sample.

(b) "It must be pure, since it chromatographed as a single spot on a TLC plate".

(c) "Examination by LC showed the sample to be a two-component mixture". – The separation was monitored using a UV absorption detector operating at a wavelength of 365 nm.

These examples pin-point the potential dangers of making assumptions, often completely unjustified, about the nature of the sample without having established the same. The first remark quoted was simply an expression of the lack of attention to practical detail that was given to the examination: One injection of the solvent used to dissolve the sample would have shown up the fault. This is the type of error which most operators are likely to make, at least once. The second remarks made above presuppose that the separation of possible impurities on the TLC plate would have occurred if they were present in the sample, a situation which needs much extra work to establish. The third remark equally presupposes some definite knowledge that any impurities will have some definite spectral characteristics. When using highly selective detectors to study an unfamiliar sample it is recommended that a non-specific detector should also be fitted, in series if possible, to ensure that no components are missed.

These examples illustrate the serious mis-interpretation that can occur through insufficient attention to the qualitative aspects of a chromatographic procedure.

Apart from the situations described in this section, perhaps one of the greatest sources of mis-interpretation of a LC analysis is the failure to establish whether or not the entire sample has eluted from the column: Although it is easy to recognise this possibility, the work necessary to confirm that all the sample has eluted is quite difficult to achieve. One approach could be to inject a known mass of sample, collect the total column effluent and remove the solvent; in principle, weighing the residue will indicate the recovery of the sample. For this result to be accurate the mass injected should be at least 1 mg and "blanks" should be run on the system to make sure that residues from the solvents or column packing do not invalidate the results. An alternative method is to study the sample by a steric exclusion technique as, in most cases, all the sample will elute within the region $k' = 0$ and 1. This approach unfortunately only confirms that the sample will elute from a steric exclusion column and indicates nothing about the degree of retention on, say, a partition column. Independent studies by TLC can also be helpful provided the coating on the TLC plate and the column packing are strictly comparable and the mobile phases are identical. The recommended approach is to remove spots from the TLC plate, extract the component, and run it in the liquid chromatograph to establish the retention behaviour of the observed spots.

METHODS OF ESTABLISHING OR CONFIRMING THE IDENTITY OF AN ELUTING PEAK

From the earlier remarks it will be apparent that careful attention to the qualitative aspects of an analysis is important in all but the most predictable chromatographic separations. There are a number of methods by which the identity of the eluting component may be checked. These rely either on comparison of the chromatographic characteristics of the components with reference materials of known identity or on the characteristics of the detection system. In this sense the "detection" may be considered as either in-line, *i.e.*, as a flow through detector, or isolated, *i.e.*, by collecting fractions for further study.

Identification methods based on comparison of retention data

In principle, it could be considered that since the retention characteristics of a sample component are dependent on its chemical and physical properties, *i.e.*, molecular size, functional groups, and solubilities, correlation of retention times (or volumes) with the type of sample should provide a means of tentatively identifying an eluting component. This rather over-simplified concept has some degree of truth but if it is to be applied to the identification of eluting components without additional data considerable caution must be exercised as to the resolving power of the chromatographic system used. Work is often reported where the resolving power of the system is not capable of separating compounds of similar structure, thus the results may be strictly invalid. Any chromatographic system must offer a high resolving power for the samples being studied. In Chapter 2 it was shown that the resolving power is directly dependent on the capacity, the selectivity, and the square root of the efficiency of the column.

The most frequently used method of establishing the identity of an eluting component is comparison of the retention data of an "unknown" peak with the retention of a similar injection made under identical operating conditions of a reference substance, which, based on other considerations, *e.g.*, a known synthesis precursor, could possibly occur in the sample. For this method to be successful a number of supplementary points must be considered.

Firstly, it is imperative that the precision of the measurement of retention time must be very good in relation to the variation of retention time due to the selectivity of the chromatographic system for compounds of similar structure.

The precision of retention times clearly is very dependent on the repeatability of operations such as recorder chart speeds, injection technique and the flow through the column during the analysis. The injection technique of an experienced operator or where a valve injection system is employed and the recorder chart speed variations can generally be neglected, however, liquid flow-rates can fluctuate for a number of reasons. Significant changes in flow can result from a change in the temperature of the chromatographic system and column permeability. A change in inlet pressure can affect the flow in a fairly straightforward manner, but in some high-pressure systems flow changes are complicated by an increase in viscosity with increasing pressure and compressibility of the mobile phase.

Secondly, retention volumes, if derived simply by the expression "retention time × flow-rate = retention volume" can suffer from the same limitations as were discussed for retention time measurements in the previous paragraph. However, in this instance alternative methods are available in that one may, with the appropriate equipment, measure retention directly in terms of the volume of mobile phase passing through the column from the moment of injection of a sample to its detection. This may be achieved by using a siphon counter, as described in Chapter 10, to aid accurate assignment of elution volumes during the characterisation of molecular weights of polymers.

In any situation where retention characteristics are being compared, there are several practical ways which can be employed to endorse the tentative identification of a component.

(1) When it is believed that the "unknown" peak has been identified, prepare a 50:50 mixture of the unknown and the anticipated reference compound and analyse the mixture. Clearly, if correctly identified only one peak will be observed.

(2) If the equipment used has recycling capability (as described in Chapter 6), an even more critical test is to recycle the mixture, prepared for the first test described above, through the chromatographic column system for as many times as it is practicable to establish whether the "unknown" and "reference" substances can be resolved.

(3) Where possible the exercise should be repeated with other chromatographic phase systems which exhibit different types of selectivity, *i.e.*, a normal partition, a reversed-phase system and a liquid–solid (adsorption) system. Confidence in the identification of a component is considerably increased if three different systems fail to separate the prepared mixture of "unknown" and "reference" compounds.

In studies of this nature, perhaps a negative result is far more decisive. Thus, if the retention characteristics of a reference substance and an unknown are different, they are most definitely not the same substance. On the other hand, if the retentions are identical, they may be the same substance.

When tabulating retention data, use of the capacity factor term (k') is to be recommended. This term is a measure of the effective retention of the compound on a column and is not influenced by column geometry and mobile phase flow-rate; thus comparison of results is simplified. Table 11.1 is an example of recording the results from studies of a

TABLE 11.1

INFLUENCE OF MOBILE PHASE COMPOSITION ON THE ELUTION ORDER OF SEVERAL SUBSTITUTED-UREA HERBICIDES

Details taken from Table 8.5 (Chapter 8).

Mobile phase★	Capacity factor of compound, k'		
	Diuron	Monuron	Neburon
Methanol–water (35:65)	4.3	1.0	25.6
Dioxane–hexane (1:99)	1.5	1.1	0.3

★The stationary phase employed was Permaphase ETH.

group of compounds on two different chromatographic systems; the order of elution and extent of retention of the compounds are clearly seen from such a format. The data in Table 11.1 are taken from Table 8.5 (Chapter 8), where details of compound structure and other phase systems are given.

The use of retention volumes or capacity factors as characteristics of a sample is perhaps more common in the field of GC. In a number of instances the relationship between the logarithm of the retention volume and the carbon number has been shown to be linear in a homologous series in liquid–liquid partition, in a similar manner to that observed in GLC.

A vast amount of information exists relating retention characteristics of samples in GC systems with chemical structure. The very considerable bulk of such data that have been published gives some indication of the large amount of work which should be performed to ensure that one is assigning the correct structure to an eluting component. Recent studies carried out on hydrocarbon samples using adsorption chromatography[1] and reversed-phase partition[2] indicate the potential of a comparable method in the liquid phase. Both of these papers, however, although quite detailed, are in reality only scratching the surface of the subject.

Identification methods using in-line selective detectors

Perhaps the most common application of this approach is the tentative identification of components which contain characteristic UV or visible absorption spectra. If a photometric detector with multi-wavelength capability is employed, the analysis of a sample may be repeated several times, each run being monitored at a different wavelength. Comparison of a number of chromatograms obtained in this manner will show the size of the peaks due to the separated components varying in accordance with their spectral characteristics, which in many instances will be known or can readily be determined.

With some photometric detectors based on the spectrophotometer, it is possible to record the spectrum of an eluting peak while it is contained in the detector flow cell. Although in principle one could rapidly scan the spectrum of the mobile phase containing the component as it elutes from the column, a more faithful representation of the spectrum is obtained by stopping the liquid flow during the time that the spectrum is recorded.

Since diffusion in the liquid phase is very slow, this method is more attractive for LC procedures than for those in the gas phase.

Various devices for the rapid scanning of column effluents have been proposed. One particularly interesting approach has been the use of a cathode ray tube display of the spectrum to enable the operator to observe any change in the absorption spectrum during the elution of a sample component[3].

The use of more than one detector, linked in series or parallel after the chromatographic column, can provide comparative information which reduces the possibility of incorrect assignment of the identity of a component. A simple example is the use of an UV absorbance detector in line with a differential refractive index (RI) detector. The latter will respond to most substances, whereas the first mentioned detector is quite selective in its response. A ratio of the peak heights will normally provide the easiest method of comparing the relative responses of a "reference" and an "unknown" sample. When recording the data, the sensitivity or attenuation settings of the detectors should not be overlooked. Comparisons of this kind are best performed on the same instrument as post-column band broadening will reduce the peak height, especially in the second detector, if they are coupled in series. The relative response of, say RI/UV, should be a characteristic of a compound provided both detectors are operating within their linear range; injections of

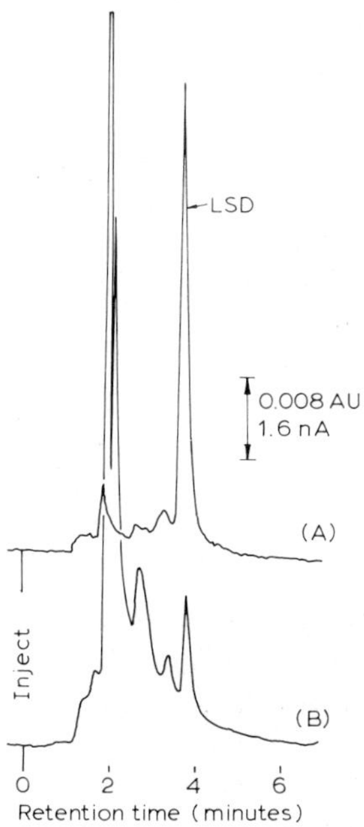

Fig.11.1. Use of combined fluorescence (A) and absorbance (B) detection for increased confidence in the identification of LSD in illicit tablets. Operating conditions: column, 0.25 m × 2.1 mm I.D.; packing, Zorbax SIL; mobile phase, methanol–dichloromethane–acetic acid (30:70:0.1); temperature, 24°C; inlet pressure, 1200 p.s.i.; flow-rate, 0.6 ml/min; detection by UV absorbance at 334 nm (0.08 a.u.f.s.) and fluorescence (16 nA full scale; excitation wavelength 334 nm; emission wavelength 408 nm and above). (Reproduced by courtesy of DuPont and from D.R. Baker, R.C. Williams and J.C. Steichen, *J. Chromatogr. Sci.*, 12 (1974) 499.)

samples of different mass will check this point. A similar combination of detectors which gives valuable information regarding the eluting component is the combination of UV absorbance with fluorescence. Fig. 11.1 illustrates the simultaneous UV/fluorescence detection of LSD (lysergic acid diethylamide) in illicit tablets. The combination of detection methods in this manner reinforces the certainty of identification of the components being sought.

Detector non-linearity can be somewhat more acute in fluorescence measurements as the absorbance of a compound will reduce the intensity of the excitation radiation, which in turn will lower the fluorescent emission. This phenomenon is known as the "inner filter effect" and can cause some non-linearity in the response of a fluorescence detector if the background absorbance (at the wavelength of the excitation radiation) is in excess of approximately 0.05 absorbance units.

Monitoring of column effluents by mass spectrometry

The combination of GC–MS and computerised data handling systems has proved to be one of the most powerful analytical methods for identifying minute components which may be present in chemical samples. Perhaps the greatest successes have been in its application in the fields of forensic science, pollution, and biochemistry.

Following the impact of the GC–MS technique, it is logical to consider a similar approach involving LC, particularly since the separation of a wider range of sample types may be studied. There are, however, several fundamental differences associated with the concept of an interfaced LC–MS system. Firstly, the sample will in general be non-volatile, polar and/or of high molecular weight: If it were volatile, then GC should have been the chromatographic method to employ. Secondly, the mobile phase is very dense, particularly when it is compared to helium, the most popular carrier gas in GC–MS. There is also the risk of corrosion within the spectrometer due to the solvent itself or dissolved solids, *e.g.*, from buffer solutions from the effluent of an ion-exchange separation. On the other hand, quantitative collection of a sample from an effluent leaving a liquid chromatograph is very easily accomplished as the component is in solution. If the identity of a compound is being sought, a portion of such a collected fraction can be evaporated on to a direct insertion probe and introduced manually into the mass spectrometer, thus avoiding the inherent problems associated with an in-line system.

Mass spectral measurements of compounds eluting from a liquid chromatograph can be made in several ways. The simplest is to collect the fraction of column effluent containing the sample and determine its spectral characteristics as an independent exercise. Alternatively, the column effluent may be "sampled" automatically, either in a discontinuous or in a continuous manner. Instrumental requirements for these methods vary considerably and are described in the following paragraphs.

Manual procedure for evaporating the collected fractions and examining the residue

This method may be very simply achieved if, as is usual, the vapour pressure of the mobile phase is considerably higher than that of the sample. Juhasz *et al.*[4] have described the application of a refinement of this method, whereby the column effluent is collected in a small sample tube containing 6 mg potassium bromide. A steady stream of dry nitrogen is fed into the small sample tube via a hypodermic needle to assist the evaporation of the solvent. In this manner the residue of the collected fraction is coated on to the potassium bromide, which is subsequently formed into a disc suitable for examination in an IR spectrophotometer fitted with beam condensing optics. The potassium bromide disc may be subsequently transferred to the direct insertion probe of a mass spectrometer. On heating the probe, the sample is vaporised into the spectrometer allowing a spectrum of good quality to be obtained.

This approach possesses the distinct advantage of simplicity and does not require a complex interfacing system. The greatest limitation is the amount of operator handling of the sample which can increase the possibility of contamination of the fraction and also lead to loss of sample, particularly during the evaporation of the solvent.

Semi-automated sample collection and insertion into a spectrometer

Lovins *et al.*[5] have described a liquid chromatograph-to-mass spectrometer interface where a motor-driven insertion probe is employed. In this approach the fraction of column effluent which is to be studied is initially collected in a small reservoir. On opening a valve, the solution passes through narrow-bore capillary tubing to the tip of the probe, where the solvent is flash evaporated in the reduced atmosphere of the fore-chamber. When this stage is completed, the valve controlling the entry of solution is closed automatically, the fore-chamber is reduced to low pressure whereon a high-vacuum valve isolating the mass spectrometer ion source from the fore-chamber is opened and the tip of the sample probe is advanced into the ion source.

All operations are accomplished by motorised components giving the interface a semi-automatic capability. By this method the complete operation from sample collection to obtaining a mass spectrum is reported as taking 3–5 min with minimal operator attention. In common with the first method described, some loss of sample has been observed when the solute has a fairly high vapour pressure at the temperature at which the mobile phase is being flash evaporated.

In-line, coupled liquid chromatograph–mass spectrometer systems

Several approaches have been described, notably by Horning *et al.*[6], Arpino *et al.*[7] and Jones and Yang[8] whereby the effluent from a liquid chromatograph is introduced into a mass spectrometer in a similar manner to that employed in GC–MS.

The interfaces described to date are generally fairly simple in design when one considers that they have to overcome the very large difference in sample environment, *i.e.*, from solution in a liquid at high pressure to a vapour in a high-vacuum system. Not surprisingly, a good deal of the emphasis in design has been given to providing a sufficiently high pumping capacity in the MS analyser to avoid the pressure within the system exceeding approximately 10^{-5} torr, at which point the spectrometer will no longer operate efficiently. With the exception of very corrosive mobile phases, *e.g.*, acids and buffer solutions, it would appear that simultaneous introduction of solvent and sample molecules can simplify the mass spectra by providing an "atmosphere" within the ion source comparable to that employed in chemical ionisation mass spectrometry. This latter variation of the technique allows simple spectra to be obtained from labile substances, often showing molecular ions. When using electron impact for ionisation, similar molecules are more completely fragmented, leading to a spectrum showing ions of much lower *m/e* value, which can be confused with ions produced from the molecules of mobile phase.

Horning *et al.*[9] have illustrated the practical utility of this approach using a quadrupole mass spectrometer fitted with a ^{63}Ni radioactive ionisation source operating at essentially atmospheric pressure. This ionisation source is situated at the end of a heated capillary from which the column effluent is vaporised into a nitrogen stream immediately adjacent to an aperture, 10–25 μm in diameter, leading into the mass spectrometer.

Some disparity exists in the sensitivity reported for LC–MS systems, but it can be anticipated that when these rather new systems become more established limits of detection approaching picogram sensitivity should be possible. One of the most interesting uses of a combined LC–MS system comes from the inherent "tuning" characteristics of the spectrometer. As in GC–MS it is possible to focus the spectrometer to any desired *m/e* value and record the variation in the concentration of that ion with time. In applications where specific chemical species are being sought, for example, drug metabolism studies and pesticide residue analysis, most of the co-extracted substances which could interfere with a conventional chromatographic analysis will be rejected by the *m/e* value set on the mass spectrometer. This arrangement leads to high sensitivity and very selective detection of the components of interest.

It is quite probable that this last-mentioned approach could well play a major role in the future analytical chemical research studies in areas of toxicology, metabolism, and pollution. The most serious drawback of the technique will no doubt be the high cost of the equipment required, particularly if, as is often the case, a computerised data handling system proves necessary.

OTHER CONSIDERATIONS WHEN SEEKING TO IDENTIFY AN ELUTED COMPONENT

While on the subject of the identification of components eluting from a liquid chromatograph it is important to bear in mind the purity of the solvents used in the separation process. Clearly, when seeking to collect a sample component for further study it is imperative to select solvents which are free from any non-volatile impurities and fairly easily vaporised. Careful distillation of all solvents prior to use will normally prevent any difficulties in respect of the mobile phase. The stationary phase, however, poses a different problem. Many of the original phase systems were based on fairly polar and particularly high-molecular-weight materials, for instance, 1,2,3-tris(2-cyanoethoxy)propane and polyethylene glycols. Chromatographic systems using these phases were described in Chapter 8, where the need to saturate mobile phases with respect to the stationary liquid was emphasised. In the event of wishing to collect pure components from a chromatographic column, clearly these systems pose considerable problems, as the sample components would be contaminated with stationary phase. In this application the use of liquid–solid chromatography with silica or alumina types of adsorbent and partition or ion-exchange column packing materials which have the stationary phase bonded chemically to the support is to be recommended. These materials will not bleed and lead to contamination of the column effluent.

In an analogous manner, the ability to select a phase system that will enable minor components which must be identified to elute before the major components of a sample will greatly facilitate the collection of pure materials. In this way the chromatographic system may be overloaded significantly with respect to the major component in order to collect a larger amount of the impurity. If the minor component elutes later in the chromatogram, contamination by residual amounts of the major component is frequently encountered.

REFERENCES

1 M. Popl, V. Dolanský and J. Mostecký, *J. Chromatogr.*, 91 (1974) 649.
2 R.B. Sleight, *J. Chromatogr.*, 83 (1973) 31.
3 A. Bylina, D. Sybilska, Z.R. Grabowski and J. Koszewski, *J. Chromatogr.*, 83 (1973) 357.
4 A.A. Juhasz, J. Omardoali and J.J. Rocchio, *Int. Lab.*, July/August (1974) 29.
5 R.E. Lovins, S.R. Ellis, G.D. Tolbert and C.R. McKinney, *Anal. Chem.*, 45 (1973) 1553.
6 E.C. Horning, D.I. Carroll, I. Dzidic, K.D. Haegele, M.G. Horning and R.N. Stillwell, *J. Chromatogr.*, 99 (1974) 13.
7 P.J. Arpino, B.D. Darokins and F.W. McLafferty, *J. Chromatogr. Sci.*, 12 (1974) 574.
8 P.R. Jones and S.K. Yang, *Anal. Chem.*, 47 (1975) 1000.
9 E.C. Horning, D.I. Carroll, I. Dzidic, K.D. Haegele, M.G. Horning and R.N. Stillwell, *J. Chromatogr. Sci.*, 12 (1974) 725.

Chapter 12

Quantitative analysis

INTRODUCTION

One of the most important features in the development of modern LC is that quantitative analysis should be possible with very much the same methodology and quality of results as are available from GC. The concept of modern LC, *i.e.*, fast separations, the ability to control the separation conditions, *e.g.*, flow-rate and temperature, and in-line detectors, makes the technique more amenable to obtaining good replication of data than the more conventional systems using either TLC or gravity flow column chromatography.

In discussing quantitative analysis the terms accuracy, precision and reproducibility have clearly defined meanings. In practice the terms are often misused, giving rise to misleading, if not incorrect, statements. To avoid any possible confusion these terms will be explained here. The accuracy of a method is its ability to measure the quantity being determined. For example, if a component is present in a sample to the extent of 50%, an accurate method would confirm this figure, whereas one which was inaccurate might have given an answer of 55%. The accuracy of a method can only be assessed by comparing the results obtained with information taken from another source, *e.g.*, a sample made up with a known (weighed) composition. The precision of a method is its ability to consistently give the same result for a number of replicate determinations. Taking the earlier example, a series of results – 45.1%, 45.0%, 45.0%, 44.9%, 45.1% and 45.0% – could be considered as very precise for there is very little variation in the results from one determination to another. The results are, however, quite inaccurate as the true value is 50%.

The reproducibility of a method is the ability for independent determinations to consistently give the same result, *i.e.*, different operators in different locations analysing the same sample by the same procedure.

In LC techniques, the precision obtainable is of paramount importance as this reflects the confidence which can be placed on the results obtained from an analysis. The precision gives some measure of the degree of control that can be exerted on the instrumental and separation conditions. The accuracy of the method relies almost exclusively on the ability to calibrate the system with standards of known composition. Reproducible results, *i.e.*, minimal variations from different operators and locations, depend considerably on the ability to accurately define the operating conditions for the particular separation. A typical example of a source of discrepancies in this last instance is control of temperature – one apparatus, set to operate at 50°C, may in fact control at 48°C, whereas another apparatus may control at say 52°C.

The ability to achieve high-quality quantitative data often depends as much on the attention to detail given by the operator as to the design of the equipment. Very often careful calibration will minimise the effect of unavoidable variations in instrumentation or technique. Careful attention must be given to all stages of the analysis, from sample collection to the final calculation of the results: a non-representative sample is of no value however carefully it has been analysed.

SOURCES OF ERROR IN CHROMATOGRAPHIC ANALYSIS

The best quantitative results in LC will be obtained by giving attention to the following aspects of the analysis:

(1) Obtaining a representative sample.

(2) Preparation of a solution of the sample for introduction into the liquid chromatograph.

(3) Injection of the sample.

(4) The chromatographic separation.

(5) Detection.

(6) Quantitation.

Obtaining a representative sample

In many applications there is no problem in obtaining a representative sample, for instance, any portion of a liquid preparation may be considered as representing the whole, provided the liquid has been thoroughly mixed immediately before the sample is taken. Similarly in the production of pharmaceutical products such as tablets, there is seldom reason to believe that any one tablet is more representative than any other.

With bulk samples, such as those of agricultural importance, *e.g.*, crops, or from biological origin, *e.g.*, urine, it is sometimes particularly difficult to ensure that the sample is representative. The most suitable procedure for solid samples is to collect the largest sample which is practicable and thoroughly mix it and "quarter it down" according to a standardised procedure. In this way a sample small enough for analysis will be obtained without bias. In all sample collection procedures it is recommended that several sub-samples are taken from each product. These should be processed individually and the results compared. Any variation of result will immediately indicate whether or not the procedure adopted was satisfactory.

Preparation of a solution suitable for injection

It should always be remembered that a true solution is the only acceptable sample which may be introduced into a liquid chromatograph and it should be dissolved in a solvent which is miscible with the mobile phase. Although this is a seemingly obvious statement, on many occasions one finds operators complaining of poor reproducibility after having injected samples which contain suspended matter.

Modern chromatographic columns, particularly those filled with microparticles of 10 μm diameter or less, must be considered as highly efficient filters. Consequently any trace of particulate matter will be retained on the column packing material, leading to premature failure of the column due to a restricted flow of mobile phase.

It is good practice to filter all sample solutions which are to be analysed by LC. One very convenient method, particularly for small samples, is to pass the solution from a hypodermic syringe through a small cartridge filter. A convenient type is available from Millipore, under the name "Swinnex filters"; these fit directly on to the syringe in place of the needle and contain a replaceable pad of filter paper. When using this method it is important to push the sample gently through the filter as excessive force will rupture the filter paper. Larger volumes of solution may be filtered by conventional methods, *e.g.*, under suction through a 0.5-μm-porosity filter paper. Centrifugation is also very acceptable as an alternative method of removing particulate matter from a suspension.

It is also important to ensure that the solvent used to dissolve the sample is compatible with the liquids used in the chromatographic system. The solvent should ideally be the same as the mobile phase used for the separation. In this way there is no possibility of precipitating the sample at the head of the column, no disturbance of the equilibrium of the phase system and, which is of considerable importance if some components are weakly retained on the column, very little "solvent response" will be seen on the resultant chromatogram. The composition of the solvent used to dissolve the sample can have a very pronounced effect on the resultant chromatogram. Fig. 12.1 shows this effect using

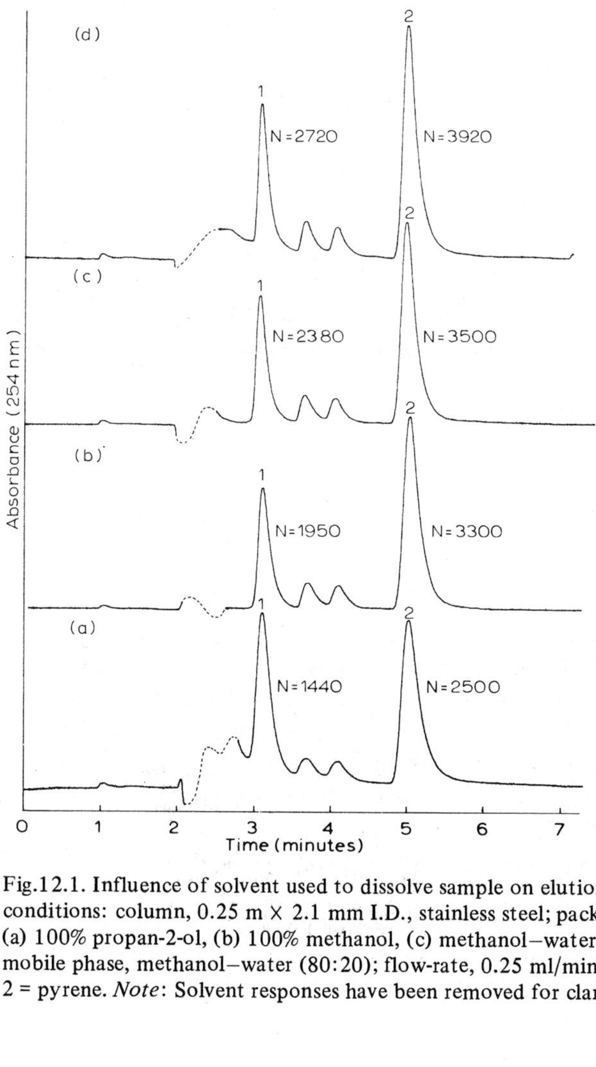

Fig. 12.1. Influence of solvent used to dissolve sample on elution behaviour of components. Operating conditions: column, 0.25 m × 2.1 mm I.D., stainless steel; packing, Zorbax ODS, 6–8 μm; solvent, (a) 100% propan-2-ol, (b) 100% methanol, (c) methanol–water (80:20), (d) methanol–water (60:40); mobile phase, methanol–water (80:20); flow-rate, 0.25 ml/min; injection volume, 10 μl. 1 = Naphthalene; 2 = pyrene. *Note*: Solvent responses have been removed for clarity sake; minor peaks are ignored.

a series of equivalent injections of a sample differing only in the solvent used to dissolve the sample. The mobile phase composition was 80% methanol in water and the samples were dissolved in (a) pure propan-2-ol, (b) pure methanol, (c) the mobile phase, and (d) 60% methanol in water. In addition to the difference in the size and shape of the solvent response, very distinct differences in the efficiency of the column can be seen. The most noticeable change is the improved efficiency of the early eluting component when dissolved in a solvent containing less methanol than the mobile phase. This improved performance is most certainly due to an initial concentration of the sample on the head of the column, followed by elution with the more powerful mobile phase. This effect can be considered comparable to step-wise gradient elution, only on a micro scale.

There are occasions where it is not practicable to dissolve the sample in the mobile phase, for instance, when a minor component is to be detected while the solubility of the major component in the mobile phase is insufficient to obtain the necessary mass of impurity in the column system for adequate detection. In these circumstances it is necessary to employ a better solvent for the sample: This should be selected with some care in order to avoid disturbing the equilibrium condition of the phase system or the generation of solvent peaks that might obscure the components of the sample.

A general feature which should be borne in mind is that the volume of sample solution injected during analytical procedures is, in general, quite small, *i.e.*, less than 100 μl. For some applications, particularly trace analysis, this means that solutions may need to be evaporated to low bulk prior to injection. The solvents used in such procedures must be absolutely free from non-volatile impurities; all solvents should be carefully redistilled before use. Samples analysed by LC are usually of low volatility, even so, if the evaporation of solutions is taken to dryness during the preparation of the sample, this requires particular care, for significant losses of microgram and nanogram quantities of sample can be incurred by evaporation, even at very moderate temperatures.

Injection of samples

This step is the most critical in obtaining good quantitative data. The various types of sample injection systems used in LC apparatus were described in Chapter 4. Basically, one either uses a microsyringe to inject the sample directly into the system or first loads a sample valve under essentially atmospheric pressure and subsequently actuates the valve to introduce the sample into the liquid chromatograph.

Regarding the relative merits of the two approaches, the use of a good sampling valve for injection gives an unquestionably higher precision. These improved results arise from

(A) The volume introduced is well defined by a fixed cavity, either as a length of capillary, as in a loop valve, or as a groove in the shaft in other types of valves.

(B) Since the valve is loaded at low pressure, the possibility of leakage of sample is very low compared with an injection port.

(C) Loss of sample can occur during a high-pressure syringe injection due to leakage between the plunger and barrel of a worn syringe.

(D) The sample solution flows into the column in a reproducible manner at a reproducible rate, unlike in the case of a syringe system, where the rate of sample injection depends on the operator's manipulations and the precise point at which the sample enters the column on the depth and angle at which the syringe needle pierces the septum.

(E) Syringe injection can also cause disturbance at the head of the chromatographic column, for example, if the syringe needle enters the column, the force of the sample being injected can disturb the uniformity of the chromatographic packing from injection to injection. Similarly, if a septum style of injection is used, fragments of septa can be torn away during injection, giving a disturbance of the flow of liquid entering the column.

(F) Larger volumes of sample solution, *i.e.*, 20–100 μl, may be introduced more easily with a loop valve than with a syringe. In many instances injection volumes larger than those commonly employed in GC may be employed without adverse effect on chromatographic resolution. In this way the errors in sample introduction may be reduced significantly, for example, the loss of 1 μl from a 50-μl injection is much less important than a similar loss from a 5-μl injection.

(G) Cross contamination of samples is perhaps one of the more common problems that can be experienced with a valve sampling system. It is common practice to load a valve at low pressure with the aid of a conventional hypodermic syringe, of approximately 1-ml capacity. Capillary tubing is used to carry the sample solution into the body of the valve, excess liquid being flushed through the sampling cavity of the valve to drain. This capillary tubing and the syringe used to load the sample must be kept scrupulously clean. In the microsyringe method of sample introduction, syringes of approximately 10 μl are easily rinsed ten or twenty times with a modest volume of fresh solvent to ensure they are clean. However, a common fault when using a larger syringe, *e.g.*, of 1-ml capacity, to load a valve is not to clean it as efficiently. This almost certainly arises from the much larger volume of rinsing solvent which must be available for the purpose. Clearly for results of good quality it is imperative that the syringe should be clean, whatever its capacity.

(H) Valves very significantly reduce variations in technique from one operator to another, thus improving the reproducibility of a method.

Data relating to the precision attainable by the various techniques may be found in the literature. It has been mentioned that the microsyringe injection technique is very much dependent on the operator and therefore some spread in reported results can be expected. Generally, however, using a syringe injection (*i.e.*, no valve) and measuring the resultant chromatogram by the peak height method will give a precision in the order ± 5%, while measurements based on the areas of peaks will result in a precision of approximately ± 1–2%. Sample introduction with valves gives a precision in the order of ± 1–2% for peak height and approximately ±1% or less for peak area measurements. The true significance of differences in reported values less than 1% becomes debatable as a number of sources derive values of precision on a very limited number of injections, *e.g.*, four to five*.

Errors arising from the chromatographic separation

Once a LC system has reached equilibrium the errors in quantitation arising from this aspect of the analysis are quite small. The main emphasis is to avoid disturbing the equilibration of the column system through allowing the operating conditions to vary in an un-

*Throughout this text, values of precision quoted are obtained statistically as the coefficient of variance, which is the root mean square of the standard deviation of a group of results expressed as a percentage of the mean result.

controlled manner, for example, the column temperature and the composition of the mobile phase: The effect of very small changes in the polar components of mobile phases used in LSC were described at length in Chapter 7. Changes of this kind produce variations in the activity of the column packing leading to a change in the retention characteristics of samples. Similarly, in these systems it is particularly important that the sample be dissolved in a solvent which will not disturb the equilibration of the column. For example, in adsorption chromatography using silica or alumina as the column packing, the injection of samples dissolved in water or methanol can create considerable problems if the normal mobile phase is, say, hexane or chloroform.

Decomposition of samples during the chromatographic separation is comparatively rare. Compounds which would give problems in GC, for example, corticosteroids, antibiotics and vitamins, show no signs of decomposition in the liquid phase, particularly as analysis times are in the order of a few minutes and the temperature seldom exceeds 75°C. Complications occur more frequently with samples which form complexes with heavy metal ions, *e.g.*, tetracyclines, as most instrumentation is constructed from stainless steel, if not the column, the injector and the detector flow cell. Additionally, the possibility of trace amounts of metal ions being present in the chromatographic support cannot be ruled out. Complexing reagents such as EDTA (ethylenediaminetetraacetic acid) often reduce this type of problem considerably; making the mobile phase say 0.01 *M* with respect to this reagent is usually adequate.

A certain amount of common sense is required when studying compounds which are sensitive to acids or bases in order that they are not subjected to conditions that might lead to decomposition. There are conflicting opinions regarding the reproducibility of LSC separations. It has been reported that partial loss or decomposition of samples can occur on surfaces due to irreversible adsorption or chemical reaction. This situation would seem understandable when it is remembered that in other areas of work finely divided silica can act as an efficient catalyst. Generally, however, the use of a trace of water in the mobile phase to partially deactivate the surface of the adsorbent minimises this effect so that it rarely needs to be considered.

In a large number of applications, programming techniques, particularly gradient elution, are used to increase the range of sample components that may be eluted within a given time. Although many of the modern instruments which are designed for this purpose are capable of generating highly reproducible gradients, the use of the technique in quantitative analysis should be considered only when absolutely necessary. The principal reasons for this cautious view originate from:

(A) The possibility of generating sharp, spurious peaks due to dehomogenisation of the mobile phase (solvent dimixing) which may be misinterpreted as sample components.

(B) An additional time is required between successive injections for the system to return to the starting composition and attain equilibrium.

(C) In many applications solvent programming will be accompanied by a baseline shift which complicates the measurement of the height or area of peaks.

Clearly the relative merits of using a solvent gradient must be assessed for the application in hand. Some detectors, notably the solvent transport to flame ionisation and fluorescence detectors, are essentially unaffected by changes in the mobile phase composition. Photometric detectors will generally give a good baseline stability during a gradient elution programme. Their stability is, however, adversely influenced by changes in the optical characteristics of the solvents, for example, differences in the absorbance at the wavelength of operation and in the refractive indices of the solvents used.

When attempting to analyse any sample, repeatable results will only be obtained within the limits of the capacity of the column. Before analysing samples of widely differing size, it is prudent to be aware of any signs of column overload. This condition can usually be diagnosed by observing the dependence of retention times and column efficiency on sample size. A marked change in the retention characteristics with a corresponding increase in plate height should be taken as an upper limit for the sample capacity of the system.

Detection

A stable, low-noise, in-line detector is no doubt a vital element in the whole process of quantitative analysis. The characteristics which make a detector attractive for quantitative work are: good stability, a response which is linear with respect to sample concentration over a wide range and a sensitivity, or gain, which does not fluctuate with environmental conditions, such as laboratory temperature.

All of the above features reflect not only the design and quality of construction of a detector, but also the other parts of the chromatographic system. The smoothness of flow of mobile phase will influence the noise level of a detector, while changing flow conditions will affect both baseline stability (drift) and possibly the concentration (weight) *versus* response characteristics of a detector. The linearity of the response of a detector with sample concentration depends on the physical property being monitored as much as on the design of the detector. In general the linearity of a LC detector is not as wide as that of an equivalent detector used in GC, the former detectors being typically linear over ranges of four to five orders of sample size, *e.g.*, for a refractive index and UV absorbance detector, respectively. With the current vogue of employing electronic digital integrators having very wide linear dynamic ranges and built-in computation capability, it has become more important to use detectors offering a wide linear range. This is particularly so as these data systems often rely on a constant "response factor" to convert raw peak areas into concentration of components; this is a feature that is discussed more fully in the next section.

It should not be forgotten that the older, perhaps somewhat more laborious, method of constructing an "area" *versus* sample mass calibration graph, as illustrated in Fig. 12.2, does enable one to perform good quantitative analysis. In this case, even though the detector response may not be perfectly linear, provided the concentrations of samples are interpolated from a calibration curve which has been constructed with data derived from a series of injections of "standard" solutions, a degree of non-linearity will not create too large an error for most practical purposes. When preparing a calibration curve it is recommended that a series of solutions of different (known) concentrations of sample are used rather than one solution and altering the injection volume to vary the mass of sample injected. To minimise inaccuracies in the calibration arising from any contamination of the injection system, it is often worthwhile injecting the most dilute solution first and follow with the other solutions in order of increasing concentration.

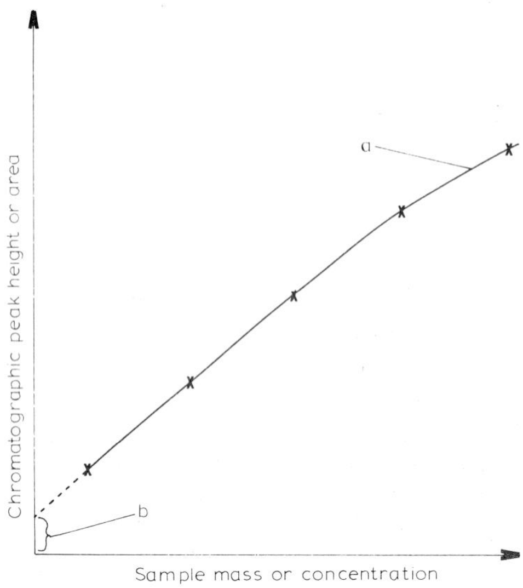

Fig.12.2. Typical calibrated detector response curve for a quantitative LC method. (a) Deviation from linear behaviour; (b) intercept, normally due to an error created by the volume of a syringe needle or the internal volume of a valve.

Quantitation

Deriving quantitative information from a chromatographic analysis is made by measuring the height or area of the respective peaks. In principle, either dimension is valid provided that the quantity measured shows a linear relationship with the mass, strictly the concentration, of the component eluting from the column. Depending on the type of analysis being undertaken, either of these methods can hold advantage.

Peak area measurements will normally be more precise than height measurements as the peak area is less dependent on minor variations in operator technique, especially the injection, or variations in the instrumental parameters. The measurement of the area of a peak is also readily interpreted by electronic data systems, *i.e.*, digital integrators and computors which eliminate the extra effort that would be required for manually measuring areas relative to the peak height.

The width of a peak, measured in units of time as distinct from liquid volume, is, however, very dependent on the velocity of the mobile phase in the chromatographic column. This leads to the area of a peak being more dependent on the rate of liquid flow than is the peak height. Fig.12.3 illustrates this point with a graph showing the height and area measurements for a single-component peak taken from a typical modern chromatogram. The greater dependence of area on flow-rate may be seen from the different slopes of the curves, suggesting that in instances where flow-rates may vary, peak height measurements may provide superior results, assuming this facet is limiting the precision, *i.e.*, injection, etc. is perfect.

A similar effect is observed when measuring overlapping peaks, particularly small peaks on the trailing edge of a larger peak. In these circumstances the width of the peak is con-

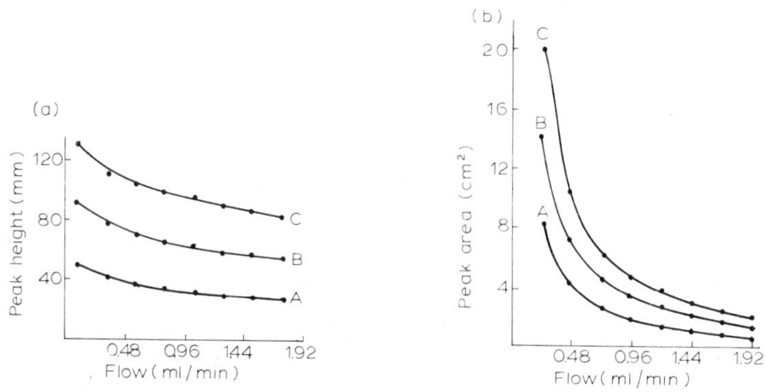

Fig.12.3. Influence of mobile phase flow-rate on peak heights (a) and areas of chromatographic peaks (b). Operating conditions: column, 1 m × 2.1 mm I.D.; packing, Zipax SAX, strong anion exchanger; mobile phase, water + 10 m*M* sodium borate + 50 m*M* sodium nitrate; temperature, 25°C; detector, UV absorbance, 254 nm; sample, phenobarbital – curve A 0.34 μg sample, curve B 0.58 μg sample, and curve C 0.82 μg sample. (Redrawn from data by R.W. Roos, *J. Pharm. Sci.*, 61 (1972) 1979, with permission of the copyright owner.)

siderably increased by the "tail" of the preceding peak. This broadening is reflected in area measurements giving a higher-than-expected value for the peak area of the later eluting component. The height of the peak, measured from a constructed baseline as shown in Fig.12.4, will generally indicate a more accurate value for the concentration of the later eluting component than the peak area, as the height is less influenced by the presence of the earlier eluting component. In many situations the decision between whether

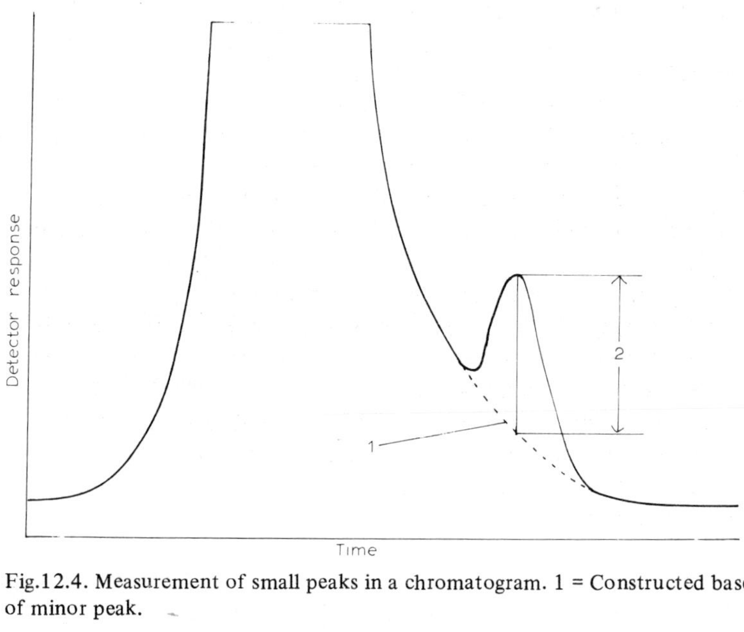

Fig.12.4. Measurement of small peaks in a chromatogram. 1 = Constructed baseline; 2 = peak height of minor peak.

to measure the height or the area of a peak is based on individual requirements, such as time and equipment available.

The measurement of the size of peaks on a chromatogram will only give the magnitude of the response of a particular detector to the components being studied. It must always be remembered that many of the detectors used in LC are very selective in their response, so that the existence of a large peak on a chromatogram may not necessarily indicate a major component.

With any integration of a chromatographic peak, the most accurate and precise value is usually obtained by the method which involves the least number of operations after the detector has responded to an eluting component. Thus, considering the normal sequence of events in a chromatographic analysis: (i) the component elutes from the column; (ii) is "seen" by the detector, (iii) an electrical signal relays this information to a potentiometric recorder, and (iv) a trace is drawn on a strip chart. Depending on the equipment available, integration of the eluting peak may be made immediately after detection, simultaneously with recording the chromatogram or after the completion of the separation. Clearly, the former method will involve the "detector signal" passing through a minimum number of operations, but this method requires quite expensive electronic equipment. The various methods of integrating peak areas are summarised below in an approximate order of increasing cost.

MANUAL METHODS OF INTEGRATION MADE AFTER COMPLETION OF THE ANALYSIS

Multiplication of the peak height by the width at half height

This method is probably the simplest approach for the measurement of peak areas and provided a standardised approach is adopted, little problems are encountered when measuring well resolved peaks which approximate closely to Gaussian shape. Problems can occur, however, when very broad, low peaks must be measured. The height of chromatographic peaks may be measured with sufficient accuracy with a good quality rule. Peak widths, being of much smaller dimension, are best measured with the aid of a magnifying glass type graticule. The use of a comparatively fast chart speed for the recording of the chromatogram can provide a larger, thus more accurately measured, peak width for an eluting component. Whichever method is adopted, a conscious decision should be made as to whether to measure widths based on the "inside" or "outside" edges of the ink line drawn by the recorder pen, in this way avoiding an error equivalent to the thickness of the pen line, which can be significant in the case of narrow peaks. The precision of measurement obtained by this method is generally considered to be in the order of 3%.

Area of a constructed triangle

This approach is similar to the first method, but as seen in Fig. 12.5, an approximation is made in that the area between the peak and the apex of the triangle is the same as the

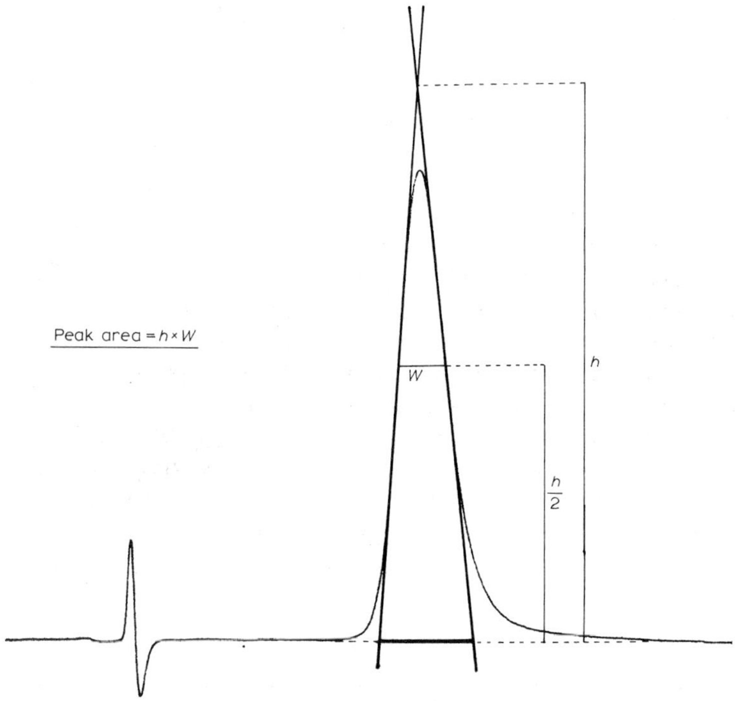

Fig.12.5. Measurement of peak areas by triangulation.

area of the leading and tailing edges of the peak. With this method, considerable error can be introduced if care is not taken in the construction of the tangents to the sides of the peak. Literature values for the precision of measurement are generally about 4%. This method holds no advantage over multiplying peak height by the width at half-height.

Counting squares

In this method, use is made of the small divisions usually printed at 1-mm intervals on the charts used in most potentiometric recorders. After constructing a suitable baseline beneath the peak, the total number of squares contained within the perimeter of the peak are counted, areas of half a square or greater are counted as unity, while those less than half are ignored. Although tedious, a precision of approximately 3% may be obtained by this method. Counting squares holds advantage only in the measurement of low, very broad peaks.

Cut and weigh the chart paper

It may be generally assumed that good quality chart paper is of uniform thickness and density. The method of integration described here is achieved by cutting out the peaks from a chromatogram with scissors and weighing the paper using an analytical balance.

The weight of a peak can be related to area by weighing known areas of chart paper. The uniformity of a particular paper may readily be assessed by cutting out equal areas of paper and comparing their weights. In favourable circumstances this method will yield a relatively high precision in the order of 2%, but unfortunately the chromatogram is destroyed by the procedure.

Use of a planimeter

A planimeter is a small, hand-held mechanical device commonly used for assessing the areas of irregular shapes, for example areas of land from the detail shown on a map. This is accomplished by using a pointer, or crossed wires in a magnifying glass, on the device which is traced round the perimeter of the area to be measured; an indication of the contained area is given on a scale on the planimeter.

A similar approach may be used to measure the area of a chromatographic peak after a baseline has been constructed. For satisfactory results, this method must be performed by an operator with good eyesight and steady hands. This method of integration yields results with a precision of approximately 4%, which is comparable with the triangulation method.

INTEGRATION MADE DURING THE COURSE OF THE ANALYSIS

Electromechanical devices

Mechanical integration and recording of the chromatogram may be accomplished concurrently using a disc integrator. This novel, yet inexpensive, device may be incorporated into a conventional potentiometric recorder and comprises a sensing element coupled to the moving "wiper" on the slide wire of the recorder. A second recording pen is provided, taking the integrator response, and spans a nominal 10% of the width of the strip chart. The remaining 90% is used by the normal recorder pen to trace the chromatogram.

While a horizontal baseline is being drawn by the recorder pen, the second "integrator" pen remains stationary, drawing a similar horizontal line. When the recorder responds to an eluting peak, the integrator pen produces a series of sweeps, backwards and forwards across the 10% span of the chart paper. The rate at which the pen sweeps the chart and hence the number of lines drawn by the pen is proportional to the height of the peak drawn by the recorder pen. Peak areas are deduced by counting the number of strokes made by the integrator pen, usually by giving one stroke a nominal value of ten area units and partially completed strokes a value depending on the number of lines on the chart paper that the pen has crossed. This system can provide a useful method of determining peak areas provided the baseline of the chromatogram is perfectly horizontal between the peaks, in which case a precision of about 1% may be obtained. Any baseline slope is, however, recorded as a pen displacement and added to the area of the peak, thus significant error may be introduced which is corrected only by manual manipulation of the data. In an analogous manner, the relative areas of partially resolved peaks must also be decided by the operator. For satisfactory operation, it should be remembered that since the device functions from the potentiometer of the recorder it is essential that the peaks being measured stay on the recorder scale.

Electronic, digital, integration

The current emphasis with all instrumental techniques is to obtain data as quickly as possible with the minimum of operator involvement. In chromatographic methods, the use of electronic digital integrators, especially those equipped with some computation capability, allows results to be presented in the form of an "analysis" report very soon after the completion of the separation. Digital instrumentation for this purpose has become very sophisticated and space in this text does not permit a detailed description of all of the capabilities offered by individual instruments. In general very precise data, in the order of 0.5% precision, may be obtained with electronic integrators.

A "basic" digital integrator is usually connected electrically immediately after the LC detector. The basic function is to accept the analogue output signal from the detector, provide a digital output of peak areas and simultaneously provide an analogue output signal to drive a strip chart recorder. Electronic integrators have the capability to "detect" variations in the output of the chromatographic detector which would correspond to changes in the slope of the baseline on a recorded chromatogram. The onset of a peak which corresponds to an increase in the baseline slope, relative to any baseline drift, is sensed by a phase-sensitive rate-of-change detector, as is the change in slope from positive to negative at the peak maximum and the return to a normal baseline at the end of the peak. A "slope sensitivity" adjustment is used so that the integrator will respond to the onset of a peak and its return to a normal baseline while disregarding baseline shifts caused by other effects which are generally of a much slower nature. The value selected for this control is usually related to the width of the broadest peak in the chromatogram, *i.e.*, that showing the lowest slope on the leading edge. Integration of short-term noise is reduced in a somewhat analogous manner by rejecting, or filtering out, responses which are faster, *i.e.*, of greater slope, than that of the sharpest peak in the chromatogram. Thus, when correctly adjusted, the only areas which are integrated are those with peak widths falling between the narrowest and the broadest peak in the chromatogram. Clearly, for optimum integration of a peak with maximum rejection of noise, the values selected for the filtering and slope sensitivity should respond to a peak just narrower and wider, respectively, than the peak to be measured. The controls on most digital integrators are capable of adjustment for peak widths from about one second to several minutes. The upper limit on selectable peak width is an important consideration when choosing integrators for LC, as many instruments designed primarily for GC do not cover the very broad peaks that can be encountered in LC.

Depending on the equipment used, the basic integration function will be complemented with facilities for mathematically correcting the integrated areas for baseline drift and differentiating areas beneath partially resolved peaks. The slope sensing circuitry also provides an additional useful role in that the inflection corresponding to the peak maximum is memorised to provide a signal (usually to a digital print out or punched tape) indicating the retention time of the component.

The principal limitations of basic integrators are that they produce only "raw data", that is they cannot provide correction for the selectivity characteristics of the detection system and also a small proportion of the area of each peak is lost from the total area of the peak corresponding to the beginning and end of the peak, since the "slope detector" can only actuate the integrator after a certain finite value of slope has been exceeded.

The "computing" integrator is a more sophisticated electronic device which offers a number of important features in addition to basic digital integration. The principle difference with the computing integrator is, as the name suggests, that it contains memory and computation facilities. By using these added features it is possible for the integrator to be programmed with the operational details necessary for a given analysis. For example, the anticipated peak widths for the components in a given analysis, once determined, hence the settings of the filtering and slope sensitivity may be stored in the memory until required on another occasion. The memory of such integrators is capable of retaining a number of sets of operational parameters, thus it is possible to reset the integrator for a given analysis by simply selecting the appropriate programme. Computing integrators tend to offer somewhat more elaborate circuitry for the sensing of a peak, for instance, in being able to "up-date" or programme the slope sensitivity and filtering during the course of the analysis in order to maintain optimum integration as successively eluting peaks become progressively broader. These integrators also tend to be better equipped to compensate, or correct, for errors which can occur when peaks are only partially resolved and with chromatograms in which the baseline shift is considerable.

One of the more important features of computing integrators is that it is also possible to apply pre-determined detector response factors to the raw peak areas to overcome detector selectivity and thus enable accurate correlation of peak areas to the analytical concentration of the components in the sample mixture. At the end of an analysis a print-out of the normalised analysis, based on raw peak areas and based on "corrected" data, incorporating the response factors, is generally possible within seconds of the elution of the final peak in the chromatogram. For this latter facility to be applicable, however, the retention times and the anticipated range of experimental variations in the same, for each peak must be programmed into the integrator by the operator. Since retention times are generally highly reproducible and the information may be stored in the integrator for subsequent re-use, it should be apparent that the technique is of great value in laboratories performing routine quantitative analysis, as a very considerable saving in operator time and effort is achieved.

Almost without exception, digital integration equipment which has been developed for GC may be used without modification for LC analysis. Most commercial detectors for LC are perfectly compatible with such integration devices. The use of computing integrators for the calculation of results from routine LC analysis is no doubt one of the most convenient and time-saving methods.

One restriction, common to nearly all integrators, is their inability to measure reversed peaks which may occur when monitoring a separation using a refractive index detector. Manual methods must be employed in such circumstances.

Integration clearly is a means of measuring the size of a detector response to a particular component in a sample. With the exception of the pre-programmed computing integrator described in the last section, all methods yield what is generally referred to as "raw data", *i.e.*, no account has been made of detector selectivity. Various methods are commonly adopted to convert these data or peak heights into analytically significant information. The relative merits of these methods, which are described below, often depend on the requirements of the operator rather than on analytical accuracy. For instance, a manufacturing plant operator, based on previous experience, may be able to decide on the acceptability of a product by simply observing the ratio of the area of two peaks on a chromatogram; in these circumstances, the need for calibration and accurate injection, etc., are of no importance. In the description of these methods it will be assumed that peak areas are the quantity being measured; the same methodology can be applied to peak height measurements.

NORMALISATION OF THE PEAKS

This method is based on measuring the area of every peak observed in the chromatogram, perhaps rejecting that due to the solvent response, and expressing the area of each component as a percentage of the sum of all of the peak areas. This will clearly give an "analysis" which adds up to 100% whether or not all of the sample is eluted from the column or is "seen" by the detector used. As no correction is applied to compensate for the selectivity characteristics of the detector, the results must be regarded as an "apparent composition" and as such are of value only for comparative purposes. This kind of data is readily obtained using digital integrators, as the result is available almost immediately after the last peak has eluted from the column.

NORMALISATION OF PEAKS WITH CORRECTION FACTORS

To overcome the inaccuracy created by the detector selectivity in the straightforward normalisation method, use can be made of detector response factors which allow for differences in the weight-to-area response for each component of the sample to be taken into account. A more accurate analysis of a sample is given by this normalisation procedure, but each peak in the chromatogram must be corrected by its own response factor.

The work involved in establishing the response factors can be considerable as each component must be studied separately. Alternatively, if a sample of accurately known composition is available, this may be analysed and the response factors determined from the area of the peaks produced. Computing integrators are normally capable of storing such information in a memory circuit so that subsequent analyses may be converted directly to a weight composition basis. Perhaps the greatest source of error with this method is that it presupposes a constant value for the response factor for each component. Clearly any non-linearity of a detector towards one or several components can give rise to misleading results. Excepting this last restriction, this method of quantitation is probably the most useful when seeking to determine all of the components in a complex sample.

CALIBRATION BY MEANS OF AN EXTERNAL STANDARD

Much of the earlier GC and LC quantitative analyses have been performed by interspersing injections of known concentrations of the compound being analysed between injections containing the solutions derived from samples. In this manner, if a fixed volume of a range of solutions of different known concentrations of the compound is chromatographed, a weight (or concentration) *versus* area calibration curve may be constructed. The concentration of the compound in the various samples is subsequently determined by interpolating their areas on the calibration curve. Even with a detection system which shows a slight deviation from true linear behaviour, provided the values for the samples are obtained by interpolation rather than extrapolation, the method will continue to provide acceptable results.

The principal source of error using this approach comes from the injection of the sample into the liquid chromatograph. Unlike the normalisation procedures, any loss of sample will directly affect the results. This source of error is reduced considerably by making replicate injections of each solution during the analysis. Valve sampling systems, which generally are less susceptible to loss of sample, also give good results by this method. Automated sampling systems using a valve injector are ideally suited to this method.

With this approach, it is necessary to prepare a calibration curve for each component to be determined. For this reason, the method finds greatest use in applications where the quantitative information is required for only a limited number of components of a sample, *e.g.*, the active ingredients in a formulated sample. The method would be tedious for a multi-component sample where all the components must be determined – for such applications the normalisation techniques are preferred.

CALIBRATION USING AN INTERNAL STANDARD

The principle limitation of the external standard method of calibration, considered to be variations in injection volume, may be overcome by the use of an internal standard. By this method an accurate concentration of an additional component, which will give a separate chromatographic peak, is added to the sample solution prior to the analysis. Similar addition of internal standard is also made to all solutions used for calibration of the system. In this manner, when the sample is introduced into the chromatograph, any loss of sample will be accompanied by the loss of an equivalent amount of internal standard. Calibration of the system, using the internal standards, is made by comparing the responses from the desired peak with the peak due to the internal standard and not against the mass of sample thought to have been injected. This approach virtually eliminates variations caused by differences in injection volume.

The choice of an internal standard can prove difficult, particularly if the sample contains many peaks with little baseline available between peaks for another component to be resolved. Although the ultimate choice depends on the separation in hand, selection should be based on the ready availability of a compound in a highly purified form at a modest price. Volatile, exotic or unstable compounds should be avoided. In certain complex analyses, where the recovery of the active component in the sample is to be determined, *e.g.*, through an extraction or clean-up process prior to chromatography, more complex internal standards may be worthwhile. The use of an isomer of the compound being assayed as the internal standard holds some advantage in this last-mentioned application provided the standard and sample may be resolved in the LC step.

Chapter 13

Practical aspects of trace analysis

INTRODUCTION

The term "trace analysis" in this context is used to indicate the separation and detection of a minor component in a sample, where its concentration is less than about 100 ppm. Types of sample analysis which enter this category include the detection of: (a) impurities in technical materials; (b) air-borne pollutants; (c) trace components in bulk water; (d) minor components in bulk solids, *e.g.*, pesticide residues in crops and foodstuffs; (e) drugs and their metabolites in body fluids.

The methodology of modern LC is proving of considerable value in all of these areas of application, as in the more favourable circumstances current limits of detection are in the order of 1 ppb. Generally, also, the extent of sample preparation in terms of clean-up and derivative formation required prior to performing the LC method is less than that required for GC methods. Described here are details of the practical aspects of LC which are most relevant to applications of this type. The nature of these applications makes it extremely improbable that the sample may be injected directly into the liquid chromatograph since the component of interest is normally too dilute and/or contains an excess of other substances that interfere in the analysis.

Very special problems can be encountered when requiring to collect and prepare a sample when a trace constituent is to be determined. In most instances a new procedure must be developed for each application; generally, however, these methods will be based on one of the following approaches.

SAMPLE PRETREATMENT

Impurities in technical materials

In this application, the nature of the impurity often resembles the major component in solubility characteristics. It is therefore improbable that selective concentration of the impurity, using simple liquid partition in a separation funnel, will be practicable. In many instances it is found that LC resolution between the components of the sample may be achieved only when the components are present in approximately equal concentrations. A disproportionate concentration of one component, as is experienced in this type of application, leads to inadequate resolution or detection due to the column system being overloaded by the major component before a detectable quantity of the trace component has been introduced. In these circumstances, a column chromatographic clean-up technique offers the most reliable approach. In this clean-up method the impurities in the sample are essentially freed from the bulk of the major components by eluting the whole sample in the normal chromatographic manner save that the effluent liquid containing the impurities is collected while discarding the column effluent containing the bulk of the major component. This method is best performed using an instrumental liquid chromatograph equipped to accept large samples, *i.e.*, preparative chromatography (as described in Chapter 14). Most frequently, particularly when collecting components present at a very low level, detection of the components of interest is not realised so that the effluent must be collected in the region of the chromatographic separation where the minor components are expected to elute. The appropriate region must be decided by previous experiment. Evaporation of the collected column effluent to low bulk when a sample is run produces a solution which may be analysed with an analytical-scale LC system.

Many preparative LC columns will accept samples of approximately 100 mg, thus 100 ng of any component which is present at the part per million level in the sample will be collected. This quantity is adequate for most analytical-scale determinations which utilise high-sensitivity detection, such as UV absorbance. In this approach the ease of collection of components from a LC column without loss of sample is in direct contrast to the comparatively inefficient task of condensing samples from the vapour phase which elute from a gas chromatograph.

Air-borne pollutants

Air pollutants can be either in the form of vapours, particulate matter or compounds adsorbed on the surface of air-borne particles. The former category, being volatile, are generally studied by techniques other than LC, the most common methods being photometry and GC. Consequently, the methods of trapping the sample will not be described in any detail in this text. Some of the more serious air-borne pollutants which have been studied by LC are typified by harmful compounds such as the carcinogenic hydrocarbons which are not carried in the vapour phase but adsorbed on the surface of dust particles.

Concentration of solids from the atmosphere is accomplished by aspirating quantities of air through low-porosity filters; the volume of gas sampled is monitored using a "gas meter" capable of recording volume corrected for pressure and temperature variations. In this manner very large volumes of air may be sampled and, provided the air intake is remote from the exhaust of the sampling system, air-borne material present only at very low levels may be effectively concentrated.

When the trace impurity is present as a vapour, or droplets of liquid, the process of trapping is more complex. Collection by absorbing the desired product in some liquid reagent, *e.g.*, acid vapours in an alkaline solution or condensation in a cold trap, can be effective in some applications. However, water vapour is simultaneously collected from the atmosphere and can create considerable problems. Selective trapping of vapours relies a good deal on the chemical nature of the compounds being sought, consequently each application must be considered as a separate problem.

Having trapped or filtered the material from the atmosphere by either of the above approaches, a considerable concentration of the desired sample component has already been effected. Dissolution of the collected material in a good solvent for the compound of interest, followed by filtration, will usually render the sample in a form directly amenable to analysis by LC.

Trace components in bulk water

The most effective method of concentrating dissolved solids from water is by the method of freeze drying. The method enables comparatively large volumes of water to be reduced to low bulk or dryness with minimal loss of solids which are present in the water, thus for analyses of this type sensitivity of detection need not necessarily be as severe a problem as one might expect, particularly as the available sample volume is usually very large.

Selective concentration of components which are dissolved in water can also be effected by using the water as the mobile phase that is passed through a column which contains a chromatographic packing which is highly retentive towards the components of interest. The use of strong cation and anion exchangers for the selective extraction of ionic species should be fairly self-evident. Clearly, if sea-water is to be sampled, the ion exchangers should show considerably greater selectivity towards the components of interest than to sodium and chloride ions.

Non-ionic species can often be concentrated in a similar manner using a column packing designed for reversed-phase chromatography, *e.g.*, one of the most widely reported is the polystyrene-based resin Amberlite XAD-2 (Rohm and Haas)[1]. These types of packing are generally highly retentive when water is used as the carrier liquid; components concentrated in this manner are only eluted when organic solvents such as alcohol or ethyl acetate are present in the mobile phase.

Solvent extraction of aqueous samples with an immiscible organic solvent, *e.g.*, chloroform or carbon disulphide, can also be employed in many applications. The ratio of sample to extraction solvent can, in many instances, be adjusted to concentrate the component of interest in the organic phase, for example a 100-ml volume of a water sample extracted with 5 ml of carbon disulphide will effect a twentyfold concentration of a component which shows greater solubility in that solvent than in water. Evaporation of the liquid extract to low bulk provides a solution directly amenable to analysis by LC.

Minor components in bulk solids

Typical of this application is the determination of the toxic residues such as pesticides in foodstuffs, crops and soil. The initial step must involve homogenising the sample to ensure that any sub-samples will be representative of the whole. If the components are readily soluble in solvents at ambient temperature, one of the most efficient methods of extraction can be to combine homogenisation and solvent extraction by employing a high-speed macerator or blender. After filtration, the slurry will yield a dilute solution containing the components of interest along with other co-extractives. An alternative approach which, in many instances, can yield at least equally efficient recovery of components is to extract the sample under reflux, for example in a 'Soxhlet' extractor. Particular advantages of this approach are that the sample is extracted with hot, freshly distilled, solvent and that the solution containing the extracted components at most requires only simple filtration, as the major proportion of the solid sample is retained in the extraction thimble. Since, in general, the compound which is to be determined by LC will have a low vapour pressure, the dilute solution produced by either of these methods may be reduced to low bulk using a rotary evaporator.

The residue remaining after evaporation may still be too complex for direct analysis by LC and further clean up of the sample might be required. In many applications the technique described for impurities in technical materials is directly applicable, in that a fairly large quantity of residue can be chromatographed to free the component of interest from the excess of co-extractives. As an example, Stalling *et al.*[2] have described a gel permeation (steric exclusion) method of cleaning up extracts of fish oil samples prior to assaying pesticide residues. These authors have also described the construction of an apparatus which permits the clean-up procedure to be performed automatically[3].

Drugs and their metabolites in body fluids

The procedures for extraction and isolation of trace components of interest in this application are very similar to those used in the analysis of pesticide residues, with one major exception in that only rarely is the starting material available in large quantities. This important consideration places considerable strain on the entire analysis, in particular, the type of detection system used to monitor the analysis. In the instance of metabolism studies, it is also all important that the speed of analysis or sample pretreatment does not influence the nature or the concentration of the products detected. Careful attention must be given to the possibility of breakdown of compounds occurring after the sample has been collected and before the sample is analysed. The complexity of the subject is such that detailed recommendations are beyond the scope of this text. However, when prolonged storage of the samples prior to analysis is unavoidable, refrigeration of samples should always be considered.

Isolation of drugs of abuse from dilute body fluids, such as urine, using non-polar (reversed-phase) column packings has been reported by several authors, the extraction of amphetamine and phenobarbitone from urine using a column packed with Amberlite XAD-2 resin[4] being a typical example of such a procedure. In the procedure reported by Kullberg *et al.* urine buffered to pH 9 was passed through a column containing the XAD-2, hydrophobic, resin. The compounds of interest were strongly retained while the excess of more polar species present in the sample passed through with little or no retention. After washing the column with water, the amphetamine was displaced using a mobile phase prepared from a mixture of dichloroethane–ethyl acetate–isopropanol (30:45:25). Phenobarbitone could be eluted from the column by using a mobile phase containing dichloroethane and ethyl acetate in the ratio 40:60.

SAMPLE INJECTION

In most cases involving the determination of minor components in a sample, the solution which must be analysed by LC will be dilute with respect to the components of interest. When considering methods available to introduce sufficient mass of sample into the chromatograph, it is important to appreciate the effect that injection volume will have on the chromatographic resolution of components obtained during the analysis. Fig. 13.1 illustrates the influence of increasing the volume of sample injected on the efficiency of

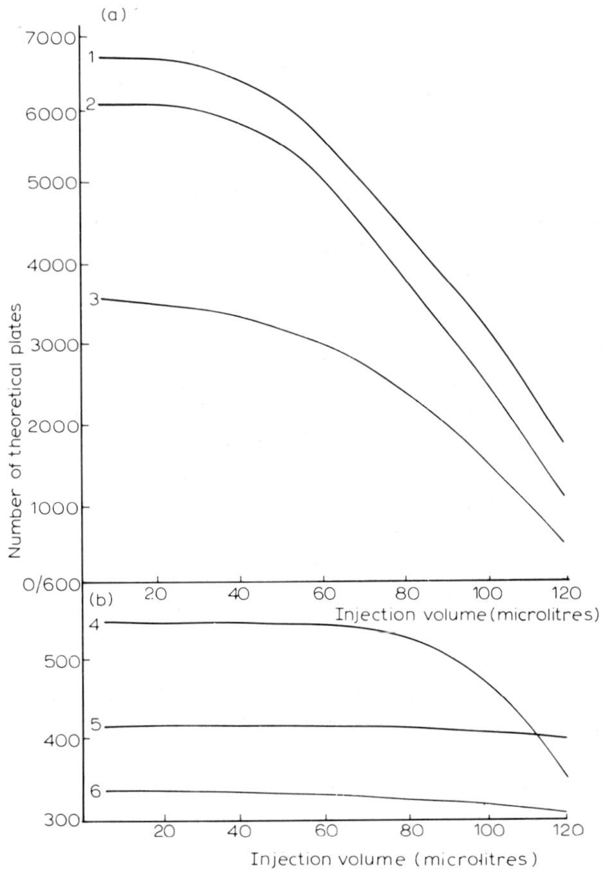

Fig.13.1. Effect of injection volume on column efficiency. Operating conditions: (a) Column, 0.25 m × 2.1 mm I.D.; packing, Zorbax ODS, 4–6 μm; mobile phase, methanol–water (3:1); column temperature, 40°C; inlet pressure, 100 bars (1500 p.s.i.); flow-rate, 0.35 ml/min. (b) Column, 0.5 m × 2.1 mm I.D.; packing, Permaphase ODS, 25–37 μm; mobile phase, methanol–water (65:35); column temperature, 40°C; inlet pressure, 40 bars (600 p.s.i.); flow-rate, 0.9 ml/min. 1 = Pyrene (k' = 2.8); 2 = anthracene (k' = 2.0); 3 = naphthalene (k' = 0.9); 4 = naphthalene (k' = 0.6); 5 = anthracene (k' = 2.9); 6 = pyrene (k' = 5.3). The sample is dissolved in a solvent mixture identical in composition to the mobile phase.

two types of chromatographic column, curve b relating the effect to a solid core material of an overall diameter of approximately 30 μm, while curve a refers to similar tests performed with a totally porous support of approximately 5-μm diameter. The chromatographic packings were chosen to have similar selectivity.

It will be immediately apparent from these curves that the volumes of sample solution used in LC may be considerably larger than those of liquid samples employed in GC. These curves indicate that, even when working with quite narrow-bore columns, *i.e.*, 2 mm I.D., a volume of 90 μl causes only a limited (approximately 9%) decrease in column efficiency in the case of the solid-core material, while a comparable decrease in efficiency is observed with the column packed with porous microparticles when the injection volume is greater than 50 μl. This result should be interpreted while remembering that resolution between successive chromatographic peaks is dependent on the square root of the column efficiency; thus a 9% decrease in efficiency will decrease the resolution by only 3% (see page 14). In this way it is often possible to offset the inherently lower sensitivity of LC detectors compared with those used in GC, to yield analytical procedures with comparable detection limits. Using other data Kirkland has reported[5] that, as a general guide, the maximum volume of sample solution that may be injected in a column is approximately 30% of the true volume of the peak which is to be measured.

When introducing large volumes of sample solution into a chromatographic column, a mechanical valve system will invariably give better results in terms of reproducibility than a manual, syringe-through-septum approach. One particular disadvantage of a valve system, however, is that some sample solution is wasted by the necessity of flushing the valve body and its associated capillary tubing with sample. In applications where the sample volume is severely limited, for example in drug metabolism and forensic studies, special design considerations must be given to the sample valve and its associated tubing to minimise sample volume requirements. In these instances a microsyringe method may represent the only method of manipulating very low volumes of sample in an efficient manner.

A considerable volume of sample solution may be introduced into a column if the nature of the mobile phase which is passing through the chromatographic column, at the time of injection, is such that strong retention of the sample components is obtained. Subsequent application of a solvent gradient will cause the components of interest to elute from the column. By using this procedure, sample volumes of several millilitres can be concentrated on the head of the column. Before considering this approach, the advantages of employing a large-sample volume should be carefully balanced against the possible disadvantages attendant with employing gradient elution over a wide range of mobile phase composition. For instance, some detectors will not give a stable baseline when monitoring a column effluent during a gradient elution run, particularly because in trace determinations the detector is most likely to be operated at high sensitivity: This situation frequently leads to increased difficulty in accurately quantifying the peaks on the chromatogram.

CHROMATOGRAPHIC CONSIDERATIONS

Resolution

Chapter 2 explained in detail that for peaks of equivalent height a resolution factor, R, having a value between 1.0 and 1.5 was adequate for quantitative measurements of peak heights or areas. This conclusion was based on the fact that closely eluting peaks, which approximate to Gaussian form, have in practice 2 and 0.03% overlap of areas when the resolution factor is 1.0 and 1.5, respectively. When considering peaks which are greatly disproportionate in size, however, this degree of overlap is unacceptable since the size of the smaller peak may be less than 2 or 0.03% of the major peak; for such situations, a higher degree of resolution must be sought.

In Chapter 3 it was shown that resolution of peaks is determined by the column efficiency (strictly the square root of this value), the selectivity of the phases and the capacity factor. With regard to the influence of column efficiency, the number of plates available for a given separation may be increased by using a longer column, a recycle technique or a column filled with particles of smaller diameter. The first-mentioned approach will usually result in an increase in the analysis time unless the velocity of the mobile phase is increased to maintain a constant retention time for the sample. This approach may be unacceptable as doubling the length of the column will require a twofold increase in the flow-rate of the mobile phase, and consequently a fourfold increase in inlet pressure, to maintain a constant retention time. Operation of extended column lengths without increasing the flow-rate of the mobile phase necessarily increases the time of analysis and hence decreases the throughput of samples. Recycle chromatography similarly increases the time of analysis, in multiples of the time taken to pass through a single column length. The most practical approach is to employ a column containing particles of smaller diameter, assuming that good column packing characteristics can be maintained. In this way, although an increase in the inlet pressure would be required, analysis times may be kept short with little, if any, increase in the volume of solvent used for each analysis.

Selectivity

Much of the success of modern LC for trace analysis depends on the appropriate choice of the many separation mechanisms available in the liquid phase. When performing any analysis where accurate determination of a minor component is required, it will always be an easier task to quantitate a minor component which elutes before the major peak. In these circumstances, the size of the sample introduced may be increased to a point when the column is overloaded with respect to the major peak, causing it to broaden and possibly tail badly; even so the leading edge is usually affected to a lesser extent, making quantitation of components present at a very low level possible.

Several approaches to the separation of a particular sample mixture may need to be investigated in order to establish one which yields the most useful order of elution for the determination of a specific impurity. Data for the elution behaviour of a group of substituted diphenylureas are presented in chromatographic form in Fig. 13.2. These clearly illustrate the different orders of elution which can be achieved by varying the nature of the stationary phase. The most common approaches to reversing the order of elution of non-ionic species is to compare the behaviour of the individual components on adsorption and normal partition systems with that obtained by reversed phase. Amphoteric ionic species, when studied on anion- or cation-exchange systems, show, in many cases, a similar reversal in their order of elution. Fig. 13.3 shows such a reversal for a group of analgesic substances. With ion-exchange systems the nature of the counter-ion will also impart a degree of selectivity to the chromatographic system, as was outlined in Chapter 9.

The most appropriate choice of mobile and stationary phase combinations is also dependent on the general nature of any co-extractives that are liable to be present in the solution to be analysed. For example, when seeking to analyse a minor component in a fatty (hydrophobic) matrix, it is preferable to try to use an adsorption or normal partition system where the major part of the co-extractives, by being largely soluble in organic solvents, will pass through the column with little or no retention. If a reversed-phase system is employed, these "fatty" co-extractives will possess a strong affinity for the stationary phase and elute fairly slowly, a situation which frequently prolongs an analysis time and

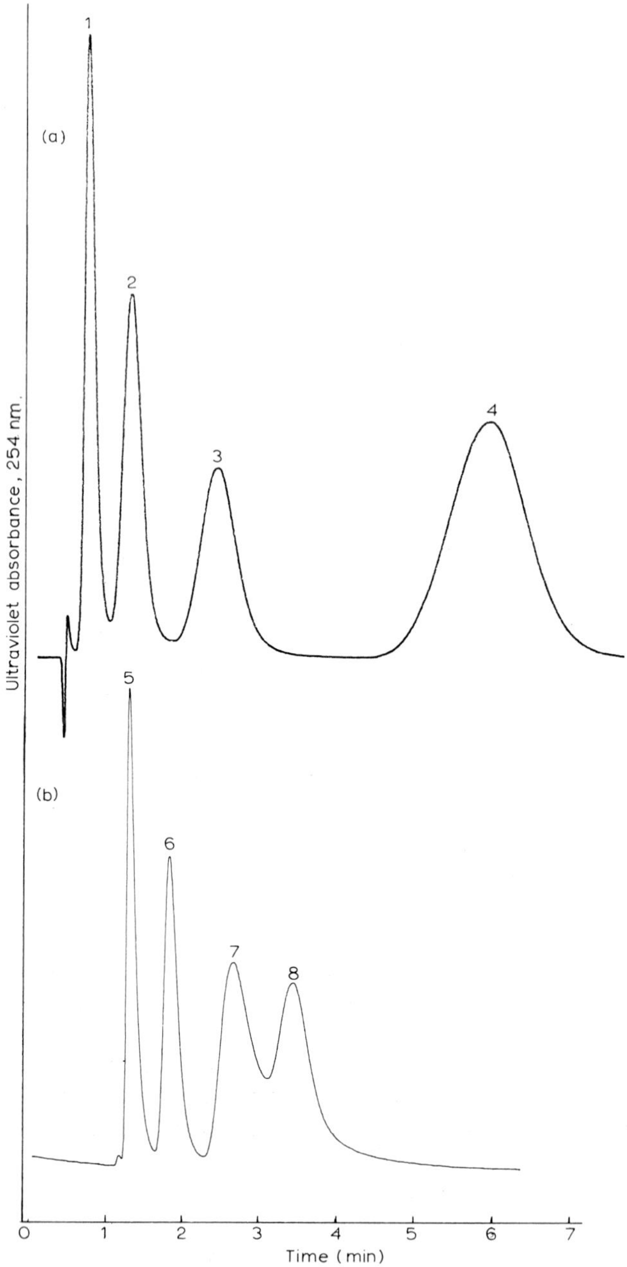

Fig.13.2. Selectivity of different stationary phases towards alkyl-substituted diphenylureas. Operating conditions: (a) Column, 0.5 m × 2.1 mm I.D.; packing, Permaphase ETH; mobile phase, methanol–water (35:65); temperature, 24°C; inlet pressure, 100 bars (1500 p.s.i.); flow-rate, 1.6 ml/min. (b) Column, 0.5 m × 2.1 mm I.D.; packing, Permaphase ODS; mobile phase, methanol–water (35:65); temperature, 24°C; inlet pressure, 40 bars (600 p.s.i.); flow-rate, 0.6 ml/min. 1 = Methyldiphenylurea; 2 = dimethyldiphenylurea; 3 = diethyldiphenylurea; 4 = diphenylurea; 5 = methyldiphenylurea; 6 = dimethyldiphenylurea; 7 = diphenylurea; 8 = diethyldiphenylurea.

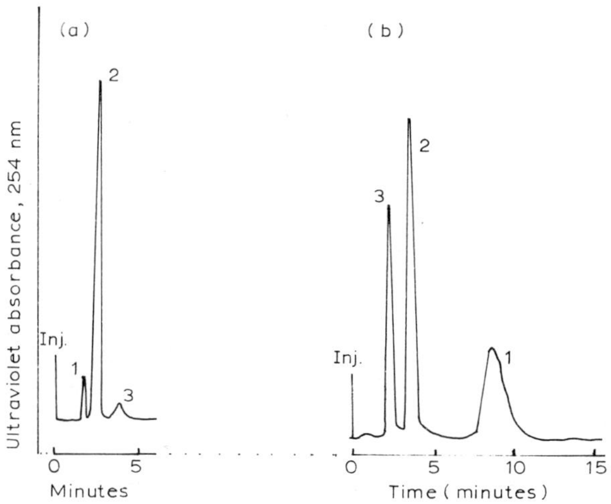

Fig.13.3. Separation characteristics of an analgesic mixture using anion- and cation-exchange packings. Operating conditions: (a) Column, 1 m × 2.1 mm I.D.; packing, Zipax SAX; temperature, ambient; mobile phase, 10 m*M* sodium borate + 5 m*M* ammonium nitrate; flow-rate, 1.5 ml/min; inlet pressure, 80 bars (1200 p.s.i.). (b) Column, 1 m × 2.1 mm I.D.; packing, Zipax SCX, sodium form; temperature, ambient; mobile phase, water; flow-rate, 1.2 ml/min; inlet pressure, 80 bars (1200 p.s.i.). 1 = Caffeine; 2 = phenacetin; 3 = aspirin. (Reproduced by courtesy of DuPont and from R.A. Henry and J.A. Schmit, *Chromatographia*, 3 (1970) 116, with permission.)

may lead to overlapping of late eluting peaks from one injection with earlier eluting peaks from subsequent injections. If reversed-phase partition is the only separation mechanism which appears to resolve the components of interest, a simple partition of the sample between, say, hexane and methanol containing a small percentage of water (*e.g.*, 5%) will often reduce the quantity of co-extractives very effectively as the component of interest, if moderately polar, will usually partition into the hydrophilic layer, leaving much of the hydrophobic material in the less polar phase. This approach has been successfully applied to formulated products, such as steroids in ointments, where the level of active steroid is in the order of 100 ppm (ref.6).

An analogous methodology may be applied to samples in a polar matrix, *i.e.*, analyse in a system where the mobile phase is water or alternatively solvent extract into an organic phase, if adsorption or normal partition chromatography is to be employed. Injection of essentially aqueous solutions containing an excess of very polar components into such columns should be avoided otherwise considerable time will be necessary before the next sample can be introduced.

In many instances, it would seem useful to employ programming techniques, such as gradient elution, to overcome the problem of the slowly eluting peaks referred to in the previous paragraph. Although, in principle, the approach should hold advantage, two factors should be borne in mind that may modify the apparent attraction of employing gradient elution. Firstly, trace analysis almost dictates that whatever detector is used to monitor the separation, it will be operated at, or close to, its maximum sensitivity. Under these conditions, maximum stability is essential and this is best achieved by maintaining a constant composition of mobile phase flowing in the system. Since, moreover, trace analysis usually involves studying many samples, the time taken to re-establish the initial starting condition for the analysis between each sample may be considered excessive.

Column switching, particularly with the use of a guard, or pre-column, as described in Chapter 6, has some merit when samples containing complex co-extractives need to be analysed. Palmer and List[7] have described an effective method employing column switching to concentrate, partially clean-up and subsequently separate organic acids present in extracts of fruit and vegetables. In their method, shown schematically in Fig. 13.4, the sample is first passed through a short pre-column where the acids are firmly retained on an anion exchanger, while the other components in the sample are flushed away with water. After actuating the lower valve the mobile phase is changed to a buffer solution; the acidic components pass from the pre-column into the principle column where resolution of the components is effected. The six-port upper valve is fitted with a 2.0-ml volume external loop which enables comparatively large volumes of dilute solutions to be loaded on the column. This approach to the separation of very complex samples has been demonstrated to be

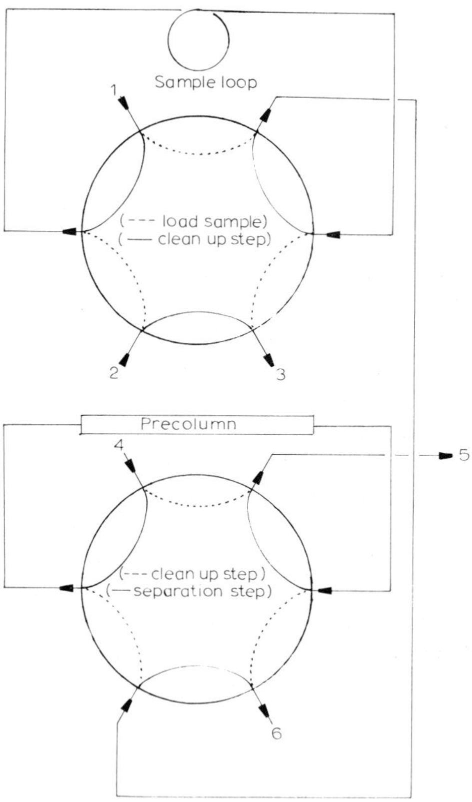

Fig. 13.4. Flow diagram for loading and purification of organic acids from fruit extracts. (1) Water pump in; (2) sample in; (3) vent to drain; (4) mobile phase pump in; (5) to separating column; (6) vent to drain. (Reproduced from J.K. Palmer and D.M. List, *J. Agr. Food Chem.*, 21 (1973) 903, with permission.)

feasible using detectors, such as the refractive index detector, which would respond unfavorably to attempts to programme the mobile phase composition.

Capacity factors

A careful balance must be made when considering the most appropriate capacity factor for a minor peak which must be detected quantitatively. A weakly retained compound, while possibly not being adequately resolved from other components in the sample, has the best characteristics for sensitive detection, that is an early eluting peak has the greatest rate of change of concentration with respect to time and is thus more readily detected. A compromise must be made between a high enough capacity factor for resolution purposes and a low capacity factor for optimum detection.

For best chromatographic resolution, the capacity factors of all peaks of interest should be kept in the range $k' = 1-10$. Below $k' = 1$ extra-column effects can seriously broaden the chromatographic peaks and above $k' = 10$ analysis times become very long. For optimum sensitivity, however, the capacity factor should be low, that is around $k' = 1$; in this region the peaks are sharpest, leading to greatest ease of detection and quantitation, particularly if measured on a peak height basis.

Efficiency

It should be fairly self-evident that if one wishes to enhance the sensitivity of a method, then a sharp, narrow peak eluting from a highly efficient column will be more easily detected than a broad, low-profile peak which is obtained when using columns of low efficiency. In addition, earlier chapters have drawn attention to the fact that the efficiency of nearly all chromatographic columns increases as the velocity of the mobile phase is decreased. This is particularly the case for columns containing moderately sized support particles, *i.e.*, in the region of 30- to 50-μm diameter. A significant reduction in the mobile phase flow-rate will thus lead to narrower peaks, improving the signal-to-noise ratio for a given mass of component. Similarly, due to a combination of effects, many detection systems show a pronounced enhancement of sensitivity as the liquid flow-rate is reduced. This phenomenon arises from decreased noise in the detector due to improvements in the flow of liquid through the cell and better thermal stability as the flow is reduced. With some detectors, the response time of the system has a less serious effect at low flow-rates. With the phase transformation detectors a reduction in the mobile phase flow will lead to an increased proportion of the column effluent being coated on the transporting wire. Separations carried out at low flow-rates will tend therefore to exhibit a somewhat better overall sensitivity compared with high flow conditions; this enhancement will unfortunately be associated with a corresponding increase in the time required to complete an analysis.

DETECTION CONSIDERATIONS

Details of the more common detectors used in LC given in Chapter 5 indicated that most detectors are selective in their response. Clearly, as the task in trace analysis is to

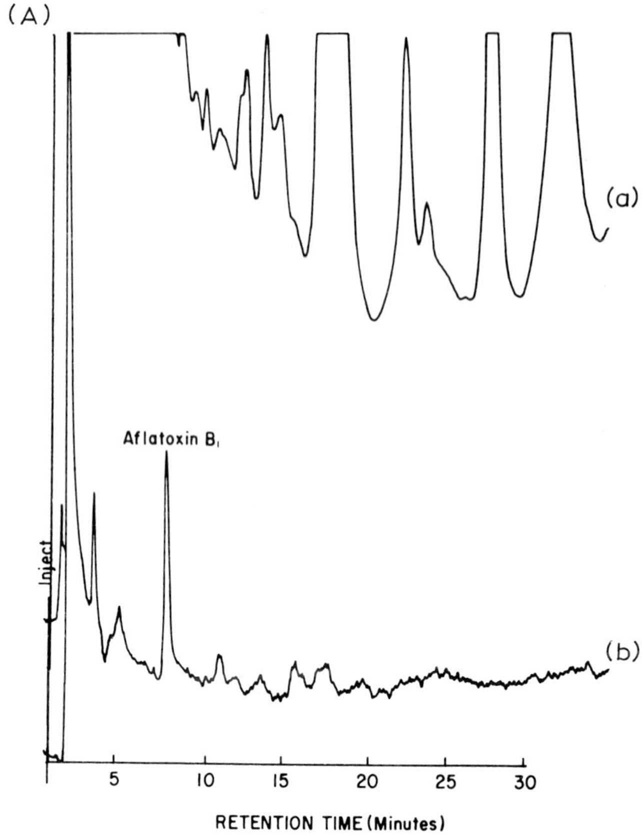

Fig.13.5. Selective detection of aflatoxins in peanut-butter extracts. (A) Comparison of UV detectors operating at different wavelengths. Operating conditions: column, 0.25 m × 2.1 mm I.D.; packing, Zorbax SIL; mobile phase, dichloromethane (50% water saturated)–chloroform (50% water saturated)–methanol (60:40:0.1); flow-rate, 0.7 ml/min; temperature, ambient; inlet pressure, 1500 p.s.i.; detector, UV photometer (a) 254 nm (× 0.02 a.u.) and (b) 365 nm (× 0.01 a.u.). The level of aflatoxin B_1 was 6 ppb. (B) Comparison of absorbance and fluorescence detection. Operating conditions: column, 0.25 m × 2.1 mm I.D.; packing, Zorbax SIL; mobile phase, dichloromethane (50% water saturated)–chloroform (50% water saturated)–methanol (60:40:0.3); flow-rate, 0.7 ml/min; temperature, ambient; inlet pressure, 2000 p.s.i.g.; detector, (c) fluorescence, excitation 365 nm, emission, Corning CS-3-72 and (d) UV photometer, 365 nm (0.02 a.u.f.s.); sample, 50 μl peanut-butter extract. 1 = Aflatoxin B_1, 5 ppb; 2 = aflatoxin G_1, 1 ppb; 3 = aflatoxin B_2, 3 ppb; 4 = aflatoxin G_2, 1 ppb. (Reproduced from D.R. Baker, R.C. Williams and J.C. Steichen, *J. Chromatogr. Sci.*, 12 (1974) 499, with permission.)

quantitatively detect components present in only minor proportions, the ideal analytical situation will be one where the detector may be "tuned" to give a maximum response to the compounds of interest and to minimise interference from other components of little or no interest. In this regard, the most popular detection system of those currently used in LC is the variable-wavelength UV/visible photometric detector, which under favourable conditions will respond to 10^{-9} g/ml of the component in the column effluent. Assuming that the compound being studied absorbs in this region of the spectrum, operation at the

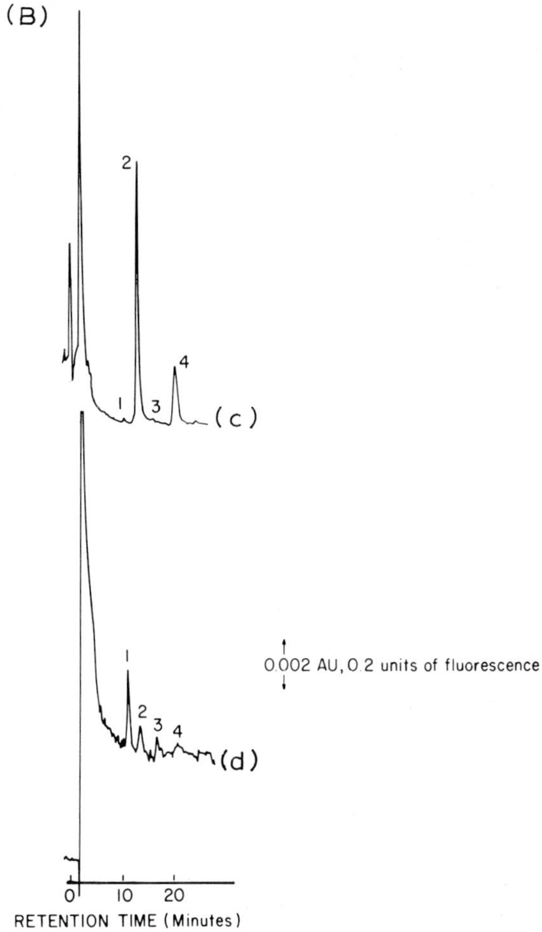

Fig. 13.5 (B).

wavelength of maximum absorption will clearly ensure the highest sensitivity towards that compound, *i.e.*, the greatest response for the least quantity of compound injected. In many applications, however, the wavelength most suitable for a particular analysis does not correspond to the wavelength of maximum absorption of the compound of interest, but rather at a wavelength where absorption from other substances in the sample mixture is at a minimum while allowing some, but not necessarily the maximum, response for the compound of interest. In a similar manner, the solvents used to form the mobile phase will absorb strongly at certain wavelengths so that the wavelength of maximum absorbance for the sample may be unattainable when working with some solvents. Clearly, selection of the optimum wavelength for monitoring a trace analysis is as much concerned with reducing the unwanted responses from interfering substances as it is in optimising the response towards the compounds of interest.

Application of fluorescence detection to trace analysis is very similar to that of UV absorption as a technique which offers the combination of high selectivity and very high sensitivity; in favourable cases it is possible to detect less than 10^{-9}g/ml of the component in the column effluent. The selectivity of response is very much greater than that of UV absorbance in that often fluorescent trace components may be detected with minimal interference, while less selective detection systems fail to discriminate between the components eluting from the column. This situation is exemplified by the chromatograms shown in Fig. 13.5, which illustrate the detection of aflatoxins, present at approximately 20 ppb, in a peanut-butter extract. The chromatograms compare the different responses obtained by monitoring the separation using in Trace A a combination of a fixed (254-nm) wavelength and a variable-wavelength absorbance detector operating at 365 nm, one of the absorption maxima of the compounds of interest where interference from the bulk of the sample is greatly reduced[8]. Trace B illustrates a similar analysis on another sample containing less toxins; here the 365-nm absorption is compared with the fluorescent emission at wavelengths above 451 nm after excitation with a medium-pressure mercury lamp. Clearly, the example was chosen to indicate selective detection where the compound of interest possessed almost ideal spectral characteristics relative to the background caused by other components in the sample. In many applications the absorbance and fluorescence characteristics will not always be as favourable as those shown. It is frequently possible, however, to chemically modify the sample being analysed in order to improve the use of selective detection. Typical of these is the detection of amino acids. One method that was described earlier was their post-column reaction with ninhydrin to produce an intense-blue coloration which is proportional to the concentration of the amino acid. The ninhydrin reaction is time consuming and more recently attention has been given to the separation of phenylthiohydantoin (PTH) derivatives of amino acids to impart a greater ease of detection by UV absorbance at 254 nm. This approach has proved particularly attractive as these derivatives are produced during the procedure according to Edman[9] for establishing the amino acid sequence in peptide chains; also, the type of detection system required, UV absorbance at 254 nm, is widely available from virtually all companies offering LC equipment.

The production of fluorescent derivatives using either "fluorescamine" (post-column) or dansyl chloride as a derivatising agent before the separation is attempted is of considerable value as more specific detection is achieved. Modern reagents such as fluorescamine (marketed by Roche under the trade name "Fluram") which yield fluorescent derivatives are particularly useful in that the reaction is completed in the cold within seconds and the reagent itself does not exhibit any appreciable fluorescence and may consequently be added in excess of the compound being determined without producing a high detector response due to the excess reagent. Dansyl chloride, another established fluorigenic reagent, has been widely used to enhance the detection of trace quantities of carbamate pesticides[10] and phenols[11]. This reagent requires that the sample is refluxed with excess reagent and thus is only applicable to a sample prior to performing the chromatographic separation.

Derivative formation to yield strongly UV absorbing compounds is generally less attractive in that the reagent is normally also a strong UV absorber and may, if present in large excess, obscure the detection of the component of interest. A number of applications of this approach have nevertheless been reported; these include the enhancement of the detection of hexachlorophene by the formation of the di-*p*-methoxybenzoate[12] and the formation of dinitrophenylhydrazones of carbonyl compounds[13], including steroid hormones from biological origin[14,15].

Electrochemical detection systems are progressively coming into more general usage. Because of the frequent need to have a conducting liquid phase these detectors are inherently more suitable for use with aqueous or semi-aqueous mobile phases. Some types of electrochemical detectors, particularly the polarographic detectors, offer a selectivity of response much different from that given by photometric or fluorescence detectors. A recently reported usage of an electrochemical device for a trace analysis application is the detection of catecholamines present in body fluids at the picogram level[16].

Detectors which are essentially non-selective, such as the refractive index detector and the solvent transport to flame ionisation detector, which currently are limited in sensitivity to about 10^{-6} g/ml of compound in the column effluent, are not ideal for trace analysis. In being non-selective, this dictates that very high resolution between components of the sample is necessary if components present at very low levels are to be resolved from components present at much higher concentrations.

The electron capture detector, a particularly selective GC detector, has been shown to have some practical value for the determination of residual quantities of chlorinated pesticides present in the effluent from a LC column[17]. This procedure is reported to be sensitive to the sub-nanogram level for favourable compounds. As little work has yet been reported on this approach, it will be interesting to find whether the technique can be considered for other applications.

From the foregoing paragraphs, it will be realised that the most appropriate detector that should be used for trace analysis will depend largely on the nature of the samples being studied. Without doubt, the photometric detector has so far proved to be the most useful detector for this purpose. Considerable need exists for the introduction of alternative, selective monitors for compounds which are not readily detected photometrically. In the near future, it would seem probable that fluorescence and electrochemical detectors will play a more important role in trace analysis, the former perhaps being coupled to the chromatographic column via some "chemical reaction" chamber which could enable, for example, a fluorigenic reagent to be mixed with the column effluent prior to its entering the detector flow cell.

QUANTITATION OF MINOR COMPONENTS

General aspects of making quantitative measurements in LC have already been described in Chapter 12. In trace analysis, the same principles of quantitation are still valid, with one exception; in this type of application, one is frequently faced with the problem of measuring a very small peak which may elute close to, and perhaps be not completely resolved from, a much larger peak, Kirkland has made a detailed study of the influence of the "neighbouring peak" on measurements of both the height and area of minor peaks falling into this category[5]. He concludes that although greatest precision is usually obtained by measuring peak areas, greater accuracy will be obtained by measuring peak heights in this special situation. The principal reason for this conclusion is that the width of a peak is influenced by a neighbouring peak to a greater extent than is the peak height.

It should be appreciated that in trace analysis accuracy is generally of greater importance than precision. For instance, if there is an impurity in a sample, it is important to know the level at which it is present: is it, say, 1, 10 or 100 ppm? An accurate method will pin-point the value, giving a direct indication as to the significance of the result. On the other hand, a method which, although precise, is inaccurate may give on replicate analysis a result indicating that the impurity is present at a level of 50 ± 5 ppm. The latter value is of little value if the true quantity is only 10 ppm. It is for this reason that peak height measurements are commonly employed when wishing to quantitate chromatograms derived from the analysis of complex mixtures where the concentration of minor components is being sought.

REFERENCES

1 G.A. Junk, J.J. Richard, M.D. Grieser, D. Witiak, J.L. Witiak, M.D. Arguello, R. Vick, H.J. Svec, J.S. Fritz and G.V. Calder, *J. Chromatogr.*, 99 (1974) 745.
2 D.L. Stalling, R.C. Tindle and J.L. Johnson, *J. Ass. Offic. Anal. Chem.*, 55 (1972) 32.
3 R.C. Tindle and D.L. Stalling, *Anal. Chem.*, 44 (1972) 1768.
4 M.P. Kullberg, W.L. Miller, F.J. McGowan and B.P. Doctor, *Biochem. Med.*, 7 (1973) 323.
5 J.J. Kirkland, *Analyst (London)*, 99 (1974) 859.
6 F. Bailey and P.N. Brittain, *J. Pharm. Pharmacol.*, 24 (1972) 425.
7 J.K. Palmer and D.M. List, *J. Agr. Food Chem.*, 21 (1973) 903.
8 R.C. Williams, *DuPont Liquid Chromatography Methods Bulletin*, May 30, 1973.
9 P. Edman, *Acta Chem. Scand.*, 4 (1950) 283.
10 R.W. Frei, J.F. Lawrence, J. Hope and R.M. Cassidy, *J. Chromatogr. Sci.*, 12 (1974) 40.
11 R.M. Cassidy, D.S. Legay and R.W. Frei, *J. Chromatogr. Sci.*, 12 (1974) 85.
12 P.J. Porcaro and P. Shubiak, *Anal. Chem.*, 44 (1972) 1865.
13 M.A. Carey and H.E. Persinger, *J. Chromatogr. Sci.*, 10 (1972) 537.
14 F.A. Fitzpatrick and S. Siggia, *Anal. Chem.*, 45 (1973) 2310.
15 R.A. Henry, J.A. Schmit and J.F. Dieckman, *J. Chromatogr. Sci.*, 9 (1971) 513.
16 P.T. Kissinger, C. Rafshauge, R. Dreiling and R.N. Adams, *Anal. Lett.*, 6 (1973) 465.
17 F.W. Willmott and R.J. Dolphin, *J. Chromatogr. Sci.*, 12 (1974) 695.

Chapter 14

Practical aspects of preparative liquid chromatography

INTRODUCTION

In this context the expression "preparative scale" is understood to mean a separation by which sufficient sample material will be resolved to enable further study to be undertaken on the individual, separate components. Clearly, all chromatographic systems can yield a separation, and could be classified as preparative if the individual components of the sample were collected after the chromatographic separation. However, it will be apparent from earlier sections of this book that analytical-scale separations are performed on samples which are frequently in the order of 1 μg in mass and therefore could not be considered productive in terms of throughput of sample.

The actual mass of sample constituting a "preparative" as distinct from an "analytical" separation depends greatly on the requirements of the purpose for which the sample is to be studied. For example, only a few micrograms are needed if the mass spectrum of the separated component is to be determined, whereas if a toxicological investigation is being considered, a minimum of several grams may be necessary.

TABLE 14.1

APPROXIMATE SAMPLE REQUIREMENTS FOR INSTRUMENTAL ANALYTICAL METHODS

Technique	Approximate sample requirements (mg)
Nuclear magnetic resonance spectroscopy (conventional)	1–10
Nuclear magnetic resonance spectroscopy (Fourier transform)	0.1–1
Infrared (conventional)	0.01–0.1
Mass spectrometry	0.001
Elemental analysis	0.1–1

The approximate quantity of pure substance required for the structural identification of an eluted component using modern instrumental methods is summarised in Table 14.1. These values should be regarded only as a guide for the actual sample requirement, for each individual instrumental technique depends very greatly on the manipulative skill of the operator and on the degree of sophistication of the equipment used.

By using these methods, complete identification of an eluted component should be possible if quantities of the sample in the order of 10 mg can be collected from a chromatographic column.

In addition to the identification of components separated in a liquid chromatograph, there is a considerable interest in the use of LC to collect samples in sufficient quantity to be used as reference grade materials. In these circumstances samples of 100 mg or even greater are generally required.

AVAILABLE METHODS FOR INCREASING THE SAMPLE THROUGHPUT OF CHROMATOGRAPHIC COLUMNS

Scaling-up a successful analytical separation to one capable of manipulating many milligrams of sample may be dealt with in several ways, the principal features of which are outlined as follows:

(1) Use a large-scale column packed with a high-performance chromatographic support and attempt to separate large quantities of sample with the degree of resolution and speed of separation comparable with the analytical-scale separation.

(2) Use a column similar to that described under (1) but overload the system and "cut" fractions from the partially resolved components and re-chromatograph the partially purified, collected fractions after concentration. Collecting a centre portion of an overloaded peak, as illustrated in Fig. 14.1, enables a fraction of higher purity to be obtained relative to collection between the onset and complete elution of the peak. The quantity of component lost by this action is small relative to the total mass collected.

(3) Use a large-scale column packed with an inexpensive support and operate the column at low linear velocity in order to improve the column efficiency, in an attempt to improve the resolution of the sample components.

(4) Use analytical-scale or moderately sized columns in any of the above methods and repeat the separation on smaller quantities of sample until the desired quantity of pure component has been accumulated.

Each of these approaches possesses some merit. However, a compromise must be made between the time taken for a separation, total sample throughput per injection, investment in expensive columns, and instrumental requirements. The most suitable method for a particular separation must be considered in terms of the practical implications which are described in detail in the following sections.

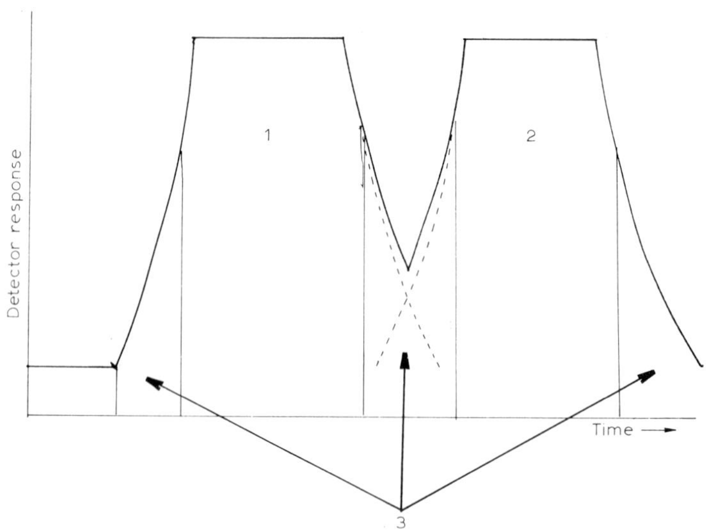

Fig.14.1. Collection of fractions from overloaded columns. (1) Collect peak A; (2) collect peak B; (3) collect, concentrate and re-chromatograph these portions.

EFFECT OF COLUMN GEOMETRY ON CHROMATOGRAPHIC RESOLUTION

Since the separation of large quantities of sample is the topic under consideration, it is important to assess how the geometry of the chromatographic column will influence its sample capacity. Clearly, in terms of geometry there are only two dimensions which may be considered: the internal diameter and the length of the column.

The influence on the performance of the internal diameter of the chromatographic column has already been described in Chapter 3. It was concluded that if columns were packed with identical chromatographic supports an increase in the internal diameter would enable the same or even higher column efficiencies to be obtained when compared under conditions of identical linear velocity of mobile phase. In other words, for columns of the same length, separation times could be comparable irrespective of the diameter of the column. The sample capacity of a column, that is the amount of sample which may be introduced, depends directly on the volume of the chromatographic packing in the column, thus on the square of the radius for columns which are identical in length. In this respect it would seem logical to use as wide a column as is practicable to allow the greatest throughput of sample.

To understand the significance of column length when considering the separation of large quantities of sample, it is necessary to clarify what happens in any given chromatographic columns as the sample size is increased. Take as a starting point a given column–mobile phase combination which provides adequate resolution when used for analytical purposes. If the same chromatographic system is subsequently used for separations where the only change is an increase in the mass of sample introduced, a point will be reached where the chromatographic resolution deteriorates. At this point the system is generally referred to as being "overloaded", that is the linear capacity of the column has been exceeded.

From details given in Chapter 3, it should be remembered that the linear (sample) capacity of column packing material is the maximum mass or concentration of sample

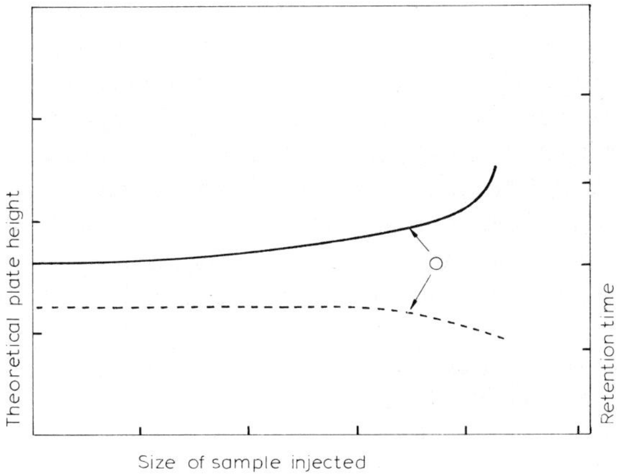

Fig.14.2. Characteristics of column efficiency measurements at the point of column overload. O = Onset of column overload. ——, Plate height curve; -----, retention time curve.

which can be retained on unit mass of chromatographic support without causing a change in the retention time. Fig. 14.2 illustrates this point. The onset of column overload is also marked by a decrease in the efficiency of the column. Since a sharp variation in the column efficiency, or retention, at a given ratio of sample to support is seldom observed, it is usual to define the linear capacity at the point where a "significant" variation of these values occurs, for example a 10% variation from values obtained with smaller samples.

For analytical separations it is good practice to work consistently within the linear sample capacity of the column; for preparative separations, some efficiency may be sacrificed, in order to increase the throughput of sample. When the size of the sample injected is such that the column efficiency deteriorates, clearly so will the resolution, since it is dependent on the square root of efficiency, *i.e.*, the column is overloading. A useful working guide when attempting preparative-scale separations is to consider overload as the mass of sample which yields only half of the resolution between two peaks of equal height, compared with the result obtained with a very small sample. This value will take into account possible changes in retention time and efficiency variations, while providing a working value to the mass of sample which may be separated in a given run.

Returning now to the effect of column length on the separation of large quantities of sample; the only improvement that will result from an increase in the length of a column is that it will off-set the decrease in efficiency when the sample capacity of the column is exceeded. It should be remembered that chromatographic resolution is a function of the square root of the column efficiency, thus if the size of the sample injected has led to a 50% decrease in resolution, it would need a fourfold increase in column length to regain the resolution. Recycle chromatography, as described in Chapter 6, is an alternative method by which the effect of a longer column may be obtained without necessarily using many columns. This can result in a considerable cost saving when separations are being carried out using wide-bore columns. Unfortunately, a significant increase in column length, however achieved, would result in a very considerable increase in the analysis time, thus decrease the throughput of sample per unit time.

The efficiency of many chromatographic columns, in terms of the number of theoretical plates available, increases linearly with the length of the column. However, there are certain instances in practice where this situation does not occur and it is important to recognise the limitations of increasing column length. Such an anomaly is sometimes found with columns containing particles of 10-μm diameter or less, where short columns may be packed very effectively by current procedures whereas longer columns tend to exhibit an efficiency lower than expected due to shortcomings in the packing method. Similarly, it is quite possible for the results obtained from a short column to be superior to those from a longer column of the same internal diameter if the former column possesses the appropriate dimensional requirements, in terms of internal diameter–column length–particle diameter, to behave in an infinite diameter manner[1].

Of the two possible dimensions which may be altered in order to improve the throughput of sample per unit time, an increase in the internal diameter of the column will be the more effective.

CONSIDERATIONS ON THE CHROMATOGRAPHIC SUPPORT

In preparative work the principal criteria for the selection of column packing materials are cost, capacity and efficiency. As the scale of separation increases, the cost and the sample capacity of the chromatographic material become of prime consideration. In analytical LC it is most desirable to have a column of high efficiency which is not markedly dependent on the linear velocity of the mobile phase. This characteristic is sometimes achieved at the expense of the capacity of the support, for example the solid core packings. Although the quality of maintained efficiency at high velocity is desirable in preparative work, the ability to increase the mass of sample per injection is of greater importance. Often a column packing material which is considered of only moderate performance for high-speed analytical LC will usually exhibit a reasonable efficiency if the velocity of the mobile phase is decreased. The characteristic of high capacity combined with modest efficiency at low carrier velocity but decreased efficiency at high velocity is typical of many chromatographic supports which were considered useful column packings for analytical separations a few years ago, but have subsequently been superseded by more sophisticated supports offering higher performance but with a correspondingly higher price. Although the ideal could be to use the packings of highest performance, their use in large preparative systems tends to be limited by cost. A reasonable alternative is to use materials of approximately 30–50 μm diameter, as these are considerably more attractive on a financial basis even though the highest efficiency may only be obtained at low mobile phase velocity.

The relative merits of carrying out multiple, high-speed separations using a small column of high performance or a single separation on a large column offering reasonable performance with an extended separation time have to be assessed for the application under consideration. An increased throughput of sample, while maintaining the chromatographic resolution per unit time, will only be achieved using large columns packed with high-performance, hence invariably high-cost, chromatographic packings.

The previous paragraphs have indicated the features which govern the selection of column packing materials with regard to cost and efficiency. The capacity of the chromatographic packing is clearly of equal importance when selecting a suitable column packing. Much of the success in analytical-scale LC has been due to the simultaneous commercial development of very sensitive detection systems and, around 1970, to the introduction of the solid core or controlled surface porosity supports, which made possible fast liquid phase separations. Full details of the characteristics of these supports have been described in Chapter 3. In this context it is important only to recall that the gain in efficiency and speed of analysis using these packings was obtained by the sacrifice of the capacity of the column packing, thus columns could be loaded only with the smallest sample sizes. While this does not place any undue limitations on most analytical-scale work, the influence in the preparative mode is considerable.

Although it is impracticable to specify the capacity for each type of chromatographic packing, a generally accepted value for a solid core support is considered to be of the order of 100 μg of sample per gram of column packing. Similarly, for a totally porous support, this figure would be approximately 1 mg per gram of column packing, reflecting the higher surface area. The nature of the retention mechanism, the solubility of the sample in the mobile phase and the surface area of packing will modify these values somewhat, but they may be regarded as fairly representative.

From these considerations, it can be seen that a totally porous support gives at least a tenfold increase in sample capacity relative to a solid core support. The comparatively recent development of totally porous microparticulate packings of high efficiency is of great importance in this area of work. Small-diameter, totally porous chromatographic packings are clearly to be preferred for preparative chromatography. Currently, however, there is a greater selection of stationary phase types available on packings of the solid core type, giving a wider choice of selectivity characteristics compared with the totally porous supports. It would seem probable that in a short period of time a similar range of selective stationary phases will be made available on totally porous supports.

It is important to remember during the selection of an appropriate column system for preparative work that the purpose of the procedure is to isolate highly purified components. This point may appear somewhat obvious, but a number of otherwise successful analytical LC systems are not capable of performing this task. The most common types of system where problems are encountered are those relying on partition chromatography where the stationary phase possesses a small, yet significant, solubility in the mobile phase. Details of these systems were described in Chapter 8. Many of the earlier analytical methods used relatively involatile stationary phases, *e.g.*, polyethylene glycols and nitriles such as 1,2,3-tris(2-cyanoethoxy)propane. In these circumstances, evaporation of a fraction of effluent collected from a liquid chromatograph will result in the isolated component being contaminated with a quantity of stationary phase. Separation of the sample from stationary phase can prove a formidable task, since as the component was retarded by the stationary phase in the column it follows that it will be readily soluble in the phase, hence difficult to separate.

For preparative partition chromatography, the use of column packings having the stationary phases chemically bonded to the support is strongly recommended. Similar remarks apply to ion-exchange, steric exclusion and adsorption chromatography, however, in these applications the support is effectively insoluble in the mobile phase, which virtually eliminates the problem.

PRACTICAL ASPECTS OF PREPARATIVE LIQUID CHROMATOGRAPHY

Description of the practical considerations is best made by sub-dividing the process. Thus, the system will be described under the following headings: (1) Sample introduction. (2) The separation process. (a) Chemical considerations. (b) Physical considerations. (3) Detection. (4) Sample collection.

Certain features of preparative chromatography are common to several categories, for example the manipulation of large volumes of mobile phase, particularly at high flow-rate. In these instances they are discussed under the heading of the separation process as the magnitude of the mobile phase flow-rate will be decided by the dimensions of the column used.

Sample introduction

The most satisfactory manner in which to introduce a large quantity of sample is with the aid of a six-port (external loop) valve. This device has been described previously (Chapter 4). The volume of sample solution introduced on activation of the valve is decided by the length and internal diameter of the capillary tubing used to form the loop. In all cases, a long, narrow capillary is to be preferred to a short, wide tube, as the latter will allow mixing of the sample and the mobile phase, leading to unnecessary band spreading and a corresponding decrease in chromatographic resolution.

One of the criticisms that has been made of the loop valve is that if a range of sample volumes is to be investigated, a series of loops of different capacity or timed delivery, as described in Chapter 4, must be used. A very simple alternative to this method is to remove the loop from the valve body and fit a simple syringe needle in the port which would have led from the loop to the column. In this manner, stopped flow injections may be made using a conventional syringe inserted into a needle fixed in one port of the valve[2]; actuation of the shaft makes it possible to transfer liquid directly from the syringe to the top of the column. This procedure is most useful for introducing large volumes of solution during preparative studies, but clearly lacks the precision desired for analytical work.

Independent studies reported in the literature confirm that, in general, better separations are obtained when a large volume of a dilute solution is injected compared to a small volume of a concentrated solution containing the same mass of sample. The principal reason for this observation is the more uniform manner in which a dilute sample is introduced, avoiding the possibility of localised overloading effects. There will understandably be a realistic limit to the volume injected; the paragraph on injection volumes in Chapter 13 clearly indicated that, as a general guide, a sample volume of up to 25–30% of the volume of the eluting peak can be introduced without serious loss in chromatographic resolution. Taking these values into account indicates that when using columns of large diameter, injection volumes may be in the order of 10 ml or even higher.

Very large volumes of sample may be employed in systems where the sample is introduced into the column dissolved in a mobile phase which causes it to be strongly retained, as shown in Chapter 12, Fig. 12.1. Subsequently a change in the composition of the mobile phase is made, either by a stepwise or a continuous gradient, to elute the sample components.

The separation process

The column geometry and the nature of the column packings have already been described. The aspects of the chromatographic process which remain to be described involve the nature of the mobile phase (Chemical considerations) and the manner in which large quantities of sample and high liquid flow-rates are manipulated (Physical considerations).

Chemical considerations

The recommended approach to the development of a preparative-scale separation is to initially obtain a good analytical-scale separation and scale up the procedure by changing, in a step-wise manner, the column geometry and the surface area (or stationary phase loading), for example, by using a totally porous support in place of a solid core material or using supports of smaller diameter. Mobile phase passed through the column which does not contain any eluting sample components can be returned to the supply reservoir so as to conserve solvent. When gradient elution is employed, this approach is clearly impracticable; fractional distillation is then the only effective method to recover solvents.

The chemical nature of the mobile phase is dependent entirely on the nature of the sample being studied and should be selected following the guidelines given in Chapters 6–10. The purity of the mobile phase is of paramount importance as the purpose of the work is to isolate highly purified components, most likely by evaporation of quite large volumes of column effluent. All mobile phases should be selected to have the lowest practicable boiling temperatures so as to avoid the sample being overheated during the evaporation of the column effluent. The mobile phase must be completely free from non-volatile matter: The existence of even minute traces of non-volatile components in the solvent will be concentrated along with the sample components. The minimum preparation for solvents intended for critical preparative work should be double distillation in efficient all-glass apparatus followed by storage in scrupulously clean containers. Even after taking these precautions, it is strongly recommended that a "blank" separation is carried out, collecting the equivalent volumes of solvent and establishing if any column bleed or other contamination is liable to interfere.

With separations based on techniques such as ion-exchange chromatography, a more difficult situation exists as invariably buffers and neutral salts must be added to the mobile phase. The removal of such "contaminants" from a collected fraction of column effluent can pose considerable problems. In some instances, after concentration, a change in the pH of the solution will cause precipitation of the desired product, permitting isolation by filtration or solvent extraction. Samples of high molecular weight have also been "de-salted" by using microcolumns containing Sephadex, a water-compatible polydextran gel which separates the high- and low-molecular-weight species by steric exclusion, *i.e.*, gel filtration[3].

Physical considerations

This second aspect of the chromatographic process is concerned with the manipulation of the relatively large volumes of mobile phase which are encountered in preparative applications, particularly when using columns of large internal diameter.

The quantities involved when scaling up a column system are best put in perspective by using an example. Take, for instance, an analytical-scale separation performed with a column of 2 mm I.D. and a mobile phase flow-rate of 0.5 ml/min. If the internal diameter of the column is increased to 24 mm, in order to achieve the same linear velocity of mobile phase, hence the same speed of analysis, it would be necessary to employ a flow-rate of 72 ml/min. Table 14.2 outlines typical values for the operational variables such as flow-rate, inlet pressure and sample size for several different types of preparative columns and chromatographic packings.

If both the small and the large columns contained the same type of chromatographic support, that is supports having the same particle diameter, the pressure drop across both

TABLE 14.2

TYPICAL VALUES FOR THE OPERATIONAL PARAMETERS IN PREPARATIVE LC

Assumes water as mobile phase; the column is operated at room temperature.

Type of column packing material	Column dimensions	Flow-rate (ml/min)	Approx. capacity (mg)	Approx. inlet pressure for 10 mm/sec linear solvent velocity (bars/p.s.i.)
Pellicular (d_p = 25–35 μm)	0.5 m × 2 mm I.D.	1	0.1	40/600
	0.5 m × 8 mm I.D.	16	1–2	40/600
	0.5 m × 24 mm I.D.	144	10–25	40/600
Porous microparticles (d_p = 5 μm)	0.25 m × 2 mm I.D.	0.5	1–3	300/4500
	0.25 m × 8 mm I.D.	8	10–100	300/4500
	0.25 m × 24 mm I.D.	72	≈500	300/4500

columns would be essentially identical. Similarly, if the overall time of separation was 15 min, the total volume of mobile phase passed through the larger column would be in excess of 1 l.

It should be immediately apparent from this approach that not all pumping systems are able to deliver liquids at this rate. Pumps must clearly be able to refill rapidly and automatically, that is be essentially continuous in their operation and connected to large-capacity reservoirs. The most suitable types of pump for this operation are the reciprocating type and pneumatic amplifier pumps.

Pumping systems having a finite volume of delivery after which the pump must be stopped for refilling are not ideally suited for preparative applications at very high flow-rates. They are, however, completely acceptable when working with columns of somewhat lower diameter, for instance, 8 mm I.D., where mobile phase volume requirements are less than 20 ml/min.

In a somewhat analogous manner, some difficulties can be encountered when passing high liquid flow-rates through narrow-bore capillaries within the chromatograph. Most analytical instruments are designed to reduce extra-column band broadening to a minimum and, understandably, use very narrow-bore capillary tubing (typically 0.25 mm I.D.) in locations such as column connectors and in the line carrying the column effluent to the detector. Tubing of this internal diameter offers no significant resistance to liquid flow at rates in the region of 1 ml/min, a typical value for an analytical separation. However, at the very high flows encountered when using large-diameter columns for high-speed separations, the resistance to flow is no longer negligible and high inlet pressures are generated when attempting to maintain the same mobile phase velocity as used in a narrower column. The use of detector flow cells and connections with a capillary of a somewhat wider internal diameter, for instance 0.5 mm I.D., reduces this back pressure considerably and, at high flow-rates, makes only an insignificant contribution to extra-column band broadening when using columns of large diameter. It should be appreciated that microbore capillary is essential when using narrower columns designed for analytical work. Consequently, if a given liquid chromatograph is required to perform a dual role of analytical and preparative capability, it is desirable to be able to easily change any flow-restricting components.

Detection

Most LC detection systems are suitable for monitoring preparative-scale separations since the concentration of the sample in the column effluent is comparatively high and therefore sensitivity of detection is not a problem. The most useful detectors are the phase transformation to flame ionisation detector, the differential refractive index detector and the photometric detector. The essentially non-specific detectors hold some advantage in this area of work, in that within the linearity of response the size of peaks observed on the chromatogram may be related fairly closely to the relative amounts of each component in a mixture. If selective detectors such as a UV photometric detector are used to monitor a separation, it is good practice to also use a refractive index detector to ensure that no other, UV-transparent, component is present in the sample which might elute at a retention time close to that of the desired component which may happen to be readily detected by UV absorption. This precaution will reduce the chances of frustration on collecting a component only to find that it is not pure.

One problem which is unique to preparative work is that of a detector being too sensitive, *i.e.*, when the concentration of components in the column effluent is sufficiently high to cause an off-scale response on the recorder even if the detector is operated at its lowest sensitivity. This is particularly the case with photometric detectors when strongly absorbing components are eluting from the column. The difficulty can be solved by replacing the flow cell with one having a shorter optical path length. Most flow cells in commercial detectors are fitted with cells of 10 mm path length to optimise sensitivity; substitution with a cell having an optical path length of 1 mm will decrease the sensitivity by one order of magnitude. An alternative and even more convenient procedure is possible when using spectrophotometric detectors which are provided with the facility for varying the wavelength of operation; in that case a wavelength can be selected at which the sample absorbs to a lesser extent.

Sample collection

Having chromatographed the sample, the remaining task is to collect the individual components. To this end the most important feature on an instrument used for preparative work is the provision of a low-volume valve immediately after the detector flow cell from which the fractions may be collected. The procedure for obtaining sample fractions is straightforward and, on a non-routine basis, may be achieved by manually collecting the column effluent as the peak is being recorded on the chromatogram. When short lengths of low-volume tubing are used to connect the detector and the collection valves, the error in taking the sample based on the appearance of the peak on the chart is virtually negligible. The most effective way of deciding if there is any significant delay time between the two events is to inject a small volume of ink into the liquid chromatograph and measure the time between the detector "seeing" the ink and its emergence at the sampling point.

When collection of sample components is to be performed on a routine basis, it is more desirable to use an automatic fraction collection device as described in Chapter 4. These devices are actuated on the basis of time elapsed since the moment of injection, on the basis of the volume of effluent, or at the onset of a chromatographic peak. Perhaps the most sophisticated procedure is to use an electronic integrator which will actuate a fraction collection as the integrator senses the onset of a peak. With any automatic collection device, elimination of all sources of band speading is of paramount importance.

APPLICATIONS OF PREPARATIVE CHROMATOGRAPHY

The number of well documented examples where high-speed LC has been used for the isolation of pure samples is rather limited, the more detailed information tending to be only available from suppliers of chromatographic equipment.

A study of the available literature can, however, provide a useful guide to the order of magnitude of sample throughput that has been realised. A summary of some of this information is produced in Table 14.3, where the sample type, mass, and other chromatographically significant data are given.

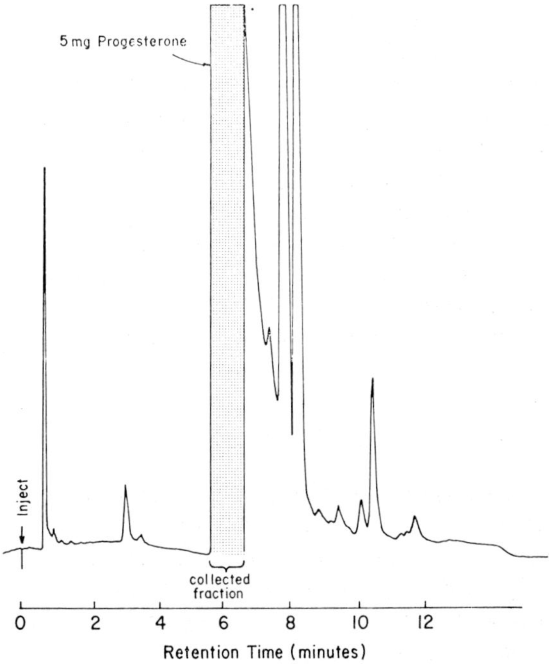

Fig.14.3. Separation of progesterone on a semi-preparative scale. Operating conditions: column, 0.25 m × 2.1 mm I.D.; packing, Zorbax SIL; temperature, ambient; inlet pressure, 100 bars (1500 p.s.i.); mobile phase, gradient from 100% dichloromethane to 90% dichloromethane–10% methanol, gradient rate 2%/min; detector, UV absorbance, 254 nm; sample size, 5 mg. (Reproduced from D.R. Baker, R.A. Henry, R.C. Williams, D.R. Hudson and N.A. Parris, *J. Chromatogr.*, 83 (1973) 233, with permission.)

TABLE 14.3

IMPORTANT FEATURES OF SOME REPORTED PREPARATIVE SEPARATIONS

Abbreviations: pell. = pellicular; por. = porous; const. = constant composition; st.g. = stepwise gradient elution; g.e. = gradient elution.

Sample		Column packing		Sample mass per unit column volume (μg/ml)	Elution type	Time for separation (min)	Flow-rate (ml/min)	Reference
Type	Mass (mg)	Volume (ml)	Type					
Pyrethrins	0.01	3.5	pell.	2.9	g.e.	15	1	4
Pyrethrins	2.0	208	pell.	9.6	const.	25	25	4
Cholesteryl phenylacetate	1000	416	por.	2400	st.g.	45	30	4
Progesterone	5.0	0.85	por.	5900	g.e.	12	0.9	4
Colchicine and isocolchicine	400	2000	por.	200	const.	100	52	5
Testosterone propionate and acetate	400	2000	por.	200	const.	55	55	5
Corticosteroids	600	2000	por.	300	const.	75	60	5
Oestradiol isomers	400	2000	por.	200	const.	60	40	5
Cholesteryl phenylacetate*	550 (75)	12	por.	46000 (6000)	const.	20	11	6

*The values in parentheses are those obtained without overloading the column.

From these tabulated data it is possible to conclude, as would be anticipated from earlier considerations, that the observed sample capacity of columns packed with solid core materials is at least an order of magnitude lower when compared to columns containing porous packings. Similarly, those systems employing gradient elution have enabled a higher throughput of sample per unit time. The latter point is illustrated in Fig. 14.3 by the separation of a 5-mg sample of technical progesterone on a 0.25 m × 2.1 mm I.D. column. The example shown, in addition to indicating the preparative capability of the system, demonstrates an alternative use in detecting impurities present at very low concentrations. The minor peaks in the chromatogram represent impurities in the progesterone present at the 0.01–0.001% level.

In situations where preparative chromatography is only of occasional interest, *e.g.*, when a new component is observed in a sample, columns of approximately 8 mm I.D. represent a useful compromise in terms of financial investment and sample throughput. Assuming such columns are 0.25 or 0.5 m in length and packed with porous particles, sample sizes in the order of 10–500 mg should be separable, the actual quantity depending entirely on the complexity of the separation. Columns of this diameter may be operated at flow-rates in the range of 2–20 ml/min, which is within the capabilities of most LC pumps.

INDUSTRIAL-SCALE CHROMATOGRAPHIC SEPARATIONS

The ability to isolate sizeable quantities of a highly purified compound from a complex mixture by LC poses the question of the viability of using preparative LC on a commercial scale. Many schemes for large-scale separations by both GC and LC have been proposed over the past two decades. Rendell[7] has reviewed these approaches to large-scale separations in terms of the sample throughput and cost. It would seem that the most promising approach for large-scale separations would be to employ a number of relatively short, wide-bore columns operated in parallel. A sample introduction device in the form of a rotating multiport valve could be used to inject pulses of sample into each column in turn. Maximum throughput of sample per column is achieved by arranging the frequency of injections such that the last peak of a previous separation just elutes before the first peak of a subsequent sample. The rate of rotation of the sample introduction valve could be synchronised with the separation time and also with the collection of the component(s) of interest at the column outlet. A "quasi-continuous" introduction of sample into the system could be achieved by employing a sufficient number of columns to ensure that after each has been sampled in turn, the first column is ready to accept another sample.

This aspect of preparative chromatography does not pose any significant new problems in terms of chemical engineering technology and could well be employed commercially in the near future.

REFERENCES

1 J.H. Knox and J.F. Parcher, *Anal. Chem.*, 41 (1969) 1599.
2 J.J. Kirkland, personal communication.
3 *Sephadex-gel filtration in theory and practice*, Pharmacia Fine Chemicals, Uppsala, Sweden.
4 D.R. Baker, R.A. Henry, R.C. Williams, D.R. Hudson and N.A. Parris, *J. Chromatogr.*, 83 (1973) 233.
5 E. Godbille and P. Devaux, *J. Chromatogr. Sci.*, 12 (1974) 564.
6 J.P. Larmann, R.C. Williams and D.R. Baker, *Chromatographia*, 8 (1975) 92.
7 M. Rendell, *Process Eng.*, April (1975) 66.

APPLICATIONS OF LIQUID CHROMATOGRAPHY

Chapter 15

Published LC applications information

Although many may feel that the renewed interest in LC is a comparatively recent event, it is quite remarkable to what extent the technique has been successfully applied. In fact, with laboratories such as those associated with the pharmaceutical industry, modern instrumental LC has brought about a complete change in the methods of routine quality control analysis, in most instances with a distinct reduction in analysis time and an increase in the analytical precision.

New and experienced chromatographers alike will be aware how often, after expending considerable efforts to develop a separation method, it is found that a similar procedure has already been achieved elsewhere. Scientific communications have proliferated in recent years, making, perhaps, more difficult the task of effectively covering all possible sources of information. In this last section of this book, a fairly comprehensive list of references relating to different areas of application are provided. The information presented is grouped into categories relating to an area of chemical application. Thus, as examples, sections are devoted to pharmaceutical, oil, food, pesticides, etc., applications. Available space does not permit full details of the experimental work to be described in each article. Data which are included, in addition to the author and literature reference, are the type of sample, individually named if of particular relevance, and the column packing used. The latter information will provide the more experienced chromatographer with an immediate idea of the mode of separation and hence possible applicability to his own studies. Likewise, information concerning the column packing, mode of separation, and possible choice of alternative chromatographic conditions may be found by referring to Chapters 6–10, where the methods of separation in the liquid phase are described. Details of all column packing materials named in this section can be quickly located by reference to the subject index located at the end of this book.

One particularly valuable source of information on LC applications is the technical literature made available by the manufacturers and suppliers of chromatographic instruments. Applications data from these sources are included in the following summaries. Only documents specifically related to applications with easily recognisable titles are quoted. A good deal of other applications information is often provided as supplementary data in sales brochures; this information is not included. In most instances, the manufacturers listed in Appendix 3 will be very willing to supply what data they may have available on a particular subject.

It is hoped that the sub-division of the information in these sections will enable the newcomer or less experienced chromatographer to quickly acquaint himself with the work which has already been reported on a particular subject. It is a certainty that during the next few years, many other useful technical papers on LC will appear in the scientific literature. Unfortunately, the number of journals in which the data are likely to appear has also increased in recent years. Faced with this situation, the most practical way in which to maintain an awareness of developments in the subject is by making use of the specialised abstracting services which are available. A few of the organisations who provide a regular LC abstract service are:

(1) *Gas and Liquid Chromatography Abstracts* – 4 issues a year
Subscription address: Applied Science Publishers Ltd., Ripple Road, Barking, Essex, Great Britain
(2) *Liquid Chromatography Literature, Abstracts and Index* – 6 issues a year
Subscription address: Preston Technical Abstracts Company, P.O. Box 312, Niles, Ill. 60648, U.S.A.
(3) *Liquid Chromatography Abstracts* – 4 issues a year
Subscription address: Science and Technology Agency, 3 Harrington Road, South Kensington, London SW7 3ES, Great Britain
(4) *Macro Profile – High-Speed Liquid Chromatography* – 26 issues a year
Subscription address: United Kingdom Chemical Information Service, The University, Nottingham, NG7 2RD, Great Britain

PHARMACEUTICAL ANALYSIS

Drugs of abuse, including tranquillizers, barbiturates and amphetamines

1 Adrenaline and noradrenaline
Spherosil XOA 400 (Rhone-Progil)
J. Merzhauser, E. Roeder and C. Hesse, *Klin. Wochenschr.*, 51 (1973) 883

2 Barbiturates
Zipax SAX (DuPont)
R.W. Roos, *J. Pharm. Sci.*, 61 (1972) 1979

3 Barbiturates, detection by use of fluorescent derivatives
ODS-SIL-X-II (Perkin-Elmer)
W. Dunges, G. Naundorf and N. Seiler, *J. Chromatogr. Sci.*, 12 (1974) 655

4 Benzodiazepines
Corasil II, Durapak OPN (Waters)
D. Weber, *J. Pharm. Sci.*, 61 (1972) 1797

5 Benzodiazepines
Durapak OPN (Waters)
C.G. Scott and P. Bommer, *J. Chromatogr. Sci.*, 8 (1970) 446

6 Butalbital in multicomponent analgesic
Corasil C_{18} (Waters)
D. Rosenbaum, *Anal. Chem.*, 46 (1974) 2226

7 Carbamazepine in blood
Perisorb A (Merck)
G. Gauchel, F.D. Gauchel and L. Birkfer, *Z. Klin. Chem. Klin. Biochem.*, 11 (1973) 459

8 Diphenylhydantoin and phenobarbital in blood serum
Micropak SI-10 (Varian)
J.E. Evans, *Anal. Chem.*, 45 (1973) 2428

9 Drugs of abuse in urine
Amberlite XAD-2 (Rohm and Haas)
M.L. Bastos, D. Jukofsky, E. Saffer, M. Chedekel and S.J. Mulé, *J. Chromatogr.*, 71 (1972) 549

10 Drugs of abuse
Corasil II (Waters); Alumina B-18 (Woelm)
M.L. Chan, C. Whetsell and J.D. McChesney, *J. Chromatogr. Sci.*, 12 (1974) 512
11 LSD
Corasil II (Waters); SIL-X (Perkin-Elmer)
J.D. Wittwer, Jr. and J.H. Kluckhohn, *J. Chromatogr. Sci.*, 11 (1973) 1
12 LSD
Zorbax SIL (DuPont)
D.R. Baker, R.C. Williams and J.C. Steichen, *J. Chromatogr. Sci.*, 12 (1974) 499
13 Phenethylamines of forensic interest
DA-X4 (Durrum); Corasil II (Waters)
P.J. Cashman, J.I. Thornton and D.L. Shelman, *J. Chromatogr. Sci.*, 11 (1973) 7
14 Phenothiazines and benzodiazepines
Ion-X-SC (Perkin-Elmer)
D.H. Rogers, *J. Chromatogr. Sci.*, 12 (1974) 742

Steroids

1 Androgen tablets
LiChrosorb SI-60 (Merck)
A.G. Butterfield, B.A. Lodge, N.J. Pound and R.W. Sears, *J. Pharm. Sci.*, 64 (1975) 441
2 Corticosteroids
5- and 10-μm silica with heavily loaded stationary phases
H. Engelhardt, J. Asshauer, U. Neue and N. Weigand, *Anal. Chem.*, 46 (1974) 336
3 Corticosteroid creams and ointments
Zipax BOP, CWT (DuPont)
J. Mollica and R.J. Strusz, *J. Pharm. Sci.*, 61 (1972) 444
4 Corticosteroids and oestrogens
Spherosil XOA-400 (Rhone-Progil)
C. Hesse and W. Hovermann, *Chromatographia*, 6 (1973) 345
5 Corticosteroids
Zorbax SIL (DuPont)
N.A. Parris, *J. Chromatogr. Sci.*, 12 (1974) 753
6 Cortisol, cortisone and aldosterone
SIL-X (Perkin-Elmer)
J.C. Touchstone and W. Wortmann, *J. Chromatogr.*, 76 (1973) 244
7 Cortisol in plasma
Zorbax SIL (DuPont)
F.K. Trefz, D.J. Byrd and W. Kochen, *J. Chromatogr.*, 107 (1975) 181
8 Derivatised non-ultraviolet absorbing hydroxysteroids
Corasil C_{18} (Waters); Permaphase ODS (DuPont)
F.A. Fitzpatrick and S. Siggia, *Anal. Chem.*, 45 (1973) 2310
9 Derivatised urinary 17-ketosteroids
Zipax BOP (DuPont); Corasil C_{18} (Waters)
F.A. Fitzpatrick, S. Siggia and J. Dingman, Sr., *Anal. Chem.*, 44 (1972) 2211

10 Equine oestrogens
Zipax with various liquid and polymer phases, Permaphase ETH and ODS (DuPont)
A.G. Butterfield, B.A. Lodge and N.J. Pound, *J. Chromatogr. Sci.*, 11 (1973) 401

11 Free and dinitrophenylhydrazone derivatives of oestrogens, androgens, progesterones, insect hormones and corticoids
Zipax coated with liquid and polymer phases (DuPont)
R.A. Henry, J.A. Schmit and J.F. Dieckman, *J. Chromatogr. Sci.*, 9 (1971) 513

12 Insect moulting hormones, ecdysones
Corasil II (Waters)
H.N. Nigg, M.J. Thompson, J.N. Kaplanis, J.A. Svoboda and W.E. Robbins, *Steroids*, 23 (1974) 507

13 Methylprednisolone residues in milk
Zipax HCP (DuPont)
L.F. Krzeminski, B.L. Cox, P.N. Perrel and R.A. Schlitz, *J. Agr. Food Chem.*, 20 (1972) 970

14 Prediction of partition coefficients in LC for many steroids
Ternary liquid partition on diatomaceous support
J.F.K. Huber and C.A.M. Meijers, *Anal. Chem.*, 44 (1972) 111

15 Progesterone preparations
Permaphase ODS (DuPont)
R.H. King, L.T. Grady and J.T. Reamer, *J. Pharm. Sci.*, 63 (1974) 1591

16 Synthetic oestrogens
Permaphase ODS, ETH (DuPont)
R.W. Roos, *J. Pharm. Sci.*, 63 (1974) 594

17 Trace oestrogens in pregnancy urine
Ternary liquid partition on diatomaceous support
J.F.K. Huber, J.A.R.J. Hulsman and C.A.M. Meijers, *J. Chromatogr.*, 62 (1971) 79

18 Quantitative assay of final dosage forms of steroids
Corasil C_{18} (Waters)
Waters Associates Leaflet, AN124, dated July, 1973

Alkaloids

1 Cinchona alkaloids, opium alkaloids and heroin
Corasil I, II (Waters); Zipax coated with polyglycols (DuPont)
C.-Y. Wu, S. Siggia, T. Robinson and R.D. Waskiewicz, *Anal. Chim. Acta*, 63 (1973) 393

2 Ergot alkaloids
Zipax (DuPont); Corasil I (Waters)
R.A. Heacock, K.R. Langille, J.D. McNeil and R.W. Frei, *J. Chromatogr.*, 77 (1973) 425

3 Morphine, heroin and methadone
Zipax SAX, SCX (DuPont)
J.H. Knox and J. Jurand, *J. Chromatogr.*, 87 (1973) 95

4 Morphine in opium
Zipax SAX (DuPont)
J.D. Wittwer, Jr., *J. Forensic Sci.*, 18 (1973) 138

5 Nicotine derivatives from tobacco
Poragel A-1 (Waters)
A. Bolt, *Phytochemistry*, 11 (1972) 2341

6 Opium alkaloids
Corasil II (Waters)
T.H. Beasley, D.W. Smith, H.W. Ziegler and R.L. Charles, *J. Ass. Offic. Anal. Chem.*, 57 (1974) 124

7 Oxindole alkaloids
Corasil C_{18} (Waters)
G.H. Jolliffe and E.J. Shellard, *J. Chromatogr.*, 81 (1973) 150

8 Strychnos alkaloids
Corasil I, II coated with polyglycols (Waters)
C.-Y. Wu and S. Siggia, *Anal. Chem.*, 44 (1972) 1499

9 Tropane alkaloids
SIL-X (Perkin-Elmer)
M.H. Knutz and S. Sass, *Anal. Chem.*, 45 (1973) 2134

Analgesics and other medicinal preparations of a general nature

1 Acetaminophen in body fluids and pharmaceutical preparations
Zipax SCX (DuPont); Pellidon (Reeve Angel)
R.M. Riggin, A.L. Schmidt and P.T. Kissinger, *J. Pharm. Sci.*, 64 (1975) 680

2 Acetazolamide in plasma
Vydac Reversed Phase (Separations Group)
W.F. Bayne, G. Rogers and N. Crisologo, *J. Pharm. Sci.*, 64 (1975) 402

3 Antihistaminic and adrenergic compounds
Zipax SCX (DuPont)
T.L. Sprieck, *J. Pharm. Sci.*, 63 (1974) 591

4 Anti-tussive preparations
Permaphase ODS (DuPont)
A. Menyharth, F.P. Mahn and J.E. Heveran, *J. Pharm. Sci.*, 63 (1974) 430

5 Anti-tussive preparations
Corasil C_{18}, Corasil-Phenyl (Waters)
I.L. Honieberg, J.T. Stewart and A.P. Smith, *J. Pharm. Sci.*, 63 (1974) 766

6 Aspirin, salicylamide, caffeine and related compounds
LFS Pellicular anion exchange (Varian)
R.L. Stevenson and C.A. Burtis, *J. Chromatogr.*, 61 (1971) 253

7 Aspirin, caffeine, phenacetin, amino-*p*-aminophen and salicylamide
Zipax-SAX (DuPont)
R.A. Henry and J.A. Schmit, *Chromatographia*, 3 (1970) 116

8 Digitoxin and digoxin
Zorbax SIL (DuPont)
D.R. Baker, R.C. Williams and J. Steichen, *J. Chromatogr. Sci.*, 12 (1974) 499

9 Hexachlorophene, enhanced detection by derivative formation
SIL-X (Perkin-Elmer)
P.J. Porcarro and P. Shubiak, *Anal. Chem.*, 44 (1972) 1865

10 Hexachlorophene
Micropak SI-10 (Varian)
C.D. Carr, *Anal. Chem.*, 46 (1974) 743

11 Hydrochlorothiazide, hydralazine and other polar pharmaceuticals
Zipax SCX (DuPont)
J.B. Smith, J.A. Mollica, H.K. Govan and I.M. Nunes, *Amer. Lab.*, 4 (1972) 13

12 *p*-Aminobenzoic acid and its metabolites
AS-Pellionex SAX (Reeve Angel)
N.D. Brown, R.T. Lofberg and T.P. Gibson, *J. Chromatogr.*, 99 (1974) 635

13 Phenothiazine derivatives with neuroleptic activity
Silica gel, 10 μm
E. Gaetani and C.F. Laureri, *Boll. Chim. Farm.*, 113 (1974) 95

14 Phenylbutazone and oxyphenbutazone in plasma
SIL-X (Perkin-Elmer)
N.J. Pound and R.W. Sears, *J. Pharm. Sci.*, 64 (1975) 284

15 Prostaglandins
SIL-X (Perkin-Elmer)
E.W. Dunham and M.W. Anders, *Prostaglandins*, 4 (1973) 85

16 Prostaglandins
AS-Pellionex SAX (Reeve Angel)
W. Morozowich, *J. Pharm. Sci.*, 63 (1974) 800

17 Reserpine chlorothiazide
Corasil C_{18} (Waters)
I.L. Honieberg, J.T. Stewart, A.P. Smith and R.D. Plunkett, *J. Pharm. Sci.*, 63 (1974) 1762

18 Sodium *o*-iodohippurate
SIL-X (Perkin-Elmer)
A.J. Falk, *J. Pharm. Sci.*, 63 (1974) 274

19 Sulphonylurea-based antidiabetic agents
Zipax HCP (DuPont)
W.F. Beyer, *Anal. Chem.*, 44 (1972) 1312

20 Warfarin and its metabolite, hydroxywarfarin
Permaphase ODS (DuPont)
E.S. Vessell and C.A. Shively, *Science*, 184 (1974) 466

Antibiotics

1 Bacitracin, a polypeptide antibiotic
Corasil C_{18} (Waters)
K. Tsuji, J.H. Robertson and J.A. Bach, *J. Chromatogr.*, 99 (1974) 597

2 Cefoxitin, cephalothin and metabolites in urine
Zipax SAX (DuPont)
R.P. Buhs, T.E. Maxim, N. Allen, T.A. Jacob and F.J. Wolf, *J. Chromatogr.*, 99 (1974) 609

3 Daunomycin antitumour agents
Micropak SI-10 (Varian)
R.E. Majors, *Liquid Chromatography at Work*, Varian Instruments, No. 7, Sep. 610G

4 Fermentation products related to griseofulvin
Permaphase ETH (DuPont)
F. Bailey and P.N. Brittain, *J. Chromatogr.*, 83 (1973) 431

5 Nalidixic and hydroxynalidixic acid in plasma and urine
Zipax SAX (DuPont)
L. Shargel, R.F. Koss, A.V.R. Crain and V.J. Boyle, *J. Pharm. Sci.*, 62 (1973) 1452

6 Sulphanilamides
Micropak SI-10 (Varian)
L. Westlie, B. Aaroe and B. Salvesen, *Medd. Nor. Farm. Selsk.*, 36 (1974) 121

7 Sulphonylureas
Zipax HCP (DuPont)
W.F. Beyer, *Anal. Chem.*, 44 (1972) 1312

8 Tetracyclines
Zipax HCP (DuPont)
K. Tsuji, J.H. Robertson and W.F. Beyer, *Anal. Chem.*, 46 (1974) 539

9 Tetracyclines and rolitetracyclines
HS-Pellionex SCX (Reeve Angel)
A.G. Butterfield, D.W. Hughes, W.L. Wilson and N.J. Pound, *J. Pharm. Sci.*, 64 (1975) 316

10 Thyroid hormones and sulphonamides
LiChrosorb SI-100 (Merck); silica gel CT (Reeve Angel) (used for ion pair partition)
B.L. Karger, S.C. Su, S. Marchesse and B.A. Persson, *J. Chromatogr. Sci.*, 12 (1974) 678

11 Trisulfapyrimidines
Zipax SAX (DuPont)
R. Poet and H. Pu, *J. Pharm. Sci.*, 62 (1973) 809

12 Polyene antifungal antibiotics
Vydac RP (Separations Group); Corasil C_{18} (Waters)
W. Mechlinski and C.P. Schaffner, *J. Chromatogr.*, 99 (1974) 619

BIOCHEMICAL ANALYSIS

Peptides and the screening of body fluids

1 Acidic, neutral and basic metabolites of tyrosine and dihydroxyphenylalanine
PA-28 (Beckman)
W. Martin and H. Cohen, *Anal. Biochem.*, 53 (1973) 177

2 Acidic and neutral catabolites from catecholamines
PA-28 (Beckman)
H.W. Lange, H.F. Mannl and K. Hempel, *Anal. Biochem.*, 38 (1970) 98

3 Amino acids and peptides
HP-AN-90, HP-B-80 (Hamilton)
J. Benson, Jr., *Anal. Biochem.*, 50 (1972) 477

4 Amino acids and amino sugars in peptidoglycans
PA-35 (Beckman)
P. Guire, *Anal. Biochem.*, 42 (1971) 1

5 Biogenic amines and their metabolites
LiChrosphere (Merck) (nearest commercial equivalent)
B.A. Persson and B.L. Karger, *J. Chromatogr. Sci.*, 12 (1974) 521

6 Biological diamines and polyamines
PA-35 (Beckman)
H. Tabor, C.W. Tabor and F. Irreverre, *Anal. Biochem.*, 55 (1973) 457

7 Brain catecholamines
Zipax SCX (DuPont)
C. Refhauge, P.T. Kissinger, R. Dreiling, L. Blank, R. Freeman and R.N. Adams, *Life Sci., II*, 14 (1974) 311

8 Phenylalanine and tyrosine in plasma or serum
Aminex A-5 (Bio-Rad)
A. Mondino and G. Biongiovanni, *J. Chromatogr.*, 67 (1972) 49

9 Porphyrins and peptides
Pellosil HC (Reeve Angel)
R.S. Ward and A. Pelter, *J. Chromatogr. Sci.*, 12 (1974) 570

10 Ultraviolet absorbing constituents in urine
Aminex A-27 (Bio-Rad)
C.D. Scott and W.W. Pitt, *J. Chromatogr. Sci.*, 10 (1972) 740

Sterols and biologically active plant components (not containing nitrogen)

1 Aflatoxins in peanuts and peanut-butter
Zorbax SIL (DuPont)
R.C. Williams, *DuPont Liquid Chromatography Bulletin*, dated May 30, 1973

2 Aflatoxins in peanut-butter extract
Zorbax SIL (DuPont)
D.R. Baker, R.C. Williams and J.C. Steichen, *J. Chromatogr. Sci.*, 12 (1974) 499

3 Bile acids in serum
Amberlite XAD-2 (Rohm and Haas)
H.P. Schwarz, K.V. Bergmann and C. Paumgartner, *Clin. Chim. Acta*, 50 (1974) 197

4 Flavonoids and flavones
LiChrosorb SI-60 (Merck); Pellosil HC, Pellidon (Reeve Angel)
R.S. Ward and A. Pelter, *J. Chromatogr. Sci.*, 12 (1974) 570

5 Furocoumarins
Corasil I (Waters)
F.R. Stermitz and R.D. Thomas, *J. Chromatogr.*, 77 (1973) 431

6 Lipid composition of soya beans
Corasil II (Waters)
O.S. Privett, K.A. Dougherty, W.L. Erdahl and A. Stolyhwo, *J. Amer. Oil Chem. Soc.*, 50 (1973) 516

7 Neutral lipids, glycolipids, phospholipids and steryl glucosides in soya
Corasil II (Waters)
W.L. Erdahl, A. Stolyhwo and O.S. Privett, *J. Amer. Oil Chem. Soc.*, 50 (1973) 513

8 Optically active diastereoisomers related to abscisic acid
Porasil T, Corasil II (Waters)
Waters Associates Leaflet, AN125, dated April, 1973

Amino acids and their derivatives

1 Amino acids (acidic and neutral)
Aminex A-5 (Bio-Rad)
G.E. Atkin and W. Ferdinand, *J. Chromatogr.*, 62 (1971) 373

2 Amino acids in protein hydrolysates
DC-4A (Durrum)
A.G. Georgiadis and J.W. Coffey, *Anal. Biochem.*, 56 (1973) 121

3 Basic amino acids and related
Aminex A-5 (Bio-Rad)
A. Mondino, G. Bongiovanni, V. Noe and I. Raffaele, *J. Chromatogr.*, 63 (1971) 411

4 Basic amino acids and related
DC-2 (Durrum)
M. Friedman, A. Noma and M. Masri, *Anal. Biochem.*, 51 (1973) 280

5 Polyfunctional amino acids (demosine, isodemosine, lysinonorleucine)
Aminex A-5 (Bio-Rad)
R. Green, J. Foster and L. Sandberg, *Anal. Biochem.*, 52 (1973) 538

6 PTH amino acids
Corasil C_{18} (Waters)
A. Haag and K. Langer, *Chromatographia*, 7 (1974) 659

7 PTH (phenylthiohydantoin) amino acids
LiChrosorb SI-60 (Merck)
G. Frank and W. Strubert, *Chromatographia*, 6 (1973) 522

Nucleotides, nucleosides and related purines and pyrimidines

1 Adenosine and guanosine nucleotides
AS-Pellionex SAX (Reeve Angel)
P.R. Brown and R.E. Parks, *Anal. Chem.*, 45 (1973) 948

2 Adenosine mono-, di-, triphosphates
Pellicular anion exchange (Varian)
H.W. Shmukler, *J. Chromatogr. Sci.*, 8 (1970) 653

3 Analysis of nucleotides
Permaphase AAX (DuPont)
DuPont Liquid Chromatography Methods Bulletin, 820M11, dated May 1, 1972

4 Arabinosyladenine 5′-formate
Zipax SCX (DuPont)
A.J. Repta, B.J. Rawson, R.D. Shaffer, K.B. Sloan, N. Bodor and T. Higuchi, *J. Pharm. Sci.*, 64 (1975) 392

5 Comparison of cell extraction procedures for nucleotides
Pellicular anion exchange (Varian)
P.R. Brown and R.P. Miech, *Anal. Chem.*, 44 (1972) 1072

6 Deoxyribonucleotides
Zipax SAX (DuPont)
T.F. Gabriel and J. Michalewsky, *J. Chromatogr.*, 67 (1972) 309

7 DNA in RNA and *vice versa*
Aminex A-7 (Bio-Rad)
D. Duch and M. Laskowski, Sr., *Anal. Biochem.*, 44 (1971) 42

8 Free nucleotides in rat brain
LFS pellicular anion exchange (Varian)
H.W. Shmukler, *J. Chromatogr. Sci.*, 10 (1972) 137

9 Nucleotides in foods, blood and tissue extracts
Permaphase AAX (DuPont)
R.A. Henry, J.A. Schmit and R.C. Williams, *J. Chromatogr. Sci.*, 11 (1973) 358

10 Nucleotides and nucleic acid bases
Zipax SAX, SCX (DuPont)
J.J. Kirkland, *J. Chromatogr. Sci.*, 8 (1970) 72

11 Nucleic acid constituents
Pellicular cation exchange, pellicular anion exchange (Varian)
C.A. Burtis and D.R. Gere, *Nucleic Acid Constituents by Liquid Chromatography*, Varian Aerograph, Walnut Creek, Calif., U.S.A., 1970

12 Oligonucleotides
Zipax WAX (DuPont)
T.F. Gabriel and J.E. Michalewsky, *J. Chromatogr.*, 80 (1973) 263

13 Oligonucleotides
LFS Pellicular anion exchange (Varian)
D. Duch, I. Borkowski, L. Stasiuk and M. Laskowski, Sr., *Anal. Biochem.*, 53 (1973) 459

14 Oligonucleotides
Pellionex WAX (Reeve Angel); Zipax WAX (DuPont)
T.F. Gabriel and J.E. Michalewsky, *Amer. Lab.*, 5, November/December (1973) 10

15 Purine and pyrimidine bases and their nucleosides
Aminex A-28 (Bio-Rad)
P.R. Brown, S. Bobick and F.L. Hanley, *J. Chromatogr.*, 99 (1974) 587

FOOD ANALYSIS

Lipids, fatty acids and simple acids

1 *cis–trans* Isomers of lipid esters
μBondapak C_{18} (Waters)
J.D. Warthen, Jr., *J. Amer. Oil Chem.*, 52 (1975) 151

2 Fatty acids as benzyl esters
Corasil II (Waters)
I.R. Politzer, G.W. Griffin, B.J. Dowty and J.L. Laseter, *Anal. Lett.*, 6 (1973) 539
3 Fatty acid methyl esters
Corasil C_{18} (Waters)
C.R. Scholfield, *J. Amer. Oil Chem.*, 52 (1975) 36
4 Frying fats
Silica gel Type 7754 (Merck); Durapak Carbowax 400 (Waters)
K. Aitzetmuller, *Fette, Seifen, Anstrichm.*, 74 (1972) 598
5 Long-chain fatty acids as 2-naphthacyl esters
Corasil C_{18} (Waters)
M.J. Cooper and M.W. Anders, *Anal. Chem.*, 46 (1974) 1849
6 Mono-, di-, hydroxy- and ketocarboxylic acids
Aminex A-14 (Bio-Rad)
U.J. Kaiser, *Chromatographia*, 6 (1973) 387
7 Organic acids in food
Aminex A-25 (Bio-Rad)
J.K. Palmer and D.M. List, *J. Agr. Food Chem.*, 21 (1973) 903
8 Polar products in frying oils
Merckogel SI-50 (Merck); Porasil A (Waters)
K. Aitzetmuller, *Fette, Seifen, Anstrichm.*, 75 (1973) 256
9 Polyglycol esters of fatty acids
Silica gel Type 7719 (Merck)
R. Wickbold, *Fette, Seifen, Anstrichm.*, 74 (1972) 578
10 Separation into lipid classes
Corasil II (Waters)
A. Stolyhwo and O.S. Privett, *J. Chromatogr. Sci.*, 11 (1973) 20

Vitamins

1 Fat-soluble vitamins (quantitative data)
Zipax HCP, Permaphase ODS (DuPont)
R.C. Williams, J.A. Schmit and R.A. Henry, *J. Chromatogr. Sci.*, 10 (1972) 494
2 Fat- and water-soluble vitamins
Permaphase ODS, Zipax SAX, SCX, HCP (DuPont)
DuPont Liquid Chromatography Methods Bulletin, 820M10, dated March 23, 1972
3 Free tocopherols in plant oils
Corasil II (Waters)
P. van Niekerk, *Anal. Biochem.*, 52 (1973) 533
4 Hydroxylated derivatives of vitamin D_3
Permaphase ODS (DuPont)
E.W. Matthews, P.G.H. Byfield, K.W. Colston, I.M.A. Evans, L.S. Galante and I. MacIntyre, *FEBS Lett.*, 48 (1974) 122

5 Oil-soluble vitamins
Micropak SI-10 (Varian)
C.D. Carr, *Anal. Chem.*, 46 (1974) 743

6 Provitamin A carotenoids in orange juice
Alumina (Woelm)
S.K. Reeder and G.L. Park, *J. Ass. Offic. Anal. Chem.*, 58 (1975) 595

7 Riboflavin in multivitamin preparations
LiChrosorb SI-60 (Merck)
D. Wittmer, and W.G. Haney Jr., *J. Pharm. Sci.*, 63 (1974) 588

8 Riboflavin in prepared food
Zipax SCX (DuPont)
D.R. Baker, R.C. Williams and J.C. Steichen, *J. Chromatogr. Sci.*, 12 (1974) 499

9 Vitamins B_1, B_2, B_6 and nicotinamide in commercial preparations
HS-Pellionex SCX (Reeve Angel)
K. Callmer and L. Davies, *Chromatographia*, 7 (1974) 644

10 Vitamin D_2 in A acetate-D_2 capsules
Zorbax SIL (DuPont)
D.F. Tomkins and R.J. Tscherne, *Anal. Chem.*, 46 (1974) 1602

11 Water-soluble vitamins
Zipax SAX, SCX (DuPont)
R.C. Williams, D.R. Baker and J.A. Schmit, *J. Chromatogr. Sci.*, 11 (1973) 618

Food additives, flavours including beverages

1 Aminobutyric acid and arginine in orange juice
Aminex Q-150-S (Bio-Rad)
C.E. Vandercook and R.L. Price, *J. Ass. Offic. Anal. Chem.*, 57 (1974) 124

2 Complex flavour mixtures
Bio-Beads SX-2 (Bio-Rad); Permaphase ETH, ODS (DuPont)
J.A. Schmit, R.C. Williams and R.A. Henry, *J. Agr. Food Chem.*, 21 (1973) 551

3 Flavour chemicals
Poragel 60 (Waters)
J.P. Walradt and C.-K. Shu, *J. Agr. Food Chem.*, 21 (1973) 547

4 Food preservatives, including benzoic and sorbic acid, ethyl- and propylhydroxybenzoates
LiChrosorb SI-60 (Merck)
W.A. Wildanger, *Chromatographia*, 61 (1973) 381

5 Hop acids
A.P. 212 pellicular anion exchanger, Northgate (now sold by Varian)
R. Vanheertum and M. Verzele, *J. Inst. Brew. London*, 79 (1973) 324

6 Hop bitter acids
Corasil II (Waters)
R.J. Molyneux and Y. Wong, *J. Agr. Food Chem.*, 21 (1973) 531

7 Hop resin acids
Vydac (Separations Group)
S.R. Palamand and J.M. Aldenhoff, *J. Agr. Food Chem.*, 21 (1973) 535

8 Vanillin and ethyl vanillin
Porasil coated with Carbowax 400 (Waters); SIL-X (Perkin-Elmer)
G.E. Martin, G.G. Guinand and D.M. Figert, *J. Agr. Food Chem.*, 21 (1973) 544

Food-colouring materials (including other non-food dyestuffs)

1 Disperse dyes and azo dyes
Zipax BOP, HCP (DuPont); Durapak OPN (Waters)
R.J. Passarelli and E.S. Jacobs, *J. Chromatogr. Sci.*, 13 (1975) 153
2 F. D. and C. Blue No. 2
Zipax SAX (DuPont)
M. Singh, *J. Ass. Offic. Anal. Chem.*, 58 (1975) 48
3 F. D. and C. Red No. 40
Zipax SAX (DuPont)
M. Singh, *J. Ass. Offic. Anal. Chem.*, 57 (1974) 219
4 F. D. and C. Yellow No. 6
Zipax SAX (DuPont)
J.A. Bailey and E.A. Cox, *J. Ass. Offic. Anal. Chem.*, 58 (1975) 609
5 F. D. and C. Yellow No. 6
Zipax SAX (DuPont)
M. Singh, *J. Ass. Offic. Anal. Chem.*, 57 (1974) 358

Sugars, saccharides and artificial sweeteners

1 Anhydroalditols, alditols and saccharides
DC-2 (Durrum)
H. Matsui, E. Paart and O. Samuelson, *Rep. Gov. Ind. Res. Inst. Nagoya*, 21 (1972) 267
2 Carbohydrates (Separations Group)
Vydac cation exchanger
W. Funaska, T. Hanai and K. Fujimura, *J. Chromatogr. Sci.*, 12 (1974) 517
3 Mono- and disaccharides
Aminex A-6 (Bio-Rad)
J.S. Hobbs and J.G. Lawrence, *J. Chromatogr.*, 72 (1972) 311
4 Metabolism of saccharin
Permaphase AAX (DuPont)
J.L. Byard, E.W. McChesney, L. Goldberg and F. Coulston, *Food Cosmet. Toxicol.*, 12 (1974) 175
5 Neutral sugars in complex carbohydrates
DA-X4 (Durrum)
Y. Lee, G. Johnson, B. White and J. Socca, *Anal. Biochem.*, 43 (1971) 640
6 Saccharin, sodium benzoate and other artificial sweeteners
Zipax SAX (DuPont)
J.J. Nelson, *J. Chromatogr. Sci.*, 11 (1973) 28

7 Simple sugars
Aminex Q-150-S (Bio-Rad)
J.K. Palmer and W.B. Brandes, *J. Agr. Food Chem.*, 22 (1974) 709

8 Sugar in barley kernels, determined as borates
Aminex A-25 (Bio-Rad)
D.E. LaBerge, A.W. MacGregor and W.O.S. Meredith, *J. Inst. Brew. London*, 79 (1973) 471

PESTICIDES AND RELATED COMPOUNDS

1 Abate, larvicide
Zipax BOP (DuPont)
R.A. Henry, J.A. Schmit, J.F. Dieckman and F.J. Murphey, *Anal. Chem.*, 43 (1971) 1053

2 Aldrin, DDT, DDD, Lindane, Endrin
Corasil I, II (Waters)
J.N. Little, D.F. Hogan and K.J. Bombaugh, *J. Chromatogr. Sci.*, 8 (1970) 625

3 Benomyl residues in soils and plant tissues
Zipax SCX (DuPont)
J.J. Kirkland, R.F. Holt and H.L. Pease, *J. Agr. Food Chem.*, 21 (1973) 368

4 Benomyl residues in cow milk, urine, faeces and tissues
Zipax SCX (DuPont)
J.J. Kirkland, *J. Agr. Food Chem.*, 21 (1973) 171

5 Carbaryl
Porasil coated with Carbowax 400 (Waters)
B.M. Colvin, B.S. Engdahl and A.R. Hanks, *J. Ass. Offic. Anal. Chem.*, 57 (1974) 648

6 Chlortoluron
LiChrosorb SI-60 (Merck)
A.E. Smith and K.A. Lord, *J. Chromatogr.*, 107 (1975) 407

7 GPC clean-up for pesticide residues
Bio-Beads SX-2 (Bio-Rad)
R.C. Tindle and D.L. Stalling, *Anal. Chem.*, 44 (1972) 1768

8 N-(4-Chlorophenyl)-N′-(2,6-difluorobenzoyl)-urea in milk
Permaphase ODS (DuPont)
C. Corley, R.W. Miller and K.R. Hill, *J. Ass. Offic. Anal. Chem.*, 57 (1974) 1269

9 Chlorinated insecticides, substituted ureas and carbamates
Zorbax SIL, porous silica microspheres (DuPont) (nearest commercial equivalent)
J.J. Kirkland in S.G. Perry (Editor), *Gas Chromatography 1972*, Applied Science Publishers, London, 1973, p. 39

10 Degradation of DTE
Corasil C_{18} (Waters)
F.A. Beland and R.D. Geer, *J. Agr. Food Chem.*, 22 (1974) 1148

11 Lannate
Zipax BOP (DuPont)
R.E. Leitch, *J. Chromatogr. Sci.*, 9 (1971) 531
12 Patulin in apple juice
Zorbax SIL (DuPont)
G.M. Ware, C.W. Thorpe and A.E. Pohland, *J. Ass. Offic. Anal. Chem.*, 57 (1974) 1111
13 Pesticides
Corasil C_{18} (Waters); AS-Pellionex SAX (Reeve Angel)
R. Stillman and T.S. Ma, *Mikrochim. Acta*, 4 (1974) 641
14 Polychlorinated biphenyls
Permaphase ODS (DuPont)
S.H. Byrne, J.A. Schmit and P.E. Johnson, *J. Chromatogr. Sci.*, 10 (1971) 592
15 Pyrethrin extracts and formulations
Bio-Beads SX-2 (Bio-Rad); Permaphase ODS (DuPont)
DuPont Liquid Chromatography Methods Bulletin, 820M12, dated September 1, 1972
16 Residues of Lindane
Corasil II (Waters)
R.H. Larose, *J. Ass. Offic. Anal. Chem.*, 57 (1974) 1046
17 Vitavax (carboxin pesticide) and decomposition products
Corasil C_{18} (Waters)
A.W. Wolkoff, F.I. Onuska, M.E. Comba and R.H. Larose, *Anal. Chem.*, 47 (1975) 754
18 Zectran
ODS-SIL-X-II (Perkin-Elmer)
G.F. Hosler, Jr., *Bull. Environ. Contam. Toxicol.*, 12 (1974) 599

OIL AND PETROLEUM ANALYSIS

Hydrocarbons

1 Aromatics in automotive exhaust condensates
Permaphase ODS (DuPont); Bio-Beads SX-2, SX-8 (Bio-Rad)
J.A. Schmit, R.A. Henry, R.C. Williams and J.F. Dieckman, *J. Chromatogr. Sci.*, 9 (1971) 645
2 Aromatics in heavy distillates
F-20 Alumina (Alcoa)
D.M. Jewell, R.G. Ruberto and B.E. Davis, *Anal. Chem.*, 44 (1972) 2318
3 Aromatic hydrocarbons, use of large-diameter columns
Permaphase ODS (DuPont)
J.P. Wolf, III, *Anal. Chem.*, 45 (1973) 1248
4 Aromatic hydrocarbons
μPorasil (Waters)
R.V. Vivilecchia, R.L. Cotter, R.J. Limpert, N.Z. Thimot and J.N. Little, *J. Chromatogr.*, 99 (1974) 407

5 Benzo[*a*]pyrene in tars and petroleum
Durapak OPN (Waters)
M. Popl, M. Stejskal and J. Mostecky, *Anal. Chem.*, 46 (1974) 1581

6 Benzo[*a*]pyrene in smoke condensate
Vydac Reversed Phase (Separations Group)
J.R. O'Hara, M.S. Chin, B. Dainius and J.H. Kilbuck, *J. Food Sci.*, 39 (1974) 38

7 Benzo[*a*]pyrene metabolites
Permaphase ODS (DuPont)
J.K. Selkirk, R.G. Croy, P.P. Roller and H.V. Gelboin, *Cancer Res.*, 34 (1974) 3474

8 Group analysis of industrial mixtures of aromatic hydrocarbons
Alumina (Woelm)
M. Martin, J. Loheac and G. Guiochon, *Chromatographia*, 5 (1972) 33

9 Metabolism of benzo[*a*]pyrene
Permaphase ODS (DuPont)
J.K. Selkirk, R.G. Croy and H.V. Gelboin, *Science*, 184 (1974) 169

10 Multiple-ring aromatic compounds, parts per billion detection
Spherosil XOA-400 (Rhone-Progil)
W. Strubert, *Chromatographia*, 6 (1973) 205

11 Organic compounds in polluted river water
Porasil T (Waters)
R.A. Hites and K. Biemann, *Science*, 178 (1972) 160

12 Polyaromatic hydrocarbons
Corasil I (Waters) (support impregnated with trinitrofluorenone)
B.L. Karger, M. Martin, J. Loheac and G. Guiochon, *Anal. Chem.*, 45 (1973) 496

13 Polycyclic aromatic hydrocarbons in automotive exhaust condensate
Porasil T (Waters)
T. Doran and N.G. McTaggart, *J. Chromatogr. Sci.*, 12 (1974) 715

14 Polynuclear aromatic hydrocarbons
Spherosil XOB-075 (Rhone-Progil); Alumina (Woelm)
J. Loheac, M. Martin and G. Guiochon, *Analysis*, 2 (1973) 168

15 Polynuclear aromatics in combustion products of fuel, tobacco smoke and food
Corasil II (Waters)
E.D. Pellizari and C.M. Sparacino, *Anal. Chem.*, 45 (1973) 378

16 Polynuclear hydrocarbons
Corasil C_{18} (Waters)
C.G. Vaughan, B.B. Wheals and M.J. Whitehouse, *J. Chromatogr.*, 78 (1973) 203

PETROCHEMICAL AND RELATED COMPOUNDS

Explosives

1 Identification of TNT byproducts
Porasil A (Waters)
C.D. Chandler, J.A. Kohlbeck and W.T. Bolleter, *J. Chromatogr.*, 67 (1972) 255

2 TNT in waste waters
Permaphase ODS (DuPont); Corasil C_{18} (Waters); Amberlite XAD-2 (Rohm and Haas)
J.T. Walsh, R.C. Chalk and C. Merritt, Jr., *Anal. Chem.*, 45 (1973) 1215

3 Propellants containing nitroglycerin
Vydac adsorbent (Separations Group)
R.W. Dalton, C.D. Chandler and W.T. Bolleter, *J. Chromatogr. Sci.*, 13 (1975) 40

Surfactants

1 Chelating agents in commercial detergents
Zipax SAX (DuPont)
J.E. Longbottom, *Anal. Chem.*, 44 (1972) 418

2 Non-ionic surfactants
Spherosil and others (Rhone-Progil)
J.F.K. Huber, F.F.M. Kolder and J.M. Miller, *Anal. Chem.*, 44 (1972) 105

3 Olefin sulphonates
Silanised silica gel Type 7719 (Merck)
H. Puschmann, *Fette, Seifen, Anstrichm.*, 75 (1973) 434

4 Triton
Corasil II (Waters)
K.J. Bombaugh, R.F. Levangie, R.N. King and L. Abrahams, *J. Chromatogr. Sci.*, 8 (1970) 657

Phenols, simple aromatic compounds and alcohols

1 Aliphatic carbonyl compounds
Zipax, coated with 1,2,3-tris(2-cyanoethoxy)propane, Permaphase ETH (DuPont)
L.J. Papa and L.P. Turner, *J. Chromatogr. Sci.*, 10 (1972) 747

2 Aliphatic carbonyl compounds as dinitrophenylhydrazones
Corasil II (Waters)
M.A. Carey and H.E. Persinger, *J. Chromatogr. Sci.*, 10 (1972) 537

3 Aromatic amines
Corasil I, II, Porasil 400, 1500, T (Waters)
J.N. Little, D.F. Horgan and K.J. Bombaugh, *J. Chromatogr. Sci.*, 8 (1970) 625

4 Aromatic amines
Zorbax SIL (DuPont)
P.R. Young and H.M. McNair, *Anal. Chem.*, 47 (1975) 756

5 Aromatic amine isomers
LiChrosorb SI-60 (Merck) (support impregnated with cadmium)
D. Kunzru and R.W. Frei, *J. Chromatogr. Sci.*, 12 (1974) 191

6 Aromatic nitrogen compounds in air samples
Zipax impregnated with silver (DuPont)
R.W. Frei, K. Beall and R.M. Cassidy, *Mikrochim. Acta*, (1974) 859

7 Bases related to pyridine
Amberlite XAD-2 (Rohm and Haas)
C.H. Chu and D.J. Pietrzyk, *Anal. Chem.*, 46 (1974) 330

8 Benzenepolycarboxylic acids
Zipax SAX (DuPont)
J. Aurenge, *J. Chromatogr.*, 84 (1973) 285

9 Benzoic and phthalic acids
Permaphase AAX (DuPont)
J.H. Knox and G. Vasvari, *J. Chromatogr. Sci.*, 12 (1974) 449

10 Monosubstituted pyridine isomers
Zipax SCX (DuPont)
C.P. Talley, *Anal. Chem.*, 11 (1971) 1513

11 Nitro- and chlorophenols
Amberlite XAD-2 (Rohm and Haas)
M.D. Grieser and D.J. Pietrzyk, *Anal. Chem.*, 45 (1973) 1348

12 *o*-,*m*-,*p*-Isomers of substituted anilines and phenols
Pellidon H (Reeve Angel)
F.M. Rabel, *Anal. Chem.*, 45 (1973) 957

13 Phenolcarboxylic acids
Merckogel SI-150 (Merck)
W. Hovermann, A. Rapp and A. Ziegler, *Chromatographia*, 6 (1973) 317

14 Phenol residues in water
Corasil C_{18} (Waters); Zipax SAX (DuPont)
K. Bhatia, *Anal. Chem.*, 45 (1973) 1344

15 Substituted anilines and phenols
Durapak/Carbowax 400, Corasil II (Waters)
R.B. Sleight, *Chromatographia*, 6 (1973) 3

16 Trace phenols in polluted waters
Corasil C_{18} (Waters); Permaphase ODS (DuPont)
A.W. Wolkoff and R.H. Larose, *J. Chromatogr.*, 99 (1974) 731

INORGANIC AND ORGANOMETALLIC COMPOUNDS

1 Acetylacetonates and trifluoroacetonates of metal ions
Ternary liquid partition on diatomaceous support
J.F.K. Huber, J.C. Kraak and H. Veening, *Anal. Chem.*, 44 (1972) 1554

2 Isomers of organo-iron complexes
Permaphase ODS (DuPont)
R.E. Graf, *Liquid Chromatography at Work*, Varian Instruments, No. 3, Sep.610C

3 Organo-iron compounds
Permaphase ODS (DuPont)
R.E. Graf and C.P. Lillya, *J. Organometal. Chem.*, 47 (1973) 413

4 Organic mercury compounds
Corasil I (Waters)
W. Funasaka, T. Hanai and K. Fujimura, *J. Chromatogr. Sci.*, 12 (1974) 517

5 Polythionates
Permaphase AAX (DuPont)
A.W. Wolkoff and R.H. Larose, *Anal. Chem.*, 47 (1975) 1003

6 Rare earth elements, uranium, plutonium and thorium
Zipax (DuPont) (support coated with tricaprylammonium chloride)
E.P. Horwitz and C.A.A. Bloomquist, *J. Chromatogr. Sci.*, 12 (1974) 200

POLYMER ANALYSIS

1 Cellulose trinitrate
Styragel (Waters)
M. Chang, *Tappi*, 55 (1972) 1253

2 Cellulosic materials
Porasil (Waters)
M. van Lancker and E. Veirman, *Ann. Sci. Text. Belg.*, 20 (1972) 98

3 Fatty acids, polystyrenes and hydrocarbons
μStyragel (Waters)
R.V. Vivilecchia, R.L. Cotter, R.J. Limpert, N.Z. Thimot and J.N. Little, *J. Chromatogr.*, 99 (1974) 407

4 High-speed GPC of polystyrenes
Vit-X (Perkin-Elmer)
W. Maclean, *Amer. Lab.*, 6, October (1974) 41

5 Nylon 6
Styragel (Waters)
P.S. Ede, *J. Chromatogr. Sci.*, 9 (1971) 275

6 Poly-(2-vinylpyridine) materials
Porasil (Waters)
A. Gourdenne, N. Hoduc and H. Daoust, *J. Chromatogr.*, 74 (1972) 225

7 Polyacrylamide
Bio-Glass (Bio-Rad); controlled porosity glass (Electronucleonics); Porasil (Waters)
A.H. Abdel-Alim and A.E. Hamielec, *J. Appl. Polym. Sci.*, 18 (1974) 297

8 Polyethylene
Styragel (Waters)
G.N. Patel and J. Stejny, *J. Appl. Polym. Sci.*, 18 (1974) 2069

9 Polyethylene terephthalate
Styragel (Waters)
J.R. Overton and S.K. Haynes, *J. Polym. Sci., Symp. No.*, 43 (1973) 9

10 Polyoxymethylene
Styragel (Waters)
I. Ishigaki, Y. Morita, K. Nishimura and A. Ito, *J. Appl. Polym. Sci.*, 18 (1974) 1927

11 Polystyrene and polydimethylsiloxane
Styragel (Waters)
J.V. Dawkins, J.W. Maddock and D. Coupe, *J. Polym. Sci.*, 8 (1970) 1803

12 Polystyrenes, using microparticulate columns
LiChrospher (Merck) (nearest commercial equivalent)
K.K. Unger, R. Kern, M.C. Ninou and K.-F. Krebs, *J. Chromatogr.*, 99 (1974) 435

13 Review of applications of the analysis of high polymers
J. Mitchell, Jr. and J. Chiu, *Anal. Chem.*, 45 (1973) 273R

14 Review of applications of the analysis of coatings
M.H. Swann, M.L. Adams and G.G. Esposito, *Anal. Chem.*, 45 (1973) 39R

15 Review of applications of the analysis of rubber
C.W. Wadelin and M.C. Morris, *Anal. Chem.*, 45 (1973) 333R

16 Size separation of small molecules
μStyragel (Waters)
Waters Associates Leaflet, AN144, August 1974

Antioxidants, plasticisers and stabilisers

1 Aromatic amine antioxidants, phthalate plasticisers and hindered phenols
Zipax BOP (DuPont); Corasil I, II, Durapak OPN (Waters)
R.E. Majors, *J. Chromatogr. Sci.*, 8 (1970) 338

2 Phthalate plasticisers
Corasil II (Waters)
Waters Associates Leaflet, AN136, December 1973

Chapter 16

The latest trends and a glimpse into the future

LC has developed considerably over the decade up to 1975. The rate of growth in interest and practical application has increased almost exponentially during that time, consequently during the preparation of this book very definite advances have been made, notably in terms of higher column performance and new, more reliable equipment.

Currently the use of microparticulate column packings, *i.e.*, those with diameters of 10 or 5 μm, is fast becoming standard practice. Although speculation can be a hazardous business, amongst the generally acknowledged leaders in the field it is considered that developments involving the use of particles of very much smaller diameter is not necessarily going to contribute to further advances in the speed of analysis. In effect, below a particle diameter of 5 μm there progressively comes a law of diminishing returns as the pressure requirements, necessary for operating a long column filled with such materials, become prohibitive in relation to currently available equipment. Indeed the important questions to ask are: "How efficient a column is required?" and "How fast an analysis is fast enough?".

In Chapter 2, the relationship between the number of effective plates in a column and the selectivity of the mobile/stationary phase system was described and demonstrated in terms of the efficiency required to give a certain resolution of components. Currently, columns of 25 cm length or shorter are generating efficiencies in the order of 10,000–20,000 theoretical plates. At this point, assuming that the number of theoretical plates is approximately the same as the effective plate number, baseline resolution will still be obtained even when using phase systems which offer a selectivity between sample components of 1.05 – a very small difference which in many applications is not difficult to realise.

It can be seen that by using more highly efficient columns it has become easier to develop a "separation" as the selectivity differences of the phases do not need to be exploited to the limit as is the case when using inefficient columns. A great many applications that are currently being performed and no doubt will be performed by LC are those where the number of components is limited, for example, in the quality control assay of a given manufactured product. In these circumstances, using columns of 5,000–10,000 plates efficiency will usually enable an analysis to be performed in a few minutes and it is considered that this speed of analysis will be adequate for very many applications.

In these circumstances, where column efficiencies of a few thousand are all that are required, it is often possible to employ a very short column, for instance only 5 or 10 cm long. Even though the column is packed with very small particles, the short length reduces the pressure requirements considerably, so that many separations, taking only 5 min or so to complete, may be effected with pressures which seldom exceed 35 bars (500 p.s.i.g.). This situation reflects the case that we are dealing with a high-performance rather than necessarily a high-pressure technique.

The use of a column or combination of columns to give very high plate counts, *i.e.*, upwards from 50,000 theoretical plates, only becomes of necessity when attempting to resolve very complex samples, for instance, those originating from body fluids, natural products, tobacco or automobile exhaust smoke. For these applications, assuming that maximum speed is all important, the pressure requirements are likely to be very high indeed. In such cases, prudence prevents a probable maximum pressure from being suggested.

The outcome of these developments to give highly efficient columns is reflected most definitely in instrumental requirements. Nothing has occurred during the developments of the past five years to drastically change the basic concept of the design of a liquid chromatograph – only the quantities have changed, in that there is a need for much less dead volume and a greater need for more effective pumps and detection systems.

Highly efficient columns of the types described dictate the use of systems with very low dead volumes. These aspects were discussed fully in Chapter 6, but the very best performance from columns which are currently being produced could well be limited by instrumental design. In many chromatographs more attention needs to be given to the internal volume of the injector, detector flow cell and associated interconnecting tubing, which must be reduced to the very minimum if the maximum column performance is to be realised.

Perhaps the greatest need at the present time is for novel, tunable detectors. The variable-wavelength photometric detectors have proved outstandingly useful in LC. What is urgently needed is a detector which can be "tuned" to a different type of compound selectivity. Many consider the need may be satisfied by electrochemical detectors or even sophisticated LC/MS systems. This latter facility will go a long way to solving problems associated with applications such as metabolic and forensic studies, but such sophistication carries a considerable penalty in terms of financial investment.

There is a growing demand for reliable automated LC systems; this demand can be met in part by automatic sampling systems, column packings with chemically bonded phases and computing integrators, all of which have already been described. Several instrument companies have introduced process liquid chromatographs which are able to operate completely unattended.

Similarly, more interest is being shown in the use of LC for purposes other than straightforward analytical work. Already many are finding that the highly reproducible systems are ideal for assessing and comparing the physical constants of chemical species. Several publications have confirmed the feasibility of relating LC retention data to partition coefficients[1] and other molecular properties[2,3]. The eventual use of the technique in helping to correlate structure with biological activity for say a series of potential drugs or pesticides would seem inevitable.

The isolation of gram quantities of components from complex natural product mixtures could be of considerable assistance in many areas of biochemical research.

It would appear that LC has a most interesting and important future role in research as well as in the improvement of the speed and accuracy of routine analytical methods. Unfortunately, no one knows exactly how far or how fast the technique will develop – we can only wait and see.

REFERENCES

1 J.F.K. Huber and C.A.M. Meijers, *Anal. Chem.*, 44 (1972) 111.
2 J.D. Kindsvater, P.H. Weiner and T.J. Klingen, *Anal. Chem.*, 46 (1974) 982.
3 P. Menheere, C. Devillez, C. Eon and G. Guiochon, *Anal. Chem.*, 46 (1974) 1375.

Appendix 1

Derivation of the general resolution equation

(Referring to Fig. 2.4 in Chapter 2)

Resolution is defined by the expression

$$R = 2\left(\frac{t_{R_b} - t_{R_a}}{W_b + W_a}\right) \qquad (1)$$

Assuming that for two peaks which are close in retention time the peak widths are approximately the same, *i.e.*, $W_a = W_b$, then eqn. 1 reduces to

$$R = \left(\frac{t_{R_b} - t_{R_a}}{W_b}\right) \qquad (2)$$

The efficiency equation relates retention times with peak widths by the expression

$$N = 16\left(\frac{t_{R_b}}{W_b}\right)^2 \qquad (3)$$

where W_b is the base width of peak b. Substituting this expression in eqn. 2 to eliminate peak widths gives

$$R = \frac{1}{4}\sqrt{N}\left(\frac{t_{R_b} - t_{R_a}}{t_{R_b}}\right) \qquad (4)$$

The capacity factor, k', relates retention time of peaks relative to the void time of a column, *i.e.*

$$k'_b = \frac{t_{R_b} - t_o}{t_o}$$

Rearranging this gives

$$t_{R_b} = t_o\,(k'_b + 1)$$

Substituting for t_{R_b} in the denominator of eqn. 4 gives

$$R = \frac{1}{4}\sqrt{N}\left(\frac{t_{R_b} - t_{R_a}}{t_o}\right) \cdot \left(\frac{1}{k'_b + 1}\right) \qquad (5)$$

Multiplying numerator and denominator by $t_{R_b} - t_o$ gives

$$R = \frac{1}{4}\sqrt{N}\left(\frac{t_{R_b} - t_{R_a}}{t_{R_b} - t_o}\right) \cdot \left(\frac{t_{R_b} - t_o}{t_o}\right) \cdot \left(\frac{1}{k'_b + 1}\right) \qquad (6)$$

This reduces to

$$R = \frac{1}{4}\sqrt{N}\left(\frac{t_{R_b} - t_{R_a}}{t_{R_b} - t_0}\right) \cdot \left(\frac{k'_b}{k'_b + 1}\right) \tag{7}$$

Rearranging gives

$$R = \frac{1}{4}\sqrt{N}\left[\frac{(t_{R_b} - t_0) - (t_{R_a} - t_0)}{t_{R_b} - t_0}\right] \cdot \left(\frac{k'_b}{k'_b + 1}\right) \tag{8}$$

The selectivity factor, α, is defined by

$$\alpha = \left(\frac{t_{R_b} - t_0}{t_{R_a} - t_0}\right)$$

then

$$R = \frac{1}{4}\sqrt{N}\left(\frac{\alpha - 1}{\alpha}\right) \cdot \left(\frac{k'_b}{k'_b + 1}\right) \tag{9}$$

i.e., resolution is a function of the square root of the column efficiency, yet is directly related to the selectivity and capacity of the chromatographic system.

Appendix 2

Comparison of the U.S. (A.S.T.M.) and B.S.S. sieve sizes in relation to aperture size in micrometres

A.S.T.M. Sieve No.	Aperture (μm)	B.S.S. Sieve No.
60	250	60
70	210	72
	180	85
80	177	
100	150	100
120	125	120
150	105	150
	75	200
200	74	
230	63	240
270	53	300
	45	350
325	44	
400	37	

Appendix 3

Suppliers of liquid chromatographic instrumentation and components

Name and address	LC complete and large units, *e.g.*, detectors and pumps	Small accessories, *e.g.*, valves and tube fittings	Columns and packings
Applied Research Laboratories, Wingate Road, Luton, Beds., Great Britain	X		X
Applied Science Laboratories, Inc., P.O. Box 440, State College, Pa. 16801, U.S.A.			X
Bio-Rad Laboratories, 32nd and Griffin Avenue, Richmond, Calif., 94804, U.S.A.			X
Carlo Erba Scientific Instruments, P.O. Box 4342, 20100 Milan, Italy	X		
Cecil Instruments, Trinity Hall Industrial Estate, Green End Road, Cambridge, Great Britain	X		
Chromatec Inc., 30 Main Street, Ashland, Ma. 01721, U.S.A.	X	X	
Disc Instruments, Ltd., Paradise, Hemel Hempstead, Herts., Great Britain		X	
E.I. DuPont de Nemours, Instrument Products Div., Wilmington, Del. 19898, U.S.A.	X	X	X
Durrum Chemical Co., 3950 Fabian Way, Palo Alto, Calif. 94303, U.S.A.	X		X
Electro-Nucleonics Inc., 368 Passaic Ave., Fairfield, N.J. 07006, U.S.A.			X

(Continued on p. 306)

Appendix 3 *(continued)*

Name and address	LC complete and large units, *e.g.*, detectors and pumps	Small accessories, *e.g.*, valves and tube fittings	Colu and pack
E.M. Laboratories, 500 Executive Boulevard, Elmsford, N.Y. 10523, U.S.A.			X
Glenco Scientific Inc., 2802 White Oak, Houston, Texas 77007, U.S.A.	X		
Hamilton Company, P.O. Box 17500, Reno, Nev. 89510, U.S.A.		X	X
Hewlett-Packard, Avondale, Pa. 19311, U.S.A.	X		
Instrumentation Specialist Inc., 4700 Superior Street, Lincoln, N.E. 68505, U.S.A.	X	X	
Japan Analytical Industry, 2165 Ishihata, Mizuho Nishitama, Tokyo 190-12, Japan	X		
Jobin-Yvon, 18 Rue Du Canal, 91160 Longjumeau, France	X		
Jobling, Laboratory Division, Stone, Staffs., Great Britain	X	X	
Kipp and Zonen, Mercuriusweg 1, Delft, P.O. Box 507, The Netherlands	X		
Dr.-Ing. H. Knauer, Adenauerallee 21, 637 Oberusel Ts., G.F.R.	X		
Laboratory Data Control, Interstate Industrial Park, Riviera Beach, Fla. 33404, U.S.A.	X	X	
E. Merck, Darmstadt, G.F.R.			X

Name and address	LC complete and large units, *e.g.*, detectors and pumps	Small accessories, *e.g.*, valves and tube fittings	Columns and packings
Micromeritics Instruments, 5680 Goshen Springs Rd., Norcross, Ga. 30071, U.S.A.	X		
Millipore, Ashby Road, Bedford, Mass. 01730, U.S.A.		X	
Molecular Separations, P.O. Drawer E, Champion, Pa. 15622, U.S.A.	X		
Orlita KG, Max-Eyth-Strasse 10, 63 Giessen, G.F.R.	X		
Packard-Becker BV, Postbus 519, Delft, The Netherlands	X		
Perkin-Elmer Co., Norwalk, Conn. 06856, U.S.A.	X		X
Phase Separations Ltd., Deeside Industrial Est., Queensferry, Flintsh., Great Britain	X	X	X
Pierce Chemical Co., P.O. Box 117, Rockford, Ill. 61105, U.S.A.		X	X
Pye Unicam Ltd., York Street, Cambridge, Great Britain	X		
Reeve Angel (now marketed under the name Whatman), 9 Bridewell Place, Clifton, N.J. 07014, U.S.A.		X	X
Rheodyne, 2809 10th Street, Berkeley, Calif. 94710, U.S.A.		X	

(Continued on p. 308)

Appendix 3 *(continued)*

Name and address	LC complete and large units, *e.g.*, detectors and pumps	Small accessories, *e.g.*, valves and tube fittings	Colu and pack
Rhone-Progil, Rhone-Poulenc Courbevoie, 25 Quai Paul Doumer, 92408 Courbevoie, France			X
Schoeffel Instrument GmbH, Celsiusstrasse 5, 2351 Trappenkamp, G.F.R.	X		
Separations Group, 8738 Oakwood Avenue, Hesperia, Calif. 92345, U.S.A.			X
Siemens AG, Karlsruhe, G.F.R.	X	X	X
SpectraPhysics, 2905 Stender Way, Santa Clara, Calif. 95051, U.S.A.	X	X	X
Supelco Inc., Supelco Park, Bellefonte, Pa. 16823, U.S.A.			X
Varian Associates, 611 Hansen Way, Palo Alto, Calif. 94303, U.S.A.	X		X
Waters Assoc., Inc., Maple Street, Milford, Mass. 01757, U.S.A.	X	X	X
Whatman Inc., 9 Bridewell Place, Clifton, N.J. 07014, U.S.A.		X	X
M. Woelm, Adsorbenzien-Abteilung, 344 Eschwege, G.F.R.			X

Appendix 4

Practical aspects of using simple liquid stationary phases

The use of simple liquids, physically coated, on a chromatographic support can present some difficulties with regard to limitations of compatible solvents for mobile phases and the need for a high degree of control over the experimental conditions. The principal factors which must be given careful attention to are described in the following paragraphs.

The pair of liquids which are to serve as the mobile and the stationary phase should be selected so that they are, as far as practicable, immiscible. Also, the liquid selected to act as the stationary phase should be the better solvent of the two for the sample being studied. An adequate volume of mobile phase is prepared by saturating the appropriate solvent with respect to stationary phase. This is achieved by addition of an excess of the stationary phase to the vessel containing mobile phase and mixing, *e.g.*, with a magnetic stirrer, for several hours, ideally overnight.

Throughout this and all subsequent operations the following precautions must be taken:

(1) Avoid any significant change in temperature of the solvents and chromatographic column. A practical guide would be to limit any temperature change to 2°C.

(2) Ensure that solvent bottles, instrument reservoirs, etc., are covered to limit any evaporation of all or a proportion of the mobile phase.

(3) In the case of air-sensitive liquids, exclude air by passage of a gentle stream of nitrogen over the surface of the mobile phase in the instrument reservoir. Oxydipropionitrile, for example, has been reported to be slowly oxidised by air dissolved in the mobile phase[1]. If this occurs, the stationary phase becomes progressively more retentive as the oxidation proceeds. If the precautions described are followed, the mobile phase will be essentially saturated with stationary phase.

In earlier work reported in the literature, the stationary phase is most commonly applied to the support prior to packing the column. Following the more general use of microparticulate column packings, precoating of the support is impractical since columns are prepared by slurry techniques. In these circumstances the stationary phase must be applied to the pre-packed column.

Three procedures are currently considered for this purpose. In the first method, described by Huber *et al.*[2], the stationary phase layer is achieved by making a series of injections of stationary phase into the column while the corresponding mobile phase is pumped through. An alternative method, that according to Kirkland[3], involves the passage of a concentrated solution of the stationary phase, dissolved in a good solvent, into the column. This solution is then replaced with the desired mobile phase which displaces the bulk of the free solution in the column while precipitating the remainder uniformly throughout the column. Using this method the highest concentration of stationary phase may be deposited on to the column packing. Engelhardt and co-workers have reported a third method, which simply utilises the small degree of mutual solubility that always exists between even some of the most immiscible pairs of liquids. Passage of mobile phase that has been saturated with stationary phase through the column for a period of time will result in a slow build-up of the stationary phase until some steady state of concentration is reached. The rate at which the stationary phase is coated on to the support depends on the rate at which the mobile phase is pumped through the column and the degree of solubility of the stationary phase in the mobile phase. This method is of particular value when only a limited concentration of stationary phase is required on the support, *i.e.*, when wishing to separate compounds which would otherwise be strongly retained.

To ensure complete saturation of the mobile phase with the stationary phase it is necessary to employ a pre-column, as described in Chapter 4, filled with a coarse support, *e.g.*, 105–125 μm diameter. This material should be coated with the highest possible level of stationary phase. The pre-column must be maintained at the same temperature as the column and should be located within the chromatograph ahead of the separating column and injection device so that mobile phase entering the system comes into intimate contact with stationary phase. These conditions will provide the greatest possible opportunity of complete saturation of the mobile phase. The initial equilibrium of mobile phase, by stirring overnight, should not be considered superfluous when using a pre-column, for if solvents which have not been prepared in this manner are passed through the pre-column continuously, the level of stationary phase may be depleted in a short period of time. If this situation continues to the limit, stationary phase will be dissolved from the separating column, leading to a decrease in retention of the sample components.

Operation of the complete chromatographic system should be at constant temperature, the columns being thermostated to within 0.1°C. With physically loaded stationary phases, as described, it is also important that the velocity of the mobile phase is not raised to such a level where its force will physically shear the stationary phase from the surface of the support. This effect has been observed at very high velocity, particularly with columns having a high level of stationary phase loading. In practice this is unlikely to occur as velocities commonly employed during separation procedures, do not exceed 5 cm/sec. Column "bleed" of this type under normal mobile phase velocity conditions is indicative of the facts that the level of stationary phase being used is too high for the support selected or that the temperature of the columns/composition of the mobile phase has changed, rendering the system super-saturated with respect to stationary phase. In a similar manner it is important not to inject solvents which could lead to dissolution of stationary phase. It is strongly recommended that samples are dissolved in a portion of the solvent being employed as a mobile phase, as this cannot disturb the stability of the phase system.

Provided these precautions in operating the chromatographic system are followed carefully, a highly reproducible and stable separation system is obtained. Details of a study of the stability of such a liquid–liquid phase system have been reported by Leitch[1], showing how good quantitative reproducibility was obtained during one year of operation in a quality control application of the technique.

If it is found by experiment that a particular choice of liquid phases and of the level of stationary phase on the support gives inadequate resolution, yet strong retention, then alternative phases should be investigated. The use of a different mobile phase with the original column may be considered. Before using a new mobile phase, however, it must be carefully pre-saturated with stationary phase following the procedure described earlier. Since this requires an extended time period, evaluation of a number of possible mobile phases in this manner can prove laborious and time consuming. If insufficient retention of the sample components is achieved, then a column containing a higher proportion of the stationary phase, or an alternative stationary phase, must be considered. These remarks are valid only when assuming that the column efficiency characteristics remain essentially constant.

In a complex mixture containing components of widely differing polarity, it is frequently found that some components will be strongly retained on the column while others elute without retention. In these circumstances it is difficult to resolve such a mixture without employing some form of programming, *e.g.*, column switching or gradient elution, as described previously in Chapter 6. Unfortunately, because of the simple manner in which the stationary phase is held on the chromatographic support, gradient elution would lead to dissolution of the stationary phase and loss of the column performance. It should be apparent that for gradient elution work the column packing material, *i.e.*, the support and the stationary phase, must be capable of withstanding a change in the chemical composition of the solvents passing through the column. For this reason packings having polymeric or chemically bonded stationary phases are required.

The amount of operator involvement and time required when changing solvents has placed this approach at some disadvantage relative to using columns where the stationary phase firmly adheres to the support, *e.g.*, either as an insoluble polymer or as a chemically bonded phase.

REFERENCES

1 R.E. Leitch, *J. Chromatogr. Sci.*, 9 (1971) 531.
2 J.F.K. Huber, E.T. Alderlieste, H. Harren and H. Poppe, *Anal. Chem.*, 45 (1973) 1337.
3 J.J. Kirkland, *J. Chromatogr. Sci.*, 10 (1972) 593.

Appendix 5

Suppliers of well characterised polymer samples forlar weight standards molecular weight standards

Polymer type	Supplier
Polystyrene, polyethylene and polyvinyl chloride	Pressure Chemical Co., 3419–3425 Smallman St., Pittsburgh, Pa. 15201, U.S.A.
Polystyrene	Dow Chemical Co., Midland, Mich. 48640, U.S.A.
Polyvinyl chloride and poly(1,2-butyleneglycol phthalate)	Ar-Ro Labs, Inc., 1107 W. Jefferson St., Joliet, Ill. 60434, U.S.A.
Linear polybutadiene and linear hydrogenated polybutadiene	Phillips Petroleum Co., P.O. Box 968, Phillips, Tex. 79071, U.S.A.
Polymethyl methacrylate	Rohm and Haas, Independence Mall, Philadelphia, Pa. 19105, U.S.A.

List of abbreviations and symbols

GC	=	Gas chromatography
H (or HETP)	=	Theoretical plate height
K	=	Distribution coefficient
k'	=	Capacity factor (relative partition coefficient)
LC	=	Liquid chromatography
M	=	Molar
mM	=	Millimolar
N	=	Number of theoretical plates
N_{eff}	=	Number of effective theoretical plates
PC	=	Paper chromatography
p.s.i. (or p.s.i.g.)	=	Pounds per square inch (gauge)
R	=	Resolution factor
t_0	=	Retention time of a non-retained component
t_R	=	Retention time of a retained component
TLC	=	Thin-layer chromatography
V_0	=	Void volume of a column
V_i	=	Pore (interstitial) volume of a column
V_R	=	Retention volume
W_b	=	Base width of a peak, strictly of the triangle constructed thereon
α	=	Selectivity factor
μm	=	Micrometre, micron

Subject Index

A

Abate 292
Abscisic acid 287
Abstracting services 279, 280
Accuracy 229
Acenaphthylene–MMA copolymer 205
Acenaphthylene–styrene acrylics 205
Acetaminophen, in body fluids 283
Acetazolamide, in plasma 283
Acetonaphthalene 138
Acetonitrile, as stationary phase 148
Acetophenone 129
Acids, in food 289
---, simple 288
---, suppression of dissociation 110
Acrylates 205
Acrylic styrene–butadiene 205
Acrylonitrile–butadiene rubber 205
Activity, variations in 234
Adenine 171
Adenosine 171
Adenosine-5′-diphosphate 170
Adenosine mono-, di-, triphosphates 287
Adenosine-5′-monophosphate 170
Adenosine nucleotides 287
Adenosine-5′-triphosphate 170
Adrenaline 280
Adrenergic compounds 283
Adsorbents, activation of 39
---, activity of 96, 105, 128
---, changing activity of 140, 141
---, chemically modified 136
---, controlling activity of 136, 139
---, modest-cost 131, 132
---, porous 131
---, porous-layer 131
Adsorbent activity, mobile phase selection and 137, 138
---, temperature effects on 115
Adsorption, in ion exchange 175
---, irreversible 135
Adsorption chromatography 96, 100, 106, 127
---, mechanism of 132–135
---, relation to thin-layer chromatography 129
---, sample applicability of 127, 128
---, solvents for 109, 110
Adsorption isotherm 135, 136
Adsorptive packings, types of 129, 130, 132
Aflatoxins, in peanut-butter extract 260, 286
---, in peanuts and peanut-butter 286
Aflatoxin B_1 258
Aflatoxin B_2 258
Aflatoxin G_1 258
Aflatoxin G_2 258
Agarose 191, 204
Agarose gels 197
Air-borne pollutants 248
Alanine 169
Alcoa F-20 132
Alditols 291
Aldosterone 281
Aldrin 292
Aliphatic carbonyl compounds 295
Alkaloids 282
Alkyd resins 205
Allopurinol 172
Alumina 127, 130–132, 290, 294
Alumina B-18 281
Alumina F-20 293
Amberlite XAD-2 249, 250, 280, 286, 295, 296
Amine antioxidants 298
Aminex A-4 180
Aminex A-5 180, 286, 287
Aminex A-6 180, 291
Aminex A-7 180, 288
Aminex A-14 180, 289
Aminex A-25 180, 289, 292
Aminex A-27 172, 180, 286
Aminex A-28 180, 288
Aminex Q-150-S 290, 292
Aminex resins 176
Amino acids 81, 168, 169, 182, 286, 287
---, basic 287
---, in protein hydrolysates 287
---, polyfunctional 287
Amino-*p*-aminophen 283
p-Aminobenzoic acid, and metabolites 284
Aminobutyric acid, in orange juice 290
γ-Aminopropyltriethylsiloxane, deactivating agent for silica surfaces 201
Amino-SIL-X-I 156
Amino sugars 286
Ammonia 169
Amobarbital 185
Amphetamine 250
Amphoteric substances 182, 253
Analgesics 280, 283
Andrenosterone 121
Androgens 281
---, derivatised 282

Androstenedione 121
Anhydroalditols 291
Aniline 138
Anilines, *o*-, *m*-, *p*-isomers of 296
---, substituted 296
Anions 167
Anion exchangers 167
Anisole 154
Anthracene 107, 145, 251
Antihistaminic compounds 283
Antioxidants for polymers 205
Anti-tussive preparations 283
A.P. 212 pellicular anion exchanger, Northgate 290
Applications information 279
Aquapak 200
Arabinosyladenine 5′-formate 287
Ar Gel 200
Arginine 169, 290
Aromatic amines 295
Aromatic compounds 293, 294
Aromatic hydrocarbons 293, 294
Aspartic acid 169
Asphalts 205
Aspirin 255, 283
Azobenzene 154

B

Bacitracin 284
Balanced density slurry 37, 38
Ballotini beads 34
Band broadening 24
---, extra-column 20, 30, 122, 125, 271
---, post-column 223
---, sources of 20
Band widths, of photometric detectors 80
Barbital 185
Barbiturates 186, 280
Bases, suppression of dissociation 110
Beckman AA-15 180
Beckman PA-28 180, 285
Beckman PA-35 180, 286
Beer's Law 76
Benomyl, residues 292
Bentonite clays 130
Benzene 145, 154, 157
Benzenepolycarboxylic acids 296
Benzodiazepines 280, 281
Benzoic acid 107, 290, 296
Benzopyrene 154
Benzo[*a*]pyrene 294
Benz[*e*]pyrene 145
Benz[α]pyrene 145
Bile acids, in serum 286
Bio-Beads S 199
Bio-Beads SX-2 214, 290, 292, 293
Bio-Beads SX-8 293
Bio-Gel A 199
Bio-Gel P 199
Biogenic amines 171, 286
Bio-Glass 202, 297
Biological diamines 286
Bio-Rad AG 132
Bio-Sil A 107, 132
Biphenyl 145
Bis-(2-ethylhexyl) phosphoric acid 188
Body fluids 171, 285
Boiling points of solvents 99
μBondapak C_{18} 288
Bondapak C_{18}/Corasil 156
μBondapak C_{18}/Porasil 156
Bondapak Phenyl/Corasil 156
Borate complexes 182
Brockman scale of activities 105
Butadiene, *cis*-polymer 205
Butalbital 280
Butyl rubber 205

C

Caffeine 255, 283
Capacity 17, 110
---, linear 10
---, linear sample 266
---, sample 10, 27, 29
Capacity factors 8, 10, 14, 16, 30, 116, 146, 221, 222
---, optimum range of 100
---, role in trace analysis 257
Carbamates 292
Carbamate pesticides 260
Carbamazepine, in blood 280
Carbaryl 292
Carbazole 129
Carbohydrates 291
Carbohydrate analysis 182
Carbon disulfide 212
Carbowaxes 205
Carbowax P-4000 212
Carboxylic acids 289
Carotenoids, in orange juice 290
Carrier 7
Catecholamines 261, 285, 286
Cation exchangers 167
Cefoxitin, in urine 284
Cellulose acetate 205
Cellulose nitrate 205
Cellulose trinitrate 297
Cellulosic materials 297
Centrifugation 231
Cephalothin, in urine 284

Charcoal 127, 130, 131
Chelating agents, in commercial detergents 295
Chemically bonded stationary phases 146
---, mobile phase selection 157
---, pH stability of 154, 155
---, preparation of 152–155
---, properties of 153–155
---, types of 156, 164, 165
Chlordiazepoxide 107
Chlorinated insecticides 292
N-(4-Chlorophenyl)-N′-(2,6-difluorobenzoyl)urea, in milk 292
Chlortoluron 292
Cholesteryl phenylacetate 274
Chromatograms, recording of 73
Chromatographs, components of 43
---, dead volume in 66
---, manufacturers of 74
---, safety of 45
Chromatographic separations, sources of error in 230, 234
Chromatographic support, design of 24
Chrysene 145, 151
Cinchona alkaloids 282
Cis/trans isomers, of lipid esters 288
---, resolution of 134
Cis/trans pairs, separation of 127
Clean-up method 247, 250
Coal tar pitch 205
Codinene 214
Colchicine 274
Colorimetric detection 81, 169
Colour reactions, post-column 81, 168
Columns 66
---, coupled 196
---, equilibration of 233
---, guard 67
---, overload 235, 264–266
---, testing of 39
---, unpacking of 39, 40
Column capacity 116, 117
Column chromatography, classical 19, 26, 31, 98, 127, 129
Column connector 31, 67
Column coupling 32, 67
Column dimensions 31, 32
Column efficiency 14, 15
Column effluents, rapid scanning of 223
Column equilibration time 100, 107, 115, 140
Column geometry 31
---, sample throughput and 265, 266
Column length, calculation of optimum 15
Column packings, silanised 201
Column packing machine 36, 37
Column packing materials, capacity 267
---, cost 267
---, efficiency 267
Column packing methods 25–27, 34–38
Column selectivity 110
Column switching 118, 119, 121, 256
---, apparatus for 119, 120
---, as clean-up method 256
---, detector choice and 122
---, in partition chromatography 122
Column type, selection of 105
Column wall 31–33
Complex mixtures, elution behaviour of 110
Constant flow pumps 47
Constant pressure pumps 45, 46, 48, 49
Controlled Porosity Glass 29, 202, 297
Controlled surface porosity supports 27, 267
CO:PELL ODS 156
Corasil 28
Corasil I 120, 148, 282, 283, 286, 292, 294–296, 298
Corasil II 120, 130, 280–283, 286, 287, 289, 290, 292–296, 298
Corasil C_{18} 280–284, 287, 289, 292–296
Corasil-Phenyl 283
Corn oil glycerides 107
Corticoids 282
Corticosteroids 163, 274, 281
Corticosterone 121, 161
Cortisol 121, 161, 281
Cortisone 121, 161, 281
Countercurrent distribution techniques 143
Counter ions, UV absorbing 188
Counter-ion concentration, retention and 185
Cyano-SIL-X-I 156
Cysteine 169
Cytidine-5′-diphosphate 170
Cytidine-5′-monophosphate 170
Cytidine-5′-triphosphate 170

D

Dansyl derivatives 83, 260
Daunomycin, antitumour agents 285
Davison Code 12 132
Davison Code 62 132
DDD 292
DDT 292
Dead volume 7, 25, 30, 90, 120, 300
---, extra-column 122
Decomposition of samples, adsorbents and 135
---, minimisation of 136
Decylbenzene 121
Dehomogenisation of the mobile phase 112, 186, 234
11-Dehydrocorticosterone 121
11-Deoxycorticosterone 161
11-Deoxycortisol 161
Deoxyribonucleotides 288

Derivatisation, to enhance detection 260
Derivatised dextran 204
Detection systems 75
Detectors, charged aerosol 90
---, connection of 69, 70
---, dead volume of 91
---, electrical conductivity 88
---, electron capture 88
---, fluorescence 81, 82
---, gas bubbles in 70
---, heat of adsorption 89
---, phase transformation 86, 107
---, photometric 77–81
---, polarographic 89
---, radioactivity 89
---, refractive index 83–86
---, requirements for 77
---, requirements for preparative chromatography 272
---, response factors 242
---, response time of 90
---, selectivity of 238, 242
---, suitability for trace analysis 260, 261
---, tunable 300
---, unblocking 71
---, vapour pressure 84
Detector drift 76, 84
Detector noise 75, 235
Detector non-linearity 76, 77, 224, 235
Detector selectivity 238
Dextrans 204, 205
---, cross-linked 191, 197
Dia-Chrom 148
Dialkyl phthalates 205
Diaphragm pumps 49, 52, 54
Diastereoisomers 287
Diatomaceous earth 148
o-Dichlorobenzene 157
1,4-Dichlorobutane 212
Diethyldiphenylurea 254
Diffusion, extra-column 20
Diffusion phenomena 10
Digitoxin 283
Digoxin 283
Dihydrocholesterol 107
6,7-Dihydroxycoumarin 6-glucoside 150
Dihydroxyphenylalanine, metabolites of 285
6,7-Dimethoxycoumarin 150
Dimethyldiphenylurea 254
N^2-Dimethylguanosine 172
1,5-Dimethylnaphthalene 154
Dimethyl polysiloxanes 205
2,4-Dinitrobenzene 129
Dinitronaphthalene 138
Dinitrophenylhydrazone derivatives, of carbonyl compounds of steroids 260
Dinitrotoluenes 128, 137
Diphenylhydantoin, in blood serum 280
Diphenylurea 254
Dipole moments 98
Disc integrator 240
Dispersion 10, 20
Dissociation constant 186
Distribution coefficient 8, 145, 146, 193
Distribution Law 145
Diuron 159, 162
DNA, in RNA 288
n-Dodecane 212
n-Dodecyl ether 212
Drugs 250, 280, 281
Dry-column chromatography 3
Drying oils 205
DTE, degradation of 292
Durapak Carbowax 400 289, 296
Durapak Carbowax 400/Corasil 156
Durapak Carbowax 400/Porasil 156
Durapak *n*-octane 156
Durapak OPN 156, 280, 291, 294, 298
Durapak OPN/Corasil 156
Durrum DA-X2 180
Durrum DA-X4 180, 281, 291
Durrum DA-X8A 180
Durrum DC-1A 180
Durrum DC-2 287, 291
Durrum DC-2A 180
Durrum DC-4A 169, 180, 287
Durrum resins 176
Dyes 291

E

Ecdysones 282
Eddy diffusion 20–22, 24
EDTA 234
Efficiency 11–13, 17, 18, 20, 25, 27, 29, 30, 34, 178, 299
---, choice of sample solvent and 232
---, column 12, 300
---, future requirements 299
---, influence of sample volume 250
---, internal diameter and 32–34
---, optimisation in trace analysis 257
Effluent 7
Electrical conductivity detectors 88
Electrochemical detectors 261
Electron capture detectors 88
Eluate 7
Eluent 7
Eluotropic series 39, 98, 99, 109
Eluting peak, identity of 220
Endrin 292
Enzymes 198
Epichlorohydrin 205

Epinephrine 188
Epoxy resins, uncured 205
Ergot alkaloids 282
Ethyl acrylate polymers 205
Ethylenediaminetetraacetic acid 234
Ethylene glycol, as stationary phase 148
Ethylene–propylene copolymer 205
Ethylene–vinyl acetate copolymer 205
Ethyl ether 212
Ethylhydroxybenzoate 290
Ethyl iodide 212
Ethyl vanillin 291
Exclusion limit 195
External sample loops 65
External standard, calibration using 244

F

Factor, selectivity 10
Fatty acids 288, 297
———, benzyl esters 289
———, long-chain, as 2-naphthacyl esters 289
———, methyl esters 289
———, polyglycol esters 289
———, and derivatives 205
F.D. and C. Blue No. 2 291
F.D. and C. Red No. 40 291
F.D. and C. Yellow No. 6 291
Fenuron 159, 162
Filters, highly efficient 230
———, interference 80
———, line 57, 58
———, low-porosity 248
———, narrow band pass 80
———, Swinnex 230
Flavones 286
Flavonoids 286
Flavour chemicals 290
Flavour mixtures 290
Florisil 127, 130
Flow-controlled pumps 49, 211
Flow controllers 49
Flow programming 125
Fluoranthene 145, 154
Fluorescamine 83, 260
Fluorescence/absorbance detectors 82
Fluorescence detection 168, 260
Fluorescence detectors 81, 82, 169, 224
———, linearity of 83
———, use in trace analysis 259
Fluorigenic reagents 83
Fluram 260
Folic acid 173
Fraction collection 272, 273
Fraction collection and identification 224
Fraction collectors 71
Freeze drying 249
Fresnel, Law of reflection 84, 85
Frits, inlet 61
———, metal 33, 38, 39
———, porous metal 33, 34
———, PTFE 34, 39
Fruit juices 213
Frying fats 289
Frying oils, polar products in 289
2-Fuorylglycine 172
Furfuryl alcohol 205
Furocoumarins 286

G

Gel filtration chromatography 104, 191, 211
Gel permeation chromatography 104, 191, 208
General elution problem 110
General resolution equation, derivation of 301
Geranial 214
Glass 204
Glucose 107
Glutamic acid 169
Glycerides 205
Glycine 169
Glycolipids, in soya 287
Gradient elution 52, 111–115, 234
———, detectors and 111
———, incremental 53
———, in trace analysis 255
———, large injection volumes and 252
———, reconditioning 114
———, step-wise 114, 232, 269
Gradient elution profiles 114
Gradient elution systems 52–55
Griseofulvin, fermentation products 285
Guanosine-5′-diphosphate 170
Guanosine-5′-monophosphate 170
Guanosine nucleotides 287
Guanosine-5′-triphosphate 170
Guard columns 67, 187, 256

H

Haemoglobin 212, 213
Hamilton AN-90 180
Hamilton B-80 180
Hamilton H-70 180
Hamilton HP-AN-90 286
Hamilton HP-B-80 286
Heat exchanger 58, 59
Heat of adsorption detectors 89
Height equivalent to a theoretical plate 12, 20, 23, 24, 28
n-Heptane 212

n-Heptanol 212
Herbicides 157
---, substituted-urea 162
Heroin 282
HETP, see height equivalent to a theoretical plate
Hexachlorophene 284
Hippuric acid 172
Histidine 169
Homologues, resolution of 134, 135
Hop acids 290
Human serum proteins 212, 213
Hydralazine 284
Hydrocarbons 293
Hydrochlorothiazide 284
Hydrodynamic volume 208, 209
Hydrogel 200
19-Hydroxy-androst-4-ene-3,17-dione 121
4-Hydroxybenzoylglycine 172
Hydroxylated aromatics 149
6-Hydroxy-7-methoxycoumarin 150
Hydroxynalidixic acid, in plasma and urine 285
p-Hydroxynorephedrine 188
16α-Hydroxy-pregn-4-ene-3,20-dione 121
Hydroxywarfarin 284
Hypoxanthine 171

I

Identification of components, purity of solvents and 226
Identity of a component, incorrect assignment of 223
Incremental gradient elution 107, 108
Infinite diameter effect 33, 34
Infrared photometric detectors 81
Infrared spectra of collected fractions 225
Injection, on-column 34, 61
Injection solution, nature of solvent used 231, 232
---, preparation of 230
Injection systems 60–66, 269
Injection volume, column efficiency and 251
Inlet pressure 26
Inner filter effect 224
Inosine 171
Insect hormones 282
Insect moulting hormones 282
Instrumentation 43
---, availability 74
---, components of 43
---, suppliers 74
Integration 238–241
Integrators 73, 236
---, computing 242
Internal standard, calibration using 244
Interstitial volume 193, 198
Intrinsic viscosity 197, 209
Iodobenzene 157
Ion exchange 168–175, 184
Ion-exchange chromatography 7, 97, 100, 102 167, 168, 182
Ion exchangers 173, 176–181, 183, 185
Ion-exchange resins, conversion of 174
Ion-exchange separations, non-ionic effects 175
Ion-pair chromatography 145
Ion-pair partition chromatography 187, 188
Ion selectivity 175
Ion-X-SA 180
Ion-X-SC 180, 281
Irreversible adsorption, minimisation of 136
Isobutyl allylbarbital 185
Isocolchicine 274
Isocyanates 206
Isoleucine 169

K

11-Keto-progesterone 107
Kinetic parameters 16

L

Laminar flow 21
Lannate 293
Large molecular species, from biological fluids 211
Larvicide 292
Leucine 169
Lexan 206
LFS Pellicular Anion Exchange 283
LiChrosorb Alox T 130, 132
LiChrosorb RP-8 156
LiChrosorb SI-60 130, 132, 281, 286, 287, 290, 292, 295
---, silanised 156
LiChrosorb SI-100 130, 132, 285
LiChrospher 202, 286, 298
LiChrospher SI-100 120, 130
LiChrospher SI-500 130
LiChrospher SI-1000 120, 130
LiChrospher SI-4000 130
Lignin sulphonates 206
Limonene 214
Lindane 292
---, residues of 293
Linuron 159, 162
Lipids 206, 288
---, of soya beans 286
Lipid classes 289

Liqua-Chrom 148
Liquid chromatographic instrumentation and components, suppliers of 305–308
Liquid chromatograph–mass spectrometer systems, in-line 225, 226
Liquid flow monitor 210
Liquid–liquid chromatography 7, 96, 97, 143
———, solvents for 108, 109
Liquid–solid chromatography 7, 96, 127
Longitudinal diffusion 20, 22–24
LSD 223, 224, 281
Lubricating oils 206
Lysergic acid diethylamide, see LSD
Lysine 169

M

Macroreticular ion exchangers 179
Magnesia 129
Magnesium silicates 130
Mass spectrometer, interface 225
Mass spectrometry, liquid chromatography and 224
Mass spectrum 225
Mass transfer 20, 22–24, 27–30, 146, 177, 178
———, stationary phase 164
Mass transfer in polymer phases 151
Melamines 206
Membrane pumps 49
Merckogel OR 199
Merckogel SI-50 289
Merckogel SI-150 296
Metabolism studies 250
Metal ions, acetylacetonates of 296
———, trifluoroacetonates of 296
Metering pumps 49–51
Methacrylates 206
Methadone 282
Methionine 169
3-Methoxy-4-hydroxyphenylacetic acid 172
Methyl benzoate 129
Methyldiphenylurea 254
1-Methylguanosine 172
Methyl iodide 212
Methyl methacrylate–styrene copolymer 206
Methylprednisolone, residues in milk 282
Methyl stearate 107
Methyltestosterone 121
1-Methylxanthine 172
Micropak A1-5, A1-10 130
Micropak C–H 156
Micropak C≡N 156
Micropak NH_2 156
Micropak SI-5 130
Micropak SI-10 130, 280, 284, 285, 290
Microreticular ion exchangers 179
Microsyringes, filling of 62
———, replaceable needle 62
———, unblocking of 61, 62
Mineral oil 206
Mobile phases 7, 8, 101
———, addition of acid or base 128, 136
———, boiling point 99
———, choice for detector compatibility 101
———, classification of 98, 99
———, degassing 56
———, detector compatibility 203
———, elution characteristics of 95
———, for preparative chromatography 270
———, nature of 95
———, optimisation for ion exchange 185, 186
———, pulsations in 50
———, refractive index 99
———, selection of 100, 101, 106
———, selectivity effects 99
———, stagnant pools of 23, 28, 30
———, UV cut off 99
———, viscosity 99
Mobile phase compressibility 47
Mobile phase flow-rate 17
———, gravimetric measurement of 72
———, in preparative chromatography 271
———, measurement of 72
———, measurement with flow meters 73
———, reproducibility 221
———, volumetric measurement of 72
———, with large columns 270
Mobile phase selection, gradient elution and 106
Mobile phase velocity 17, 19, 21–24, 26, 28, 30
———, inlet pressure and 26, 27, 29
———, reduced 18
Molecular association, elution volume and 203
Molecular sieves 130
Molecular size, elution volume and 195
Molecular weight determinations 197
Molecular weight distribution 191
———, calculation of 210, 211
———, experimental errors in 211
———, of polymers 208, 209
Monochlorobenzene 157
Mononitrotoluenes 128
Monuron 159, 162
Morphine 282, 283

N

Nalidixic acid, in plasma and urine 285
Naphthalene 118, 145, 158, 231, 251
Neburon 159, 162
Neoprene 206

Neral 214
Neutral lipids, in soya 287
Nicotinamide 290
Nicotine, derivatives from tobacco 283
Nicotinic acid 173
Ninhydrin 169, 260
Ninhydrin reaction 81, 168
Nitrobenzene 129, 154
Nitroglycerin, propellants containing 295
p-Nitrotoluene 154
n-Nonane 154
Non-ionic surfactants 206, 295
Noradrenaline 280
Normalisation of peak areas 243
Normalisation of peaks with correction factors 243
Normal partition systems 144
Norphenephrine 188
Nucleic acids 170
Nucleic acid bases 288
Nucleic acid constituents 288
Nucleosides 287
Nucleotides 287, 288
Nucleotide bases 170
Nylon 130
Nylon 6 297
Nylons (4, 6, 66, etc.) 206

O

Octadecanol 107
n-Octadecyl ether 212
n-Octyl ether 212
ODS-SIL-X-I 156
ODS-SIL-X-II 156, 280, 293
Oestradiol isomers 274
Oestrogens 281, 282
Olefin sulphonates 295
Oligonucleotides 288
Opium alkaloids 282, 283
Orange juice 215
Orange oil 214
Organic mercury compounds 296
Organo-iron complexes, isomers of 296
Organo-iron compounds 296
Orotic acid 172
Orotidine 172
Overlapping peaks 255
———, measurement of 237, 241
Oxindole alkaloids 283
β,β'-Oxydipropionitrile 148, 149
Oxyphenbutazone, in plasma 284
Oxypurinol 172

P

Paper chromatography 3
Particle size 24, 25, 27, 35
Partisil 5, 10, 20 130
Partisil-10-SAX 180
Partisil-10-SCX 180
Partition chromatography 96, 97, 100, 106
———, elution order and 144
———, merits of 163, 164
———, methodology 148
———, sample applicability of 143, 144
———, selectivity in 162, 164
———, solvents for 108, 109
———, supports for 148
———, ternary liquid systems, see ternary liquid partition systems
———, theoretical basis of 146
Partition column chromatography 143
Patulin, in apple juice 293
Peaks, leading edge 10
———, spurious 112, 114, 186, 234
———, trailing edge 10
Peak area, flow dependence 236
Peak area measurements 236, 238–240
Peak broadening 211
Peak dispersion 10
Peak height, flow dependence 236
Peak height measurements 236, 261
Peak overlap 14, 16
Pellicular Anion Exchange 180, 287, 288
Pellicular Cation Exchange 180, 288
Pellicular supports, see also support materials
Pellidon 150, 283, 286
Pellidon H 296
Pellionex AS 172
Pellionex WAX 288
AE-Pellionex SAX 180
AL-Pellionex WAX 180
AS-Pellionex SAX 180, 284, 287, 293
HC-Pellionex SCX 180
HS-Pellionex SCX 180, 285, 290
Pellosil HC 120, 130, 286
Pellosil HS 120, 130
Pellumina HC 130
Pellumina HS 130
n-Pentane 212
Peptides 285, 286
Perisorb 28
Perisorb A 130, 280
Perisorb AN 180
Perisorb KAT 180
Perisorb PA-6 150
Perisorb RP 156
Permaphase AAX 170, 173, 180, 287, 288, 291, 296, 297

Permaphase ABX 180
Permaphase ETH 156, 159, 162, 254, 282, 285, 290, 295
Permaphase ODS 113, 118, 145, 156, 157, 214, 254, 281–284, 289, 290, 292–296
Permeability 28, 29, 176, 178
Pesticides 249, 292, 293
Pesticide residues, clean-up of fish lipids 215
———, gel permeation chromatographic clean-up for 292
Phase transformation detectors 86, 101, 108
Phase transformation to flame ionisation detector 87
Phenacetin 255, 283
Phenanthrene 145
Phenetole 129
Phenethylamines, of forensic interest 281
Phenobarbital 185, 237
———, in blood serum 280
Phenobarbitone 250
Phenols, hindered 298
———, in polluted waters 296
———, residues in water 296
———, substituted 296
Phenolcarboxylic acids 296
Phenol formaldehyde 206
Phenolic resins 206
Phenothiazines 281
———, derivatives with neuroleptic activity 284
Phenylalanine 107, 169, 286
Phenylbutazone, in plasma 284
Phenyl-SIL-X-I 156
Phenylthiohydantoin 260
Phospholipids, in soya 287
Phosphors 80
Photometric detectors 77–81, 101, 222, 261
Phthalate plasticisers 298
Phthalic acid 296
Planimeter 240
Plasticizers, various 206
Plate height, see also height equivalent to a theoretical plate 12, 13, 21, 22
———, calculation of 16
———, reduced 18, 26
Pinene 214
Plutonium 297
Pneumatic amplifier pumps 47–49, 54, 55
Pneumatic pumps 45, 46, 57
Polarity 98, 103, 104, 108–110
Polarographic detectors 89, 261
Polyacrylamide 204, 297
Polyalkylene glycols 206
Polyamines 286
Polybutadiene 206, 313
Polybutene-1 205
Poly (1,2-butyleneglycol phthalate) 313
Polycaprolactam 206
Polycarbonates 206
Polychlorinated biphenyls 293
Polydimethylsiloxane 297
Polyelectrolytes 206
Polyene antibiotics 285
Polyesters, non-linear and unsaturated 206
Polyethers 206
Polyethylene 206, 297, 313
Polyethylene glycols, as stationary phase 148
Polyethylene oxide 206
Polyethylene terephthalate 206, 297
Polyisobutylene 206
Polyisobutylene copolymers 206
Polyisoprene 206
Polymers, water-soluble 198
Polymeric stationary phases 96, 147, 149–152
Polymer reference compounds, suppliers of 310
Polymethyl methacrylate 313
Polynuclear aromatics 206
Polyols 206
Polyoxymethylene 297
Polyphenylene oxide 207
Polypropylene 207
Polystyrene 204, 207, 297, 298, 313
———, as sample 202
Polystyrene gel packings 199, 200
Polysulphonates 207
Polysulphones 207
Polythionates 297
Polyurethanes 207
Polyvinyl acetate 204, 207
Polyvinyl acetate copolymers 207
Polyvinyl acetate gel packings 199, 200
Polyvinyl alcohol 207
Polyvinyl butyral 207
Polyvinyl chloride 207, 313
Polyvinyl fluoride 207
Polyvinyl methyl ether 207
Poly-(2-vinylpyridine) 297
Poragel 60 290
Poragel A 200
Poragel A-1 283
Porasil 29, 202, 297
μPorasil 130, 202, 293
Porasil 400 295
Porasil 1500 295
Porasil A 132, 289, 294
Porasil B 132
Porasil C 132
Porasil Carbowax 400 291, 292
Porasil D 132
Porasil E 132
Porasil F 132
Porasil T 130, 287, 294, 295
Pore volume 193, 195, 196, 198
Porous silica microspheres 127, 149, 162, 202
Porphyrins 286

Positional isomers, separation of 127, 134
Precision 229
Pre-column 59, 256, 310
Prednisone 161
Preparative chromatography 248, 263
---, applications of 273
---, bonded phases and 158
---, features of supports for 267
Preparative separations, industrial-scale 275
---, operational parameters of 274
Pressure indication 56, 57
Pressure programming 125
Process liquid chromatographs 300
Progesterone 121, 161, 273, 274
Progesterones 282
Progesterone preparations 282
Propylene–(butene-1) copolymers 207
Propylhydroxybenzoate 290
Prostaglandins 284
Proteins 198
---, adsorption of 201
Pseudouridine 172
PTFE 130
PTFE fibre 34, 39
PTH amino acids 287
Pulse damper 50, 51
---, dead volume associated with 123
Pumps, constant-flow 47
---, constant-pressure 38, 45, 46, 48, 49
---, diaphragm 49, 52, 54
---, flow controlled 49
---, membrane 49
---, metering 49–51
---, pneumatic 57
---, pneumatic amplifier 47–49, 54, 55
---, positive displacement 38
---, reciprocating 49–52, 54
---, reciprocating twin piston 51
---, simple pneumatic 45, 46
---, syringe 46, 47
Pumping systems, for preparative chromatography 271
Purines 287
Purine bases 170, 171
---, and their nucleosides 288
Pyrene 118, 145, 158, 231, 251
Pyrethrins 213, 274
---, extracts 293
Pyridine bases 296
Pyridine isomers, monosubstituted 296
Pyrimidines 287
Pyrimidine bases 170
---, and their nucleosides 288

Q

Qualitative analysis 219, 220
Quantitation, in trace analysis 261
Quantitative analysis 229, 235
Quinoline 138

R

Radioactivity detectors 89
Rare earth elements 297
Reciprocating pumps 49–52, 54, 123
Recorder 73
Recycle chromatography 122–125, 221, 253, 266
Refractive index, temperature coefficient of 83
Refractive index detectors 83–86, 101, 223
Refractive indices of solvents 99
Relative partition coefficient 8
Representative sample 230
Reproducibility 229
Reserpine chlorothiazide 284
Resolution 15, 17
---, optimisation in trace analysis 252, 253
---, optimisation of 16
Resolution equation, general 14, 15
Resolution factor 13, 14
Resolving power 13, 110
Retention 7
Retention characteristics, and chemical structure 222
Retention data, sample identification using 220, 221
Retention volume 7, 8
Reversed-phase chromatography 97, 102, 106, 108, 157
Reversed-phase systems 144
Riboflavin 290
Riboflavin monophosphate 173
Rolitetracyclines 285
Rubber, acrylonitrile–butadiene 207
---, butyl 207
---, natural 207
---, neoprene 207
---, styrene–butadiene 207

S

Saccharides 291
Saccharin 291
---, metabolism of 291
Safety 45, 57, 204
Salicylamide 283
Samples, decomposition of 234

Sample capacity 275
Sample introduction, in preparative chromatography 269
Sample introduction devices, see also injection systems 59–66
---, sources of error in 232, 233
Sample throughput, methods for increasing 264
Saturation of the mobile phase 310
Secobarbital 185
Selective detectors, see also detectors 77
---, sample identification using 222
Selective permeation range 195–197
Selectivity 10, 11, 14, 17, 18, 110, 116, 117
---, mobile phase composition and 138, 139
---, optimisation in trace analysis 253, 255
Selectivity factor 13, 16
Separation method, deciding the 102–106
Sephadex dextran 270
Sephadex G 199
Sephadex G-200 212, 213
Sephadex LH-20 199
Sepharose 199
Septum, injection port 62
Septumless injectors 65
Serine 169
Sieve sizes, A.S.T.M. 303
---, B.S.S. 303
Silanised silica gel Type 7719 295
Silica 121, 130–132, 204, 281
Silica A 130
Silica gel 127, 129, 284
Silica gel CT 285
Silica gel Type 7719 289
Silica gel Type 7754 289
Silica microspheres, porous 29
Silicate-ester 152
Silicic acid 129
Silicones 207
SIL-X 281, 283, 284, 291
SIL-X-I 130
SIL-X-I-FE 156
SIL-X-II 130
Simple liquid partition chromatography 164
Simple liquid stationary phases, operation details of 309–311
---, types of 147, 148
Siphon counter 72, 210, 211
Sodium benzoate 291
Sodium *o*-iodohippurate 284
Soft gels 211
---, bacterial attack 198, 199
---, flow control in 198
Solution evaporation, losses 232
Solvents, purity of 232
Solvent degassing 56
Solvent demixing 112, 114, 186, 234
Solvent extraction 249, 255
Solvents of known water content, preparation of 139, 140
Solvent programming 52
Sorbic acid 290
Soxhlet extractor 249
Spectrophotometric detectors 80, 222
---, use in trace analysis 258
Spherisorb 29
Spherisorb A5W, A10W, A20W 130
Spherisorb A5Y 129
Spherisorb ODS 156
Spherisorb S5W, S10W, S20W 130
Spherosil 29, 202, 295
Spherosil XOA-075 132
Spherosil XOA-200 132
Spherosil XOA-400 132, 280, 281, 294
Spherosil XOB-015 132
Spherosil XOB-030 132
Spherosil XOB-075 294
Spherosil XOC-005 132
Spray Impact Detector 90
Spurious peaks 112, 114, 186, 234
Squalane 107, 148
Stagnant pools of mobile phase 176, 192, 194
Stationary phases 7, 8
---, chemically bonded 96, 97, 144, 146, 147
---, coating procedure 309, 310
---, for preparative chromatography 268
---, liquid 96
---, polymeric 96, 147, 149–152
---, simple liquid, see simple liquid stationary phases
---, viscosity of 22
Steric exclusion chromatography 7, 28, 29, 97, 98, 102, 191
---, applicability of 191
---, as a clean-up technique 213, 215
---, calibration curve for 196
---, column packings for 194
---, differential 209
---, for molecular weight determination 104
---, low-molecular-weight samples and 213
---, mechanism of 192–194
---, mobile phases for 202, 203
---, rigid packings for 201, 202
---, semi-rigid packings for 199, 200
---, soft gels for 197–199
---, solvent compatibility of packings 204
Steric exclusion columns, features of 195, 196
Steric exclusion packings, inorganic, solvent compatibility 201
Steroids 281, 282
Steryl glucosides, in soya 287
Structure of ion exchangers 176
Strychnos alkaloids 283

Styragel 200, 212, 297
μStyragel 297
Styrene–acrylonitrile copolymer 207
Styrene divinylbenzene 204
Styrene–isoprene copolymer 207
Substituted ureas 292
Sugars 291, 292
Sulphacyanamide 112
Sulphaguanidine 112
Sulphanilamides 112, 285
Sulphanilic acid 112
Sulphanilylurea 112
Sulphonamides 285
Sulphonylureas 285
Sulphonylurea-based antidiabetic agents 284
Support, surface area of 8
Support materials 26–30
---, capacity of 268
---, characteristics of 40
Surface area, capacity and 117, 118
Synephrine 188
Syringe pumps 46, 47, 54

T

Tailing peaks, suppression of 144
Technical materials, impurities in 247
Teflon® fibre 34
Temperature, column efficiency and 102
Temperature control, methods of 68, 69
Temperature control of detector 84, 89
Temperature control of mobile phase 58, 59
Temperature control of separating column 67, 68
Temperature programming 115, 116
Terephthalate mixture, complex 113
Ternary liquid partition 282, 296
Ternary liquid partition systems 160, 161, 163
Testosterone 121
Testosterone acetate 274
Testosterone propionate 274
Tetrabutylammonium ions 188
Tetracyclines 285
Theoretical plates, effective 17, 18, 30, 31
---, number of 12
Theoretical plate number, effective 17
Thermodynamic parameters 17
Thin-layer chromatography 3, 19, 31, 105, 127, 128
Thorium 297
Threonine 169
Thyroid hormones 285
TNT, in waste waters 295
TNT byproducts, identification of 294
Tocopherols, in plant oils 289
Total exclusion 196
Total permeation 195, 196
Trace analysis 247
Trace components, concentration of 249
Trifluorostyrene 207
Trimethylene glycol, as stationary phase 148
1,2,3-Tris(2-cyanoethoxy)propane, as stationary phase 148
Trisulfapyrimidines 285
Triton 295
Tropane alkaloids 283
Tube fittings 44, 45
Tubing 44
---, blockage in 70, 71
---, stainless steel, corrosion of 44
Turbulent flow 21
Tyrosine 169, 285, 286

U

Universal calibration, for steric exclusion 208, 209
Universal detectors 77
Uranium 297
Urethane prepolymers 207
Uric acid 172
Uridine-5′-diphosphate 170
Uridine-5′-monophosphate 170
Uridine-5′-triphosphate 170
Urine 171, 172
---, UV absorbing constituents in 286
Uses of liquid chromatographic procedures 217
UV cut off of solvents 99
UV stabilizers for polymers 207

V

Vacancy effect 209
Valine 169
Vanillic acid 172
Vanillin 291
Vinyl chloride–vinyl acetate–maleic acid terpolymer 207
Viscosity 101
---, temperature and 101, 102
Viscosity of solvents 99
Vitamins 173
---, fat-soluble 289
---, oil-soluble 290
---, water-soluble 289, 290
Vitamin A acetate 107, 151
Vitamin B_1 290
Vitamin B_2 290
Vitamin B_6 290
Vitamin D_2 151
---, in A acetate–D_2 capsules 290

Vitamin D_3, hydroxylated derivatives of 289
Vitamin E succinate 151
Vitavax, carboxin pesticide 293
Vit-X 202, 297
Void volume 7, 193, 195, 196
Vydac 290
Vydac adsorbent 130, 295
Vydac Anion Exchanger 180
Vydac Cation Exchanger 180, 291
Vydac Polar 156
Vydac RP 156, 283, 285, 294
Vydac TP 156
Vydac TP Anion Exchange 180
Vydac TP Cation Exchange 180
Vydac TP Polar Bonded Phase 156

W

Warfarin 284
Water, as stationary phase 148
Waxes (hydrocarbon) 207

Z

Zectran 293
Zipax 27, 28, 148, 162, 282, 295, 297
Zipax ANH 150
Zipax BOP 281, 291–293, 298
Zipax CWT 281
Zipax HCP 150, 151, 282, 284, 285, 289, 291
Zipax PAM 150
Zipax SAX 112, 180, 185, 237, 255, 280, 282–285, 288–291, 295, 296
Zipax SCX 171, 180, 255, 282–284, 286–290, 292, 296
Zipax WAX 180, 288
Zorbax 29
Zorbax ODS 118, 156, 158, 231, 251
Zorbax SIL 128, 130, 137, 161, 223, 258, 273, 281, 283, 286, 290, 292, 293, 295
Zwitterions 182